The Structural Basis of Membrane Function

Members of the Symposium Committee
University of Tehran

Simine Atighetchi
Fereydoun Djavadi
Lisa Djavadi-Ohaniance
Ali Estilai
Esmail Hosseini-Shokrai
Khashayar Javaherian (Chairman)
Esmail Meisami
Mahnaz Moradi
Jamshid Naghizadeh
Mohammad Sabour

Proceedings of the International Symposium
(May 5-7, 1975, Tehran)

Sponsored by

The University of Tehran
The International Union of Biochemistry
The International Union of Pure and Applied Biophysics

Number 72 in the Series of IUB-sponsored symposia

The Structural Basis of Membrane Function

Edited by

Youssef Hatefi
Department of Biochemistry
Scripps Clinic and Research Foundation
La Jolla, California 92037

Lisa Djavadi-Ohaniance
Department of Cellular and Molecular Biology
Faculty of Science
University of Tehran
Tehran, Iran

Academic Press, Inc. *New York San Francisco London 1976*
A Subsidiary of Harcourt Brace Jovanovich, Publishers

Academic Press Rapid Manuscript Reproduction

ACADEMIC PRESS, INC.
111 Fifth Avenue, New York, New York 10003

United Kingdom Edition published by
ACADEMIC PRESS, INC. (LONDON) LTD.
24/28 Oval Road, London NW1

Library of Congress Cataloging in Publication Data
Main entry under title:

The Structural basis of membrane function.

(Series of IUB-sponsored symposia ; no. 72)
Bibliography: p.
1. Membranes (Biology)--Congresses. I. Hatefi,
Youssef, (date)   II. Djavadi-Ohaniance, Lisa.
III. Teheran. Dānishgāh. IV. International Union of
Biochemistry. V. International Union for Pure and
Applied Biophysics. VI. Series: International Union
of Biochemistry. Symposium series ; no. 72.
[DNLM: 1. Cell membrane--Physiology--Congresses.
2. Cell membrane--Ultrastructure--Congresses.
3. Biological transport, Active--Congresses. 4. Energy
transfer--Congresses. 5. Mitochondria--Ultrastructure
--Congresses. QH601 I64s / W3 IN952 v. 72 1975]
QH601.S76   574.8'75   75-44119
ISBN   0-12-332450-5

Contents

CONTENTS

CONTENTS

List of Contributors

Aquila, H., Institut für Physiologische Chemie und Physikalische Biochemie, Universität München, München, Germany

Avron, M., Biochemistry Department, Weizmann Institute of Sciences, Rehovot, Israel

Azzone, G.F., Institute of General Pathology, University of Padova, Padova, Italy

Bamberg, E., Department of Biology, University of Konstanz, D-775 Konstanz, Germany

Bennett, J.P., Department of Biochemistry, University of Cambridge, Cambridge CB2 1QW, England

Bennett, V., Department of Pharmacology and Experimental Therapeutics, and Department of Medicine, The Johns Hopkins University School of Medicine, Baltimore, Maryland 21205

Boyer, P.D., Molecular Biology Institute, and Department of Chemistry, University of California, Los Angeles, California 90024

Brand, M.D., Department of Physiological Chemistry, The Johns Hopkins University School of Medicine, Baltimore, Maryland 20205

Buchanan, B.B., Institut für Physiologische Chemie and Physikalische Biochemie, Universität München, München, Germany

Capdevila, J., Department of Biochemistry, and Karolinska Institute, University of Stockholm, Stockholm, Sweden

Chang, T.-M., Department of Biochemistry, The Public Health Research Institute of the City of New York, New York, New York 10016

Chu, J.-W., Department of Chemistry, Wayne State University, Detroit, Michigan 48202

Churchill, P.F., Department of Chemistry, Wayne State University, Detroit, Michigan 48202

Colonna, R., Institute of General Pathology, University of Padova, Padova, Italy

Coon, M.J., Department of Biological Chemistry, Medical School, The University of Michigan, Ann Arbor, Michigan 48104

Craig, S., Department of Pharmacology and Experimental Therapeutics, and Department of Medicine, The Johns Hopkins University School of Medicine, Baltimore, Maryland 21205

Cuatrecasas, P., Department of Pharmacology and Experimental Therapeutics, and Department of Medicine, The Johns Hopkins University School of Medicine, Baltimore, Maryland 21205

Dallner, G., Department of Biochemistry, and Karolinska Institute, University of Stockholm, Stockholm, Sweden

Dean, W.L., Department of Biological Chemistry, Medical School, The University of Michigan, Ann Arbor, Michigan 48104

Demel, R.A., Laboratory of Biochemistry, University of Utrecht, Utrecht, The Netherlands

DePierre, J.W., Department òf Biochemistry, and Karolinska Institute, University of Stockholm, Stockholm, Sweden

DeRobertis, E., Instituto de Biología Celular, Facultad de Medicina, Universidad de Buenos Aires, Buenos Aires, Argentina

Djavadi-Ohaniance, L., Department of Cellular and Molecular Biology, Faculty of Science, University of Tehran, Tehran, Iran

Ebel, R., Department of Biochemistry, Southwestern Medical School, University of Texas Health Science Center, Dallas, Texas 75235

Ernster, L., Department of Biochemistry, Arrhenius Laboratory, University of Stockholm, Stockholm, Sweden

Estabrook, R.W., Department of Biochemistry, Southwestern Medical School, University of Texas Health Science Center, Dallas, Texas 75235

Galante, Y.M., Department of Biochemistry, Scripps Clinic and Research Foundation, La Jolla, California 92037

Garrett, N.E., Department of Biochemistry, The Public Health Research Institute of the City of New York, New York, New York 10016

Geurtz van Kessel, W.S.M., Laboratory of Biochemistry, University of Utrecht, Utrecht, The Netherlands

Goodman, B.P., Departments of Biochemistry and Pediatrics, University of Pennsylvania, Philadelphia, Pennsylvania 19174

Grebe, K., Institut für Physiologische Chemie und Physikalische Biochemie, Universität München, München, Germany

Green, D.E., Institute for Enzyme Research, University of Wisconsin, Madison, Wisconsin 53706

Griffiths, D.E., Department of Molecular Sciences, University of Warwick, Coventry CV4 7AL, England

Guengerich, F.P., Department of Biological Chemistry, Medical School, The University of Michigan, Ann Arbor, Michigan 48104

Guerrieri, F., Institute of Biological Chemistry, Faculty of Medicine, and Institute of Biological Chemistry, Faculty of Science, University of Bari, Bari, Italy

Gunsalus, I.C., School of Chemical Sciences, University of Illinois, Urbana, Illinois 61801

Handler, P., National Academy of Sciences, Washington, D.C. 20418

Hatefi, Y., Department of Biochemistry, Scripps Clinic and Research Foundation, La Jolla, California 92037

Haugen, D.A., Department of Biological Chemistry, Medical School, The University of Michigan, Ann Arbor, Michigan 48104

Hesketh, T.R., Department of Biochemistry, University of Cambridge, Cambridge CB2 1QW, England

Houslay, M.D., Department of Biochemistry, University of Cambridge, Cambridge CB2 1QW, England

Huennekens, F.M., Department of Biochemistry, Scripps Clinic and Research Foundation, La Jolla, California 92037

Huxley, H.E., M.R.C. Laboratory of Molecular Biology, Cambridge CB2 2QH, England

Jakobssen, S., Department of Biochemistry, and Karolinska Institute, University of Stockholm, Stockholm, Sweden

Jonen, H., Department of Biochemistry, Southwestern Medical School, University of Texas Health Science Center, Dallas, Texas 75235

Kaback, H.R., The Roche Institute of Molecular Biology, Nutley, New Jersey 07110

Kamp, H.H., Laboratory of Biochemistry, University of Utrecht, Utrecht, The Netherlands

Kayalar, C., Molecular Biology Institute, and Department of Chemistry, University of California, Los Angeles, California 90024

Keynes, R.D., Physiological Laboratory, Cambridge, England

Kimura, T., Department of Chemistry, Wayne State University, Detroit, Michigan 48202

Klingenberg, M., Institut für Physiologische Chemie und Physikalische Biochemie, Universität München, München, Germany

Klob, H.-A., Department of Biology, University of Konstanz, D-775 Konstanz, Germany

Läuger, P., Department of Biology, University of Konstanz, D-775 Konstanz, Germany

Lehninger, A.L., Department of Physiological Chemistry, The Johns Hopkins University School of Medicine, Baltimore, Maryland 21205

Lien, E., Departments of Biochemistry and Pediatrics, University of Pennsylvania, Philadelphia, Pennsylvania 19174

Lorusso, M., Institute of Biological Chemistry, Faculty of Medicine, and Institute of Biological Chemistry, Faculty of Science, University of Bari, Bari, Italy

Lynen, F., Max-Planck-Institut für Biochemie, D-8033 Martinsried, Germany

Maelicke, A., The Rockefeller University, New York, New York 10021

Massari, S., Institute of General Pathology, University of Padova, Padova, Italy

Masters, B.S.S., Department of Biochemistry, Southwestern Medical School, University of Texas Health Science Center, Dallas, Texas 75235

Matsubara, T., Department of Biochemistry, Southwestern Medical School, University of Texas Health Science Center, Dallas, Texas 75235

Meister, A., Department of Biochemistry, Cornell University Medical College, New York, New York 10021

Metcalfe, J.C., Department of Biochemistry, University of Cambridge, Cambridge, CB2 1QW, England

O'Keefe, E., Department of Pharmacology and Experimental Therapeutics, and Department of Medicine, The Johns Hopkins University School of Medicine, Baltimore, Maryland 21205

O'Keeffe, D., Department of Biochemistry, Southwestern Medical School, University of Texas Health Science Center, Dallas, Texas 75235

Orrenius, S., Karolinska Institute, University of Stockholm, Stockholm, Sweden

Papa, S., Institute of Biological Chemistry, Faculty of Medicine, and Institute of Biological Chemistry, Faculty of Science, University of Bari, Bari, Italy

Parcells, J., Department of Chemistry, Wayne State University, Detroit, Michigan 48202

Penefsky, H.S., Department of Biochemistry, The Public Health Research Institute of the City of New York, New York, New York 10016

Peterson, J.A., Department of Biochemistry, Southwestern Medical School, University of Texas Health Science Center, Dallas, Texas 75235

Pozzan, T., Institute of General Pathology, University of Padova, Padova, Italy

Rasmussen, H., Departments of Biochemistry and Pediatrics, University of Pennsylvania, Philadelphia, Pennsylvania 19174

Reich, E., The Rockefeller University, New York, New York 10021

Riccio, P., Institut für Physiologische Chemie und Physikalische Biochemie, Universität München, München, Germany

Roelofsen, B., Laboratory of Biochemistry, University of Utrecht, Utrecht, The Netherlands

Rosing, J., Molecular Biology Institute, and Department of Chemistry, University of California, Los Angeles, California 90024

Rudnick, G., The Roche Institute of Molecular Biology, Nutley, New Jersey 07110

Sahyoun, N., Department of Pharmacology and Experimental Therapeutics, and Department of Medicine, The Johns Hopkins University School of Medicine, Baltimore, Maryland 21205

Schatz, G., Biocenter, University of Basel, CH-4056, Basel, Switzerland

Schuldiner, S., The Roche Institute of Molecular Biology, Nutley, New Jersey 07110

Short, S.A., The Roche Institute of Molecular Biology, Nutley, New Jersey 07110

Slater, E.C., Laboratory of Biochemistry, B.C.P. Jansen Institute, University of Amsterdam, Amsterdam, The Netherlands

Smith, G.A., Department of Biochemistry, University of Cambridge, Cambridge CB2 1QW, England

Stiggall, D.L., Department of Biochemistry, Scripps Clinic and Research Foundation, La Jolla, California 92037

Stoeckenius, W., Cardiovascular Research Institute, and Department of Biochemistry and Biophysics, University of California, San Francisco, California 94143

Stroobant, P., The Roche Institute of Molecular Biology, Nutley, New Jersey 07110

van Deenen, L.L.M., Laboratory of Biochemistry, University of Utrecht, Utrecht, The Netherlands

Vermilion, J.L., Department of Biological Chemistry, Medical School, The University of Michigan, Ann Arbor, Michigan 48104

Verkleij, A.J., Laboratory of Biochemistry, University of Utrecht, Utrecht, The Netherlands

Wang, H.-P., Department of Chemistry, Wayne State University, Detroit, Michigan 48202

Warren, G.B., Department of Biochemistry, University of Cambridge, Cambridge CB2 1QW, England

Werringloer, J., Department of Chemistry, Wayne State University, Detroit, Michigan 48202

Wirtz, K.W.A., Laboratory of Biochemistry, University of Utrecht, Utrecht, The Netherlands

Zwaal, R.F.A., Laboratory of Biochemistry, University of Utrecht, Utrecht, The Netherlands

FOREWORD

It is ten years since I last had the privilege of contributing to the advancement of science and higher education in my native country. The idea of organizing an international symposium in Iran was suggested by Dr. Fereydoun Djavadi, whom I met in Stockholm in July, 1973, on the occasion of the 9th International Congress of Biochemistry. Dr. Djavadi was the leader of a group of dedicated, young Iranian scientists, who had gathered together at the University of Tehran and developed a Department of Cellular and Molecular Biology. He and his colleagues felt that an international symposium sponsored by their Department would be of considerable value. It would serve as a means of introducing the new Department and its excellent facilities to the international scientific community, and of establishing contact with major research centers around the world. More significantly, such a symposium would draw the attention of the government, the Iranian scientists abroad, and the research centers in the area to the value that the international academic community and local universities place on the development of biomedical sciences in Iran. As a result, needed research and study in this field would be stimulated within the country.

Accordingly, the possibility of an international symposium at the University of Tehran was explored with key officials of the IUB, IUPAB, and the American Society of Biological Chemists. Everyone was enthusiastic and promised full support. At a planning session in Tehran the following Spring, it was decided that the subject matter of the symposium should be sufficiently broad to coincide with the interests of the Iranian scientists at Tehran and other universities and, for the reasons given above, the speakers should represent major laboratories from as many countries and universities as could be fitted into the program. Dr. Djavadi and his colleagues appreciated the fact that I could construct such a program only in the area with which I was best familiar, and pointed out that, in fact, the field of membranes encompassed the research interests of several members of the group. Thus, six major aspects of membrane research were selected, with the relationship between structure and function as the connecting theme. The final program included 35 speakers from 31 institutions and 11 countries.

The symposium was convened by an auspicious message from His Imperial Majesty, which rewarded the efforts of the Tehran group with an order for construction of a biochemistry research institute at the University of Tehran. Following this message and the welcoming address by the Chancellor of the University, I made the following introductory remarks at the request of my Iranian colleagues.

"On behalf of the Symposium Committee of the University of Tehran, I should like to express our gratitude to His Imperial Majesty, the Shahanshah of Iran, for his valuable message; to His Excellency Mr. Assodollah Alam, the illustrious Minister of the Imperial Court of Iran, for bringing us His Majesty's words; and to Dr. Houshang Nahavandi, the enlightened Chancellor of the University of Tehran, for his continued support and encouragement. Also on behalf of the Symposium Committee, I should like to thank many friends and colleagues who have helped with advice and various arrangements for this symposium, especially P.N. Campbell (Chairman of the IUB Symposium Committee), B. Chance (Vice-President of IUPAB), L. Ernster, I.C. Gunsalus (President of the American Society of Biological Chemists), F.M. Huennekens, R.D. Keynes (Secretary General of IUPAB), J. Kirchner, E.C. Slater (Treasurer of IUB), and W.J. Whelan (General Secretary of IUB).

"The subject matter of this symposium is extremely important. The significance of biomembranes becomes instantly clear by asking 'what if Nature had *not* devised the biological membrane?' And the answer is chaos. Without membranes, the biological system would have been unable to maintain a low entropy, might not have differentiated and evolved, could not have developed individuality and identity, and would have been overwhelmed by its vast surroundings. Truly the biological membrane is a remarkable accident which probably took place not because the elements cooperated in its construction. Quite the contrary. Membranes were formed, I expect, because the polar molecules of water in the primordial pools preferred to interact with their own kind rather than with the nonpolar molecules. Consequently, the hydrophobic molecules were forced out onto the surface, where probably the first membrane monolayers were formed.

"The immiscibility of membranes seems to have imparted a similar characteristic to research in its various aspects. Investigators in the various facets of membrane research do not often mix together under the same roof and exchange ideas. Introduction of a nonelectrolyte into water requires the expenditure of 2-6 Kcal/mole of free energy. And according to the Symposium Committee's ledger, bringing together an international group of immiscible investigators in various aspects of membrane research also appears to have required some energy expenditure. However, this is a land of abundant energy and hospitality, and everyone is delighted that you have come.

"The six sessions of our program represent six important facets of membrane research today. One might think of them as the six faces of a cube, and we hope that those who explore each face will gather around the edges and the corners of the cube, cast glances at other facets, and exchange ideas. This hall, the hotel lobby and restaurants, the Chancellor's dinner reception on Wednesday night, and later the group excursions to Isfahan, Shiraz and Persepolis should provide ample opportunity for such exchange. Therefore, we hope that this and future meetings of this kind, to be planned by the University of Tehran, will provide an atmosphere in which the frame of knowledge in various facets of membrane research can be welded together at the edges and around the corners. Only by constructing this frame can we expect to have a clear understanding of the structural basis of membrane function."

The proceedings of the Symposium were masterfully summarized by Dr. Frank M. Huennekens. The concluding remarks of Dr. Huennekens provided an incisive analysis of the individual presentations, focused attention on their imaginative experiments and provocative ideas, and delineated the significant accomplishments of the Symposium in relation to the future of membrane research. Later that evening the participants were honored by a glittering dinner reception at the invitation of Chancellor and Mrs. Nahavandi. Following dinner, Dr. Philip Handler, President of the National Academy of Sciences, U.S.A., delivered the IUB Lecture. The audience was composed of Iranian and international scientists, prominent Iranian statesmen and academicians, diplomatic representatives of all the countries whose scientists had participated in the Symposium, and spouses. The main theme of the lecture was the global problem of overpopulation which threatens the future of mankind. With sharp statistics and eloquent logic Dr. Handler drove home the shattering truth that the world population, now at 4 billion, is increasing at the alarming annual rate of 2%, while food production is near the peak of feasibility and the critical minerals of the earth's crust are becoming rapidly exhausted. This remarkable lecture is reproduced as a special chapter at the beginning of this volume, and is recommended to everyone, both scientist and nonscientist.

As the readers of these proceedings will see, this was a symposium of sterling quality, which was made possible by the generous hospitality of the University of Tehran; the support of IUB and IUPAB; the tireless logistical efforts of the members of the Symposium Committee, especially its chairman, Dr. K. Javaherian; and the elegant scientific contributions of the speakers. Thus, it was a thrilling experience for me to help organize this symposium, and I am grateful for the opportunity, and for the help and support I received from every one concerned. Thanks are also due to all my friends and colleagues who graciously put up with my impatient entreaties for abstracts, manuscripts, etc. The present volume was edited jointly by Dr. L. Djavadi-Ohaniance and myself, but our editorial prerogatives were kept under tight control in order to preserve the individuality and the international flavor of the presentations. The decision to retype all the manuscripts on an IBM Composer was mine. Therefore, any error in restyling of the manuscripts which might be detected in this volume is strictly my fault. My coeditor spent long hours reading the typescripts, and corrected many typographical errors that I had missed. I am duly grateful for her help. Finally a word of thanks must be recorded for my family who supported and encouraged me in the more trying moments which inevitably arise during the course of organizing an international symposium *in absentia.*

Youssef Hatefi

Preface

This book is based on the proceedings of an international symposium on *The Structural Basis of Membrane Function*, which was held May 5-7, 1975, at the University of Tehran. The symposium was composed of six sessions, entitled *Membrane Structure, Transport in Membranes, Energy Transduction in Membranes, Regulatory Functions of Membranes, Excitable Membranes*, and *Microsomal and Related Membranes*.

This volume contains a collection of valuable articles in the frontiers of membrane research. The authors are eminent international scientists, renowned for their significant contributions to our current knowledge of biological membranes. The topics discussed are as follows: structure, function, and biosynthesis of membrane-bound enzymes and enzyme complexes; role and bilayer distribution of phospholipids; structure and mechanism of ion pumps in bacteria and mitochondria; mechanisms of amino acid, sugar, ATP/ADP, and ion transport, and the role of ionophores; structure, thermodynamics, and regulatory features of the mitochondrial energy transduction system; mechanism of energy transduction in mitochondria and chloroplasts as studied by electron microscopy, genetic mutation in yeast, oxygen exchange reactions of the ATP synthesizing system, photoaffinity labeling and resolution–reconstitution of the component enzyme complexes of mitochondria, acid–base transition experiments and determination of $\rightarrow H^+/\sim$ stoichiometry in ATP synthesis and in respiration–coupled calcium ion uptake; regulation of muscle function by the troponin–tropomyosin system and the role of the myosin cross–bridges as studied by rapid X-ray diffraction measurements; a new concept of hormone action as revealed by the effect of aldosterone on membrane lipid metabolism; mechanism of membrane–bound adenylate cyclase stimulation by cholera toxin involving certain novel features of direct interaction of a toxin subunit with adenylate cyclase; the organization of the sodium channels in excitable membranes; purification and properties of various receptor proteins; the mechanisms of agonist–antagonist binding as studied by kinetic measurements and various physicochemical techniques; the liver, adrenal, and bacterial monooxygenase systems, their composition, organization, mechanism, and role in chemical carcinogenesis and lipid peroxidation.

In addition, the volume contains a remarkable article by Philip Handler, President of the National Academy of Sciences (U.S.A.), entitled *Science and the Developing Nations*, and a trenchant *Summary of the Symposium by* Frank M. Huennekens.

It is a pleasure to acknowledge the contributions of Feodor Lynen, Albert L. Lehninger, Edward C. Slater, David E. Green, Humberto Fernández-Morán, and Irwin C. Gunsalus as session chairmen, and of Henry A. Lardy as the chairman of the discussion period.

Youssef Hatefi
Lisa Djavadi-Ohaniance

I
IUB Lecture

SCIENCE AND THE DEVELOPING NATIONS

Philip Handler

President
National Academy of Sciences
U.S.A.

Those of us who have been attending the Symposium have had the high privilege of participating, for three days, in a review of one of the most dynamic frontiers of modern science. For us, for whom the understanding of the molecular basis of life has been a life-long preoccupation, this has been an extraordinarily thrilling intellectual experience. The superb papers to which we have listened have borne witness to the dynamics of our science in a manner which would evoke stirring emotions in any of the world capitals in which this company has been accustomed to meet. That we should be gathered to partake of such rich intellectual fare in the capital of Persia, so close to the cradle of human civilization, where men were inventing agriculture and commerce and poetry when the inhabitants of Northern Europe and the Americas were but savages, has enhanced our appreciation of these modern truths in a most special way. And that appreciation is given an added zest as we observe this ancient land vigorously preparing to take its place in the vanguard of the future of our species.

For three days we have feasted on the latest understandings gained in but one small corner of the scientific endeavor, that addressed to the understanding of life in molecular terms, an endeavor that has been enormously successful in our time, while offering great potential for prevention and alleviation of human disease and for enhancement of the productivity of agriculture, the most ancient and most vital of organized human endeavors.

The dominant impression gained in these few days is the immense progress which our branch of science has made in the 40 years since, as a student, I had my first course in biochemistry. The extent of that progress is evident

Dr. Philip Handler was IUB Lecturer at the reception given by the Chancellor of the University of Tehran in honor of the Symposium participants.

not so much in the sophistication of the answers now available to the most important questions concerning the nature of life — but in the realization that, 40 years ago, we could not even properly formulate the questions themselves. Our program was admirably designed to portray this progress by those who organized this symposium. This is an opportune moment for me, on behalf of all of the visiting scientists gathered here, to convey our thanks to Chancellor Nahavandi for the generous hospitality of the University and to the organizers of the symposium for their logistical efforts and for their great good taste in science.

It is important that we also recognize that the pace of our own area of science is rivaled by the pace of science generally, that our program is but symbolic of the penetrating power of science to reveal the nature of man and his universe. Here on the plains of Persia, men were mapping the stars millenia ago. For four centuries we have been observing those stars with the aid of optical telescopes, while only for two decades have the heavens been observed with sensors that can operate over the full spectrum of electromagnetic radiation. And what a magnificent panorama stands revealed! With the use of X-ray, ultraviolet, radio and infrared detectors, pulsars, quasars, black holes and other unimaginable celestial objects have made themselves evident. The general theory of relativity has received successful experimental tests and both the origin and the future of the universe have been given what appear to be rational explanation — even if the state of the universe at time zero and at some ultimate date in the future defy understanding. The observation of 33 different species of polyatomic molecules in interstellar space render it, once and for all, unnecessary to postulate the necessity of living forms in order to account for the presence of the simple organic chemical molecules characteristic of life.

Physical chemistry and chemical physics have been blended into a single subject. The fine structure of organic chemical molecules and the rules governing their behavior are now understood in such detail that it is possible to predict the nature of the interactions between molecules and the rates at which they will occur, thereby permitting the synthesis, at will, of extraordinarily complex chemical structures. But there is still far to go, e.g., it is not yet possible to predict crystal structure from elementary considerations. The interior of the atomic nucleus has been explored with incredibly powerful tools, revealing an unexpectedly complex and as yet uncomprehended structure in which the operational forces and the particles upon which they operate blend into each other, perhaps to become one and the same.

The voyage of the research ship, Glomar Challenger, climaxed a revolution in the earth sciences which has revealed the nature of huge movements which give rise to the structure of the earth's crust, phenomena now known as continental drift and plate tectonics. That voyage will surely prove historically comparable to that of the HMS Beagle when it carried Darwin to the Galapagos. Advances in solid state physics combined with a new form of applied mathematics made possible the modern digital computer. So profound is its impact that one may reasonably wonder whether, in so doing, man guided his own evolution — in the sense that a human being combined with a computer represents almost a new species.

Truly the accomplishments of the scientific endeavor during the working lifetimes of those gathered in this hall constitute one of the most glorious chapters in the saga of the human mind.

But we must also acknowledge that the current state of our understanding, so immensely richer than in any time past, is the lineal descendant of the threads of learning as these began to form in many times and many places. The torch of learning has alternately burned and dimmed and brightened again in many centers of civilization. Many nations can number their heroes in this great progression. Since the invention of printing, 500 years ago, what is learned cannot again be unlearned. That is both the glory and the curse of the human brain. What we have learned best in this saga is that in our searching, rarely do we find the answers to the questions we pose. Rather do we tug at the veil of ignorance only to glimpse yet another layer of questions and again acknowledge that the work of the scientific endeavor is never finished. Nevertheless, we can say with reasonable conviction that we have convened here during that unique moment in history when scientific understanding of the bold architecture of the animate and inanimate universes has been grasped, even though there remains an immense fabric of detail to be gathered by our intellectual descendants.

Marching in step with this parade of scientific progress has been the growth of a wealth of technologies which have dramatically affected almost every aspect of the lives of those who dwell in what are usually called the "developed" or "industrialized" nations. And, in turn, some of those technologies have been employed to enhance the power of science itself while, at the same time, mitigating the brute condition of mankind, alleviating pain, preventing disease, relieving us of the burden of the back-breaking toil by which mankind has, for eons, fed, clothed and sheltered itself and vastly enriching the realm of the personal experience of each of us.

It is important that, this evening, we recognize that the principal institution which has made this progress possible has been the university as it evolved from its origins in medieval Italy. The genius of the university derives from the invention of the scholarly and scientific disciplines into which knowledge is organized, from firm adherence to the ideal of the university, viz., the dispassionate concern for truth, and from the zealous defense of the two principles of academic freedom as these have emerged over the centuries. Let me remind you of those two canons. The first is the freedom and obligation of the individual professor to seek to extend the bounds of human knowledge and to transmit to his students that knowledge and understanding which has been hard won by objective, rigorous scholarship — however uncomfortable that knowledge may be to the established order. The second canon is the right of the professor to engage in those civil activities that are legal for all citizens without threat to his position within the university. Where those canons have been breached, the torch of learning has dimmed or been extinguished; scientific progress has withered and humanity been denied the advantages of progress.

On the other hand, although science has been a truly international endeavor, neither the pursuit nor the benefits of science have spread equally rapidly across the earth. While science-based technology enriched the lives of peoples in those lands where science flourished, it was slow to find its way to many other nations — for diverse historical reasons upon which we need not dwell this evening.

In recent times, various less developed nations have deliberately embarked upon vigorous programs of national development, attempting to utilize their resources to accelerate the rate of industrialization, accelerate the diffusion of knowledge and of technology, and increase the standard of living of their peoples. Certainly, the scale and pace of this effort in Iran, led by His Imperial Majesty Mohammad Reza Pahlavi, Shahanshah of Iran, is more impressive than in any other so-called developing nation. We are most pleased that His Majesty has utilized this symposium as the occasion on which to express his intention to support the scientific endeavor in Iran and to announce his intention to found a Biochemistry Institute at the University.

From my talks with officials of the Ministry of Science and Higher Education, and especially from my enlightening conversation with Chancellor Nahavandi, it is evident that those planning this effort in Iran begin to understand well the basic nature of the task of development. In considerable degree, technology and technical advice can be purchased abroad. But if "development" is to be sustained and successful, the basic requirements are universal literacy, an adequate general educational system, a lively system of higher education, and an adequate cadre of scientists of your own nation engaged in first-line research. The latter are important not only to contribute to the ever-growing pool of knowledge as an essential ingredient of your culture, but to create a climate in which scientific problem solving is a normal way of life, thereby promoting continuing advance in industry, agriculture, medicine, and government. Such a society will continue to advance toward all of its goals, whereas attempts merely to import technology must necessarily fail in due course. Thus the university, seat of the fundamental research endeavor and repository and transmitter of learning, has become the most critical of all societal institutions, the key to the success of "development."

I cannot pose as an expert in the special problems of development in Iran, but there are some thoughts concerning developing nations generally that I would like to share with you. The principal thought is that the circumstances of the planet, as we now understand them, are such as to make it impossible for each developing nation to pursue its own course, oblivious of the others, or to repeat the historical development of the now industrialized nations. Let me explain.

Of the almost four billion human beings, about one billion live in developed nations with annual average per capita GNP of $2700 and energy consumption equivalent to 5000 kilograms of coal per year. About 2.5 billion people live in poorer nations with an average per capita GNP of $170 and energy consumption equivalent to 300 kilograms of coal. Another half billion people live in nations of more or less intermediate character. However, it is obviously inappropriate to consider the poorer, developing nations in a single category. About 450 million people reside in 20 nations which are now benefitting signally from the export sale of their nonrenewable resources. Obviously, Iran is in that company. Another 60 million people are in small developing nations which, like Japan, are integrating into the world economy by their manufacturing industries. Presumably, the nations of these two classes will, hereafter, be able to purchase food and the products of industrial technology on the world market as they require; their incomes should, in considerable part, sustain their economic development. Perhaps a billion people, including China, live in relatively poor nations which, however, are reasonably self-sufficient with respect to natural resources and are unaffected by the increased world price of petroleum. Although their agriculture is not always sufficient to their needs, these peoples may be expected to purchase food on the world market only in dire emergency situations.

Finally, there are almost a billion people in the fourth world, forty desperately poor countries which are in serious trouble with respect to raw materials, energy and food. These include the Indian subcontinent, Egypt and the Philippines. Whereas malnutrition remains serious in other developing nations which also exhibit rapid population growth, it is this last category of nations which truly occasions alarm, since it seems less than likely that they will be able to manage their problems with their own resources. And it is for some of them, notably India, Bangladesh, and some of the sub-Sahara nations, that this winter has been so disastrous.

One conclusion stands out above all others. The growth of human populations is the principal threat to the survival of our species. If we had started with a single couple the present population could have been attained by about 30 consecutive doublings, which need only have taken one thousand years. In fact, the growth of the world population, proceeded at less than 0.02% per year for most of the few million years since first Homo sapiens trod this planet. The growth rate began to rise noticeably during the 18th Century, increased sharply during the 19th, and became startling only with our own lifetime. For the planet as a whole, that rate now is about 2% annually, being as high as 3.5% in some countries and zero or even negative in only a few. Most of the developed countries have gone through what is called the demographic transition. As industrialization proceeded, then, as a result of a relatively assured food supply and the introduction of only modest hygienic and sanitation measures, there was a precipitate decline in the mortality rate of infants and very young children, with little decline in fertility, for quite some while. During that period, populations in Europe, the United States and Japan grew at an average rate of about 0.9% per year; since then, the fertility rate has declined virtually to match the death rate and these countries are on the way to population stabilization.

In other huge geographic areas, the introduction of the same hygienic measures was not accompanied by the development of industry. The mortality rate declined in varying degree, but there was not initiated that economic development which, by providing jobs, income, better diets, health care, education and improving the status of women, also generates the incentive and motivation for the spontaneous decline in fertility that accompanied development in Europe, the United States and Japan. And so these populations grew. Today, with already swollen populations, continued population growth at national rates of 2 to 3.5% per year constitutes the principal obstacle to true development in most developing nations, probably including Iran. The implications of what may misleadingly seem to be low growth rates – when compared to the more familiar interest rates for the use of money – are not readily appreciated. At 2% growth per year, a population increases sevenfold in a century. At 3.5% the population is doubling with each generation and would increase 32-fold in a century. The current rate in Iran is said to be about 3.0%; if it were to continue, by the year 2000, the population of Iran would be 70-75 million. And there is grave danger that all the potential benefits of development could be lost to the requirements thus posed for housing, jobs, health care and food.

The large mass of the developing world has yet to go through the demographic transition. The human population, now at four billions, is driving toward six to seven billions at the turn of the century and is unlikely to stabilize before world population exceeds ten or twelve billions unless other phenomena intervene – as well they may. Stated most simply, if mankind is to live in the state of material well-being that technology can make possible, then, given the finite size and resources of this planet, there are just too many

of us already. If today all of mankind were to experience the material standard of living of the United States, the rate of extraction of the critical minerals in the earth's crust would have to increase by a factor of seven. That is
impossible today, much less when the world has 12 billion people.

Five years ago, in *Biology and the Future of Man*, I said that "Many of
the most tragic ills of human existence find their origin in population growth.
Hunger, pollution, crime, despoliation of the natural beauty of the planet,
irreversible extermination of countless species of plants and animals, over-
large, dirty, overcrowded cities with their paradoxical loneliness, continual
erosion of limited natural resources and the seething unrest which creates the
political instability that leads to international conflict and war, all derive from
the unbridled growth of human populations...As population growth continues,
inevitably there will be instances in which the food supply to one or another
region becomes, even if temporarily, grossly inadequate and on so large a
scale that organized world food relief will be woefully incommensurate. And
the resultant political instability could have gigantic consequences for all mankind." That statement seems even more appropriate today. All of us have a
stake in reducing population growth, everywhere.

Allow me, then, briefly to consider some of the factors operative in shap-
ing our future. First, some thoughts about world nutrition.

Historically, the food eaten by the people of a given nation has been raised
within its own borders. Food supplies and population grew in parallel by
expansion of the area under cultivation. Until quite recently, agricultural
productivity – that is to say, yield per acre – did not differ greatly among
nations. The adequacy of the food supply, then, depended upon the ratio of
the population to the area under cultivation, and upon the vicissitudes of the
weather. Drought, floods, early rains or freezes, or crop destruction by a
virus or insect, have periodically rendered food supply inadequate in almost
every country with consequent human tragedy. Today, predictions abound
that more people will starve to death in the 20th Century than in any previous century in the history of man. There were about 2 million such deaths
in the 17th Century, 10 million in the 18th Century, perhaps 25 million in
the 19th Century. Despite the remarkable worldwide transportation system
and the prolific yields of modern agriculture – where modern agriculture is
practiced – available indications suggest that the death toll due to starvation
in this century, indeed in the next few years, will set an all-time high. At
this moment, the worldwide death rate from starvation is greater than 10,000
persons per week. More people will starve to death largely because there are
so many more people.

It is estimated that the lives of about 500 million individuals are today
limited by insufficient dietary calories, protein, vitamins or minerals. As pop-

ulations grow that company will surely expand. For them, life is a succession of diseases and an apathetic struggle for survival. It is noteworthy that the character of malnutrition has changed in the last 40-50 years. Apart from occasional famine, classic malnutrition was the consequence of some dietary imbalance which gave rise to a specific deficiency disease, for example, beri beri, scurvy, pellagra, xerophthalmia, rickets, sprue or goiter. Of *that* list, only xerophthalmia remains as a truly major health problem, causing blindness in thousands of children in the Asiatic tropics, although there remain pockets in which almost each of the classic deficiency diseases may be observed. Instead, the major forms of current malnutrition are iron deficiency anemia — widely prevalent in this country — marasmus and kwashiorkor. The latter diseases are the consequence of protein insufficiency in individuals also deprived of an adequate caloric intake, viz., semi-starvation. Whereas the classical forms of malnutrition were in large part the consequence of ignorance, malnutrition today generally reflects lack of food, rather than lack of scientific understanding. In tropical countries these same populations are afflicted with a host of infectious and parasitic diseases to which they are made more susceptible by their malnutrition and which, in turn, frequently deprive them of the nutritional benefit of such food as they may ingest. The reported death rates for malaria, schistosomiasis, hookworm, filariasis, bilharzia, etc., conceal much of the death rate from chronic malnutrition.

A word about food production. Worldwide, primary food production, now about 1.2 billion metric tons of cereal grain annually, has continued to grow more rapidly than has the total human population, roughly 2.5% per year, as against 2%. But the large increases in production have not occurred where populations are growing most rapidly. The great gains in cereal production have occurred where modern energy-intensive agriculture has combined irrigation, pesticides, herbicides, fertilizer, genetics and mechanization to the increase of yields. In effect, modern agriculture utilizes sunlight to transmute fossil fuels into edible crops. Yields per acre on an Iowa farm can be more than six times what they are generally in Pakistan or India; at least three times what they are in Iran.

This great disparity has developed largely during the last four decades. The principal technical basis for this difference will be evident from the fact that, whereas application of mixed fertilizer in the countries of Western Europe frequently exceeds 200 kilograms per hectare, and in the United States averages about 100 kilograms per hectare, the mean for the developing countries is less than 25 kilograms per hectare. In a general way, one ton of fertilizer applied to unfertilized land yields sufficient additional grain to feed 50 to 100 people for one year. But it is in those very countries where yields are still low that there has been such intense pressure on agriculture as the result of continuing population growth.

Meanwhile, the flourishing agriculture of the United States, Canada and Australia have made them the breadbasket of the world. North America now exports an amount of grain equivalent to about 8% of annual world production, and the United States is also the unique exporter of soybeans in quantity. Indeed, last year the United States exported two-thirds of its wheat crop, half of the soybean crop, and 40% of the corn crop. Other industrialized nations with growing economies but without an equivalent agriculture have increasingly developed their appetites for animal protein. Hence, 60% of North American agricultural sales have been to nations whose peoples are already reasonably well fed. Purchases by more affluent nations largely involve coarse grains and soybeans for their livestock, while the developing nations seek wheat, rice and soybeans for their people. The availability of massive quantities of food for shipment across international boundaries constitutes a new historic phenomenon without precedent. In 1961, the grain in storage in the United States plus 50 million acres held out of production in the American soil bank constituted the principal food reserve of the planet, about 95 days' worth in 1961. As we entered this winter, that reserve was down to less than a 30-day supply. But 1974 witnessed another worldwide crop shortfall due to adverse climatic conditions in various countries, including the United States. Last year's crop was 45 million tons less than had been projected, of which the Indian shortfall alone was about 8 million tons. Iran found it necessary to import about one million tons of grain. For lack of reserves, consumption this year, worldwide, must be almost entirely from this year's insufficient production. The result is acute famine in a number of countries, with the prospect of hundreds of thousands of deaths from starvation. What, then, can we say of the future?

1. Essentially all of the readily arable land on the planet is now under cultivation. There are opportunities for opening new lands, but at very high cost, frequently as much as $2000 per hectare. Large areas may seem potentially to be available, but the central plain of Africa remains dominated by the tsetse fly and the lateritic soils of South America convert to desert within a few years after the jungle is cleared. Hence, the principal opportunity for increased agricultural production to keep pace with population growth must be increased production on land already in cultivation.

2. To achieve significant gains in agricultural productivity, agriculture must be viewed as a complex industry which can flourish only with an adequate social and industrial infrastructure. There is required a marketing system, a transportation system, an industry that can supply fertilizer, pesticides and herbicides, large-scale irrigation, mechanical equipment, and a sophisticated program of applied genetics to breed new strains of major crops which can

benefit from the application of the other inputs while yielding maximally under specific local climatic conditions and resisting indigenous pests. This cannot be done by the simple purchase of all of the inputs from outside of the country; each nation must be essentially self-sufficient at all levels. Importantly, none of this infrastructure can be successful in the absence of an adequate educational system to train the diverse skilled individuals required and a continuing relevant research enterprise.

3. As compared to classical agriculture, modern agriculture is heavily dependent upon the input of energy, not merely to run the mechanical equipment, but for the synthesis of the fertilizer and other required chemicals, the operation of irrigation pumps, hauling to and from the market, etc. A significant cause of the failure of Indian agriculture last year was the lack of both fertilizer and fuel for the more than one million diesel powered pumps required for their irrigation arrangements.

But it is abundantly evident that the world will not again know cheap energy. The petroleum and natural gas supplies of the planet are seriously limited. A child born in 1960 will live through that brief historic period when 90% of all the petroleum and natural gas of the planet will have been consumed. Peak production will occur before the year 2000 and inevitably will decline thereafter. Moreover, it is extremely unlikely that the rate of coal production, worldwide, will increase sufficiently rapidly to compensate for declining production of liquid and gaseous fossil fuels in the next century. During that period solar power may find significant use for space heating and cooling, but for little else. Fusion at this time is not even a successful laboratory accomplishment, and one cannot predict when or even whether it will go from the research to the development stage, much less when it can become a significant source of commercial power. Hence, early in the next century, the only truly significant source of energy which might compensate for the disappearing fossil fuels will be fission in conventional or breeder reactors. That these offer a series of environmental and other hazards is well known. But even ignoring all the other associated problems, it is self-evident that the price of energy must continue to rise until that day when – if ever – fusion becomes an industrial reality. And if energy costs rise, food costs must necessarily rise as well, with catastrophic effect upon those populations that already spend 80-90% of personal income for the acquisition of food. If the energy supply is inadequate, then increased food production on the requisite scale becomes impossible in any case. And again with predictable consequences. Should breeder reactors in due course be deemed to be too hazardous for their deployment, then the uranium supply will become limiting and conventional reactors will become incapacitated for lack of fuel within half a century.

4. Even if all the other inputs to the agriculture system are adequate, then water supply must limit agricultural production, a problem no better ever evident anywhere in the world than in Iran. Again, most of the relatively easily managed major irrigation schemes have already been implemented. To increase total irrigated land in a major degree will require enormously extensive and expensive engineering feats. Declining water tables already threaten the water supply to many existing systems. To gather access to major additional water supplies, as by river diversion or desalination of brackish waters in order to make semi-arid lands available for production, will necessitate a huge increase in annual energy expenditure, thereby again driving up the cost of food. The penalty for clearing hillside forests without meticulous terracing can be devastating downstream, flooding land which is already in successful tillage; carefully maintained pasture and range can become a dustbowl when cultivated; irrigation can result in waterlogging, salinization or the leaching of nutrients, depending on the soil and its substrata. And again, there are very few such opportunities in the fourth world.

5. Looming in the background is the possibility of a significant change in the climate of the northern hemisphere. There is a distinct possibility that the recent period of extraordinary proliferation of human populations has coincided with both the availability of cheap energy and the accident of an unusual historic period when the climate of the northern hemisphere was more equable, more suited to supporting human life, than in many centuries, perhaps millenia, before. For whatever reason, the northern hemisphere has been cooling down at the rate of $0.1°$ per decade for four decades. Perhaps that fluctuation represents only the maximum fluctuation from a normal mean. But if the trend continues for another decade or two, we shall know that a pronounced shift in climate is in the making. Were the mean temperature to drop by $2°$ C, the northern hemisphere would enter an ice age. In the early stages this would have different effects in different countries, depending upon the consequences to the rainfall. Canadian surplus wheat production would be abolished; agriculture in all of northern Europe would be most limited. In the United States and in Iran, the initial effect might well be an increase in rainfall with an associated increase in agricultural productivity. If, however, the process were to continue, then agriculture here, as in the United States,.would be drastically limited.

Despite these reasons for grave concern for the future of world agriculture, however, the question is not whether the total potentially arable land of the planet, appropriately managed, could reasonably feed a significantly larger population than at present. Undoubtedly it could. The question is whether the multiplicity of resources required will be gathered and brought to bear within the developing countries, particularly in South Asia, in sufficient time.

The answer to that question largely depends on whether the growth of the world population can be slowed down sufficiently soon. At this time, the peoples of Africa, Asia and Latin America are growing at 2.4, 2.6 and 2.7% per year, respectively. Those figures conceal malevolent growth rates of the order of 3.5% in such countries as Morocco, Algeria, Rhodesia, Syria, Iraq, Colombia, El Salvador, Venezuela, Pakistan, the Philippines and Thailand, as well as numerous instances, such as Iran and Mexico, where growth is about 3% per year. The magnitude of the task of braking this process will be evident from the fact that on those three continents, about 40% of the population is under 15 years of age, so that there is an enormous built-in pressure for future growth, and from the fact that the average human female worldwide experiences 4.7 births. In India alone that figure is 5.6. That pattern will persist until the peoples of those nations believe that virtually all of their children will survive to adulthood and that there will be reasonable means for their own support in their older years. Otherwise they will live with the ancient traditions that dictate that two sons are required to support you in your old age and that since half the children die, eight births are required to assure your own welfare in later years. Even a country such as Iran, with its improving economic prospects, could find its hopes for progress lost to the new mouths.

It has been amply demonstrated that it is difficult to sustain a program of family planning among a population that is hungry, ill, poor or without hope. Representatives of several nations argued at the Bucharest World Population Conference that emphasis need only be given to economic development and that the population problem would then take care of itself. Were the populations in question much smaller and growing at lesser rates, perhaps such an approach would be acceptable today. Certainly achieving the demographic transition in the developing world would then be attended with far less misery than must now be the case. But the history of the European nations and of Japan can no longer be an acceptable model for the future of the developing nations. Populations are already immense and ultimate populations – assuming that there will be food to feed them – will be horrendous and the quality of life inversely proportional. Accordingly, one cannot await the spontaneous attainment of demographic transition as a consequence of economic development. While agricultural, economic and social development are pursued, a simultaneous, aggressive population control program is imperative. Delay in addressing these problems magnifies the task; sufficient delay may render that task essentially impossible.

What is required is the full effort necessary to bring each nation to an acceptable level of self-sufficiency, to develop its educational and health care systems and to develop the full infrastructure of industry, transportation, research facilities, etc., necessary both to upgrade agricultural productivity and increase real per capita income for the poorer citizens of each country. Only by so doing can one expect that decline in fertility rather than increase in the death rate will limit the growth of population.

There are those who believe that there is a burden upon the United States, which is the major producer of exportable grain, to mount a markedly enlarged program of food aid to the developing nations, particularly those of the fourth world. But because of the nature of modern agriculture, that places the United States in the position of importing petroleum in order to donate food, an arrangement which we might undertake briefly, but not for long. Food aid, as an emergency measure, will surely be provided, in keeping with the American tradition. But it cannot be the cornerstone of either American or world policy. We must return to the time when, in large measure, each nation can grow enough food to provide for the basic food needs of its own people; the surplus production of a few nations such as the United States could then be utilized for emergencies whenever these may occur – and they will occur ever more frequently due to the inevitability of population growth – and whatever is left over can then be used for enrichment of the diets of those nations which can so afford. But, in any case, all such food must somehow be paid for by international arrangements. The United States will gladly participate in such arrangements, as well as in a massive effort to assist the agricultural development of the developing nations.

If such a program were successful, and if mankind can manage to avoid war, then perhaps in the middle of the next century, worldwide agreement to negative population growth might be contemplated. I cannot believe that the principal objective of humanity is to conduct an experiment to establish exactly how many human beings this planet can just barely sustain. But I can imagine a better world in which a limited population can live in abundance, free to explore the full extent of man's imagination and spirit. It is not clear that such a program as I propose can succeed. But if it is not attempted, if the peoples of the developing nations do not begin to achieve a decent standard of living, it is hard to imagine a stable worldwide peace. International agreement to a world plan for global development has become imperative.

Unfortunately, the political world is not yet organized to chart a response to mankind's changing circumstances. The forces of nationalism grow ever more powerful while the sense of a worldwide community of man seems to be disintegrating. Yet it has been said that "No society can long flourish on a course divergent with reality." Instead of the imperative of international cooperation and understanding, the condition of man becomes ever more perilous as our numbers increase, food and energy supplies become less certain, our own activities threaten our life-sustaining environment, and the world arsenal of nuclear and conventional weapons proliferates.

Hence, I have difficulty facing the future with equanimity. This is bitter gall, indeed. With my fellow scientists I was enraptured by the beautiful panorama of understanding offered by science in our time. Science-based technology has provided a remarkable cornucopia of ideas, materials, and machines that offered to free man to be human. In the developed nations, advances in medicine have rendered most of life free of pain and infirmity, and one may be confident that progress in the capabilities of medicine will continue. The vision of a truly brave new world is there before us. But the planet is small, and there are too many of us.

I continue to be a technological optimist, to believe that research and development will go far to mitigate our circumstances — if there is time enough. For the worldwide research-performing community there are a thousand challenges. None are new, perhaps, but they take on added significance in the light of our understanding of the human condition. Needed, *inter alia*, is the knowledge with which to devise a truly effective, safe, cheap, reversible means of contraception; to alter climate beneficially; to expand and conserve water resources; to recycle minerals and reduce waste; to increase the productivity of agriculture while minimizing the use of energy; to expand the total area of arable land; to maximize the nutritional value of crops in each region; to store grains and foodstuffs with adequate protection; to control agricultural pests and the parasites to which man is subject; to increase the number of plant and animal species useful for human consumption; to attempt diverse innovative means of expanding food production; to conserve, generate and more efficiently utilize energy; to reduce and contain human conflict; to protect the environment, including the diverse species with which we cohabit this planet. And all that will be possible only if fundamental research in all scientific disciplines is pursued vigorously. It is difficult, however, to believe that all of this will happen sufficiently soon to manage the immense problems now evident.

As we face this future, it would be well to be mindful of a statement made in quite another context: "To look for solutions to these difficult questions is profoundly to misunderstand their nature. The quest is not to solve but to diminish, not to cure but to manage; and it is this hard truth that makes so many frustrated, for it takes great courage to surrender a belief in the existence of total solutions without also surrendering the ability to care."

The fruits of science did much to make our civilization worthwhile; only enlightened political leadership can preserve that civilization and permit science to provide a richer tomorrow for mankind. Withal, the quality of life for our descendants will be determined primarily by how many there will be.

II
Membrane Structure

MEMBRANE STRUCTURE: INTRODUCTORY REMARKS

F. Lynen

Max-Planck-Institut für Biochemie
D-8033 Martinsried
Federal Republic of Germany

The first session of our Symposium is devoted to Membrane Structure, in other words to the architecture of membranes. It seems to be very logical to have this discussion at the beginning because the many functions of membranes, for example the separation of the cellular interior from the outside or the compartmentation of the cellular interior, are most intimately connected to the membrane structure and its composition of lipids and proteins. The former constituents being essential as a barrier between two aqueous phases and the latter being involved in the specific functions of the membranes, such as membrane transport or membrane receptor functions.

The first model of membrane structure, proposed by Davson and Danielli, already had the lipid bilayer with attached proteins as the basic element. Later this model was modified to indicate that proteins are not only attached to the lipid bilayer but are also embedded into it. At present the proteins associated with membranes may be classified into two broad categories, termed peripheral and integral. Peripheral proteins are those that appear to be only weakly bound to their respective membranes and do not appear to interact with the membrane lipids, whereas the integral proteins are ordinarily more strongly bound to the membrane and exhibit functionally important interactions with the membrane lipids. From this classification it appears that peripheral and integral proteins are attached to the membrane in distinctly different ways.

It was calculated that about 70-80% of membrane proteins are characterized as integral and those include most membrane-associated enzymes, antigenic proteins, transport proteins, and finally drug and hormone receptors. One of the characteristics of enzymes that are integral proteins can be seen in the fact that they usually require lipids, sometimes even specific lipids, for their activities to be expressed.

The molecules of integral proteins are assumed to be more or less globular and amphipathic; that is, their folded three-dimensional structures are segmented into hydrophilic and hydrophobic ends, with the hydrophilic ends protruding from the membrane into the aqueous phase and the hydrophobic ends embedded within the nonpolar interior of the lipid bilayer. Within this picture the possibility also exists that a molecule of an integral protein might span the membrane protruding from both surfaces if it had two hydrophilic ends separated by an appropriate hydrophobic middle segment. If integral proteins, by analogy with multisubunit soluble proteins, exist as specific subunit aggregates in membranes and span the membranes, water-filled pores through the membrane may be generated, one of the prerequisites for some transport processes through membranes.

Many of these structural proposals are supported by electron microscopy, utilizing the technique of freeze-fracture-etching, by chemical labeling methods including the iodine-125-lactoperoxidase technique, and by proteolytic cleavage experiments.

Turning now to the lipid constituents of membranes, it should be emphasized that most biological membranes can be considered as two-dimensional fluid structures. One direct argument supporting this notion comes from the comparative study of the so-called order-disorder lipid phase transitions in both model systems of defined lipid components and several biological membranes. As a function of temperature the lipids undergo a highly cooperative transition from a "fluid," disordered state to a "crystalline," ordered state. The temperature, at which the transition takes place, is dependent on the chain length and type of functional groups of the hydrocarbon chains, the structure of the polar groups and the water content. In biological membranes the lipid molecules are normally in the disordered state and they rapidly move by lateral diffusion. As a direct consequence, membrane associated proteins can rotate or diffuse in the plane of the membrane.

At this point I would like to finish my introductory remarks about the structure of membranes, and to call on the first speaker of our session.

PHOSPHOLIPASES AND MONOLAYERS AS TOOLS IN STUDIES ON MEMBRANE STRUCTURE

L.L.M. van Deenen, R.A. Demel, W.S.M. Geurts van Kessel, H.H. Kamp,
B. Roelofsen, A.J. Verkleij, K.W.A. Wirtz and R.F.A. Zwaal

Laboratory of Biochemistry
University of Utrecht
Utrecht, The Netherlands

INTRODUCTION

Presently, it is generally accepted that in most biological membranes a significant part of the lipids is arranged in a bilayer. A lipid bilayer structure was suggested as early as 1925 by Gorter and Grendel (1) for the erythrocyte membrane. The first sentence of their paper reads: "We propose to demonstrate in this paper that the chromocytes of different animals are covered by a layer of lipoids just two molecules thick." This model was derived by measuring the surface occupied at the air/water interface by a monomolecular film of the erythrocyte lipids (Fig. 1). The lipids were extracted from erythrocytes by large quantities of acetone and were then spread from a solution of benzene onto a Langmuir-Adam trough. The surface area occupied by the red cell lipids was measured at a surface pressure of about 2 dynes per cm. The results were found: "...to fit in well with the supposition that the chromocytes are covered by a layer of fatty substances that is two molecules thick." The merits and possible pitfalls of this ingeneous and daring approach have been discussed previously (2,3). It is not our intention to review the large body of new evidence which favours the presence of lipid bilayers in erythrocytes and other membranes. In order to commemorate the pioneering studies published 50 years ago by our countrymen, we will summarize some recent work concerning the topography of phospholipids in the erythrocyte membrane.

Action of certain pure phospholipases on the outside of human erythrocytes demonstrated that about 50% of the phospholipids is localized in the outer (mono) layer of the membrane. By combining action of pure phospho-

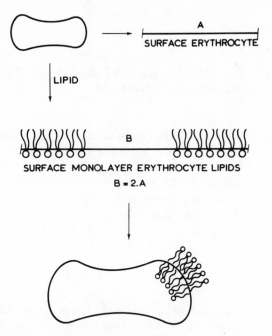

Fig. 1. Experiment of Gorter and Grendel (1925).

lipases with freeze-etch electron microscopy, it could be demonstrated that an asymmetric distribution of phospholipids exists in this membrane. The choline-containing phospholipids, lecithin and sphingomyelin, appear to be concentrated mostly in the outer half of the bilayer, whereas phosphatidylethanolamine and phosphatidylserine prefer a localization at the inner surface of the membrane. A number of phospholipases can act on the phospholipids on the exterior of the intact erythrocytes, but other phospholipases fail to attack their substrates in the outer lipid layer of the intact membrane. To understand this difference in behaviour, a comparative study was made on the action of a series of phospholipases on monomolecular films of phospholipids adjusted at various surface pressures. The distribution in the action of phospholipases on intact erythrocytes appears to be related to the ability of the enzymes to hydrolyse phospholipids in monomolecular films above a given surface pressure. This relationship permits one to deduce a value for the tightness of packing of the lipid molecules in the outer lipid bilayer of the erythrocyte membrane, which parameter is of interest with respect to the classic experiments of Gorter and Grendel.

In order to give an additional demonstration of the usefulness of mono-molecular films and phospholipases in membrane research, some studies are included on a protein which catalyses the transfer of phospholipids between membranes. This pure protein from beef liver was demonstrated to function as a specific carrier for phosphatidylcholine between two separated monomolecular films of phospholipid. One molecule of phosphatidylcholine is bound to one molecule of protein, and treatment with phospholipases under different conditions gave information about the nature of association between phospholipid and protein.

TOPOGRAPHY OF PHOSPHOLIPIDS IN THE HUMAN ERYTHROCYTE MEMBRANE AS DETERMINED BY PHOSPHOLIPASE ACTION

The lipids of the human erythrocyte membrane have been studied in great detail (4). The major phosphoglycerides are phosphatidylcholine (PC), phosphatidylethanolamine (PE) and phosphatidylserine (PS); in addition, sphingomyelin (SM) is abundant. Using labelling of intact erythrocytes and ghosts with relatively non-permeant reagents, it was concluded that phosphatidylserine and phosphatidylethanolamine are mainly located at the inside of the human erythrocyte membrane (5,6). Because certain phospholipases can act on the outside of the membrane without causing lysis of the cells, one can obtain information about a possible non-uniform distribution of phospholipid classes between the exterior and interior of the lipid core, perhaps in a more direct way (7-10).

Experiments with phospholipase A_2 (phosphatidylacyl hydrolase) from different sources are compiled in Table I. These four pure or partially purified phospholipases do not cause any haemolysis. The enzymes from pig pancreas and *Crotalus adamanteus* do not catalyze any hydrolysis of phospholipids in the intact erythrocytes, but a complete hydrolysis of the phosphoglycerides occurs in unsealed ghosts permitting the phospholipases to attack the substrates from both sides of the (modified) membrane (cf. Fig. 2). By contrast, phospholipase A_2 from *Naja naja* and bee venom can hydrolyse phospholipids present in intact human erythrocytes. It is important to note that primarily phosphatidylcholine is attacked, while in ghosts preparations all phosphoglycerides are fully degraded (Table I). It has to be mentioned that the cells after degradation of some 20% of the total phospholipid fraction revealed significant decrease in osmotic resistance. The preferential degradation of phosphatidylcholine by phospholipase A_2 from *Naja naja* and bee venom in intact erythrocytes, when compared with the complete breakdown of all phosphoglycerides in non-sealed erythrocyte ghosts, already suggest a non-random distribution of the various types of phosphoglycerides in the membrane.

In the experiments described above the phospholipase attack on the intact cells probably is confined to hydrolysis of phosphoglycerides located at the outside of the membranes, whereas in the open ghosts the phosphoglycerides at

TABLE I

Action of Phospholipase A$_2$ from Different Sources on Intact Human Erythrocytes and Ghosts

Source (phospholipase)	% Degradation					
	Total PL	PC	PE	PS	SM	Haemolysis
Intact Erythrocytes						
Pig pancreas	0	0	0	0	0	–
Crotalus adamanteus	0	0	0	0	0	–
Naja naja	20	68	0	0	0	–
Bee venom	19	55	9	0	0	–
Ghosts						
Pig pancreas	70	100	100	100	0	
Crotalus adamanteus	70	100	100	100	0	
Naja naja	70	100	100	100	0	
Bee venom	70	100	100	100	0	

PL: phospholipid; PC, phosphatidylcholine; PE, phosphatidylethanolamine; PS, phosphatidylserine; SM, sphingomyelin.

both sides of the membrane can be degraded (Fig. 2). Additional information on a possible asymmetric phospholipid distribution may be obtained by studying the action of a phospholipase at the inside of the red cell membrane. The technique of resealing ghosts (11,12) makes it possible to trap a phospholipase and to start the enzymatic reaction by the addition of a cofactor (10). It is essential that no phospholipid hydrolysis occurs during the resealing process. Pancreatic phospholipase A$_2$ was trapped in the presence of EDTA, without phospholipid hydrolysis occurring during the resealing. In a control experiment it was established that pancreatic phospholipase A$_2$ is unable to give phospholipid hydrolysis when added to the ghosts after the resealing procedure. Activation of the trapped phospholipase A$_2$ was achieved by the addition of Ca^{++}. Although this pancreatic phospholipase A$_2$ is not capable of acting on the phosphoglycerides at the exterior of the membrane, the enzyme attacked the phosphoglycerides at the interior side. About 30-35% of the total red cell phospholipase of the membrane was degraded by the pancreatic phospholipase

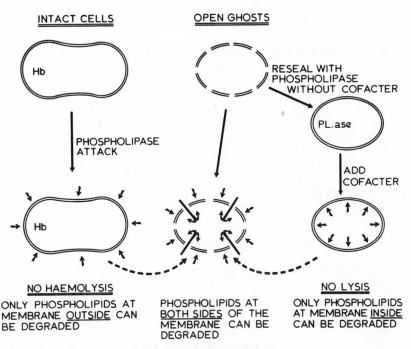

INTACT CELLS

OPEN GHOSTS

Hb

RESEAL WITH
PHOSPHOLIPASE
WITHOUT COFACTER

PL.ase

PHOSPHOLIPASE
ATTACK

ADD
COFACTER

Hb

NO HAEMOLYSIS
ONLY PHOSPHOLIPIDS AT
MEMBRANE OUTSIDE CAN
BE DEGRADED

PHOSPHOLIPIDS AT
BOTH SIDES OF THE
MEMBRANE CAN BE
DEGRADED

NO LYSIS
ONLY PHOSPHOLIPIDS
AT MEMBRANE INSIDE
CAN BE DEGRADED

Fig. 2. Action of phospholipases on erythrocytes and ghosts.

A_2 trapped within the resealed ghosts before the cells started to lyse. At this stage about 65% of phosphatidylserine, 50% of phosphatidylethanolamine, and 25% of phosphatidylcholine were found to be converted into their respective lyso-derivatives (10). This result is compatible with the concept of a non-random phosphoglyceride distribution in the human erythrocyte membrane. When considering the results obtained with phospholipase A_2 acting on the outside and on the inside, the most simple explanation appears to be that phosphatidylcholine is located dominantly at the exterior, whereas phosphatidylserine and to a somewhat lesser extent phosphatidylethanolamine are present at the interior side of the membrane. More quantitative data on the phospholipid distribution were obtained through the use of (combinations of) other phospholipases.

Intriguing differences were observed in the behaviour of various phospholipases which act on the polar headgroups of phospholipids. Pure phospholipase C from *Bacillus cereus* (hydrolyzing the phosphoryl-glycerol linkage) produces no haemolysis and no degradation of phosphoglycerides in intact erythrocytes (Table II). That the failure of this enzyme to hydrolyse the erythrocyte phospholipids in intact cells is not due to limitations in substrate specificity is demonstrated by the extensive degradation of phosphoglycerides

TABLE II

Action of Phospholipase C Sphingomyelinase and Phospholipase D
on Intact Human Erythrocytes and Ghosts

	% Degradation					
	Total PL	PC	PE	PS	SM	Haemolysis
Intact Erythrocytes						
Phospholipase C						
Bacillus cereus	0	0	0	0	0	–
*Clostridium welchii**	80	100	90	0	100	+
Phospholipase D	0	0	0	0	0	–
Sphingomyelinase	20	0	0	0	85	–
Ghosts						
Phospholipase C						
Bacillus cereus	71	100	100	95	0	
Clostridium welchii	80	100	90	0	100	
Phospholipase D	70	100	90	85	0	
Sphingomyelinase	25	0	0	0	100	

* **Formation of ghosts.**

in ghost preparations. Highly purified phospholipase C from *Clostridium welchii* appears to be the only single lipolytic enzyme used to date in our studies which haemolyses human erythrocytes. This action is believed not to be caused by possible contaminations (10). The enzyme catalyses hydrolysis of sphingomyelin, phosphatidylcholine and phosphatidylethanolamine, but not of phosphatidylserine of the human erythrocyte membrane (Table II).

Phospholipase D purified from cabbage (which enzyme splits off the nitrogenous moiety from phosphoglycerides) when incubated with intact cells, turned out to be incapable of giving any detectable hydrolysis of the membrane phosphoglycerides (10)(Table II). Action of this enzyme on erythrocyte ghosts resulted in a considerable breakdown of the phosphoglycerides.

Sphingomyelinase from *Staphylococcus aureus* appeared to give important information about the phospholipid distribution in erythrocytes (Table II). Action of this enzyme on intact erythrocytes does not cause lysis, although 85%

of sphingomyelin is hydrolyzed (7). Action of this sphingomyelinase on erythrocyte ghosts gives 100% hydrolysis of its substrate under the formation of ceramide and phosphorylcholine. The conclusion that about 85% of the sphingomyelin is located in the outer monolayer of the membrane was supported by freeze-etching studies (8).

Sphingomyelinase action on intact erythrocytes produced the formation of small spheres, probably ceramide droplets, with diameters of 75 Å and 200 Å, which were localized on the outer fracture face with corresponding pits on the inner fracture face (Fig. 3).

Of particular interest are the results obtained by incubation with combination of phospholipases, or with consecutive addition of the various enzymes (Table III). The combined action of sphingomyelinase (*S. aureus*) and phospholipase C (*B. cereus*) appears to be lytic for human erythrocytes, by contrast to the action of the individual enzymes. A nearly complete hydrolysis of all phospholipids occurs (Table III). The haemolytic and phospholipolytic action of this combination is rather similar to that of phospholipase C from *Cl. welchii* (Table II). The simultaneous formation of both ceramides and diglycerides in the outer layer of the membrane apparently triggers a haemolytic process. The deformations involve the formation of both diglyceride and ceramide droplets, and the phospholipids at the inner surface of the membrane become consecutively accessible to the enzyme(s). In this context it is of interest to note that when cells are first incubated with phospholipase A_2 and then treated with sphingomyelinase and phospholipase C, no significant haemolysis occurs (8).

Omitting a complete discussion of the various other phospholipase combinations applied (9,10), it is important to consider the results obtained by the combination of phospholipase A_2 from *Naja naja* and sphingomyelinase from *S. aureus*. As described above (Table I) the phospholipase A_2 alone attacks phosphatidylcholine only, thereby converting 68% of this compound into free fatty acid and monoacyl glycerophosphorylcholine. This treatment accounts for a breakdown of 20% of the total phospholipid without cell lysis. Sphingomyelinase treatment of intact cells gives (without lysis) a degradation of 85% of the sphingomyelin, which represents again some 20% of the total phospholipid of the membrane. Phospholipase A_2 incubated erythrocytes are not lysed by the subsequent addition of sphingomyelinase (8). The chemical analysis (Table III) of the osmotically fragile, but intact cells treated in this manner, revealed that the addition of sphingomyelinase not only results in the hydrolysis of 82% of sphingomyelin, but also allows phospholipase A_2 (which is still present in the incubation mixture) to hydrolyse 20% of phosphatidylethanolamine and to attack another 8% of phosphatidylcholine. The total quantity of phospholipids degraded in intact cells is equal to about 48% of the hydrolysis caused by this treatment of erythrocyte ghosts. In the intact cells no phosphatidylserine is degraded although this compound is rapidly hydrolyzed when ghosts are treated

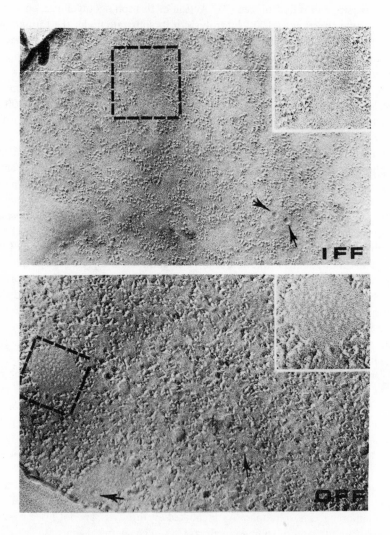

Fig. 3. Inner fracture face (IFF) and outer fracture face (OFF) of the intact red cell membrane after treatment with sphingomyelinase. Notice the characteristic areas with pits (75 Å diameter) on the IFF and corresponding spheres on the OFF (see also insets). Arrows indicate pits and spheres of 200 Å in diameter.

with phospholipase A_2. Electron microscopy, after incubation of intact red cells with phospholipase A_2 and sphingomyelinase successively demonstrated

TABLE III

Combined Action of Some Phospholipases
on Intact Human Erythrocytes

	% Degradation					
	Total PL	PC	PE	PS	SM	Haemolysis
Intact Erythrocytes						
Phospholipase C *Bacillus cereus* plus Sphingomyelinase *Staphylococcus aureus*	98-100	100	100	95	100	+
Phospholipase A$_2$ *Naja naja* plus Sphingomyelinase *Staphylococcus aureus*	48	76	20	0	82	−
Ghosts						
Phospholipase A$_2$ *Naja naja* plus Sphingomyelinase *Staphylococcus aureus*	100	100	100	100	100	

the occurrence of spheres and pits on the outer and inner fracture face, respectively, which are comparable to the results shown in Fig. 3.

Some conclusions

By selecting particular combinations of phospholipases, it is possible to achieve a non-lytic hydrolysis of nearly 50% of the phospholipids of the intact human erythrocyte. This action of phospholipases on the exterior of the erythrocyte causes the formation of spheres in the outer fracture face. Hence, it is plausible that about half of the membrane phospholipids are located in the outer (mono) layer of the (lipoprotein) membrane. The action of a series of

phospholipases on intact erythrocytes, when compared to enzymatic hydrolysis of various phospholipid classes in open ghosts demonstrated that a non-
random distribution of phospholipids in this membrane exists. This conclusion
was supported by results obtained on the action of a phospholipase trapped inside resealed ghosts. The results obtained to date suggest the following phospholipid distribution for human erythrocytes:

     Outer layer: 75% phosphatidylcholine; 20% phosphatidylethanolamine;
                   85% sphingomyelin.
     Inner layer: 25% phosphatidylcholine; 80% phosphatidylethanolamine;
                   15% sphingomyelin; 100% phosphatidylserine.

Some comments and questions

Even if one avoids the complications generated by the use of impure phospholipases the action of these enzymes on erythrocytes produces a rather complex picture. Some phospholipases do not hydrolyse any phospholipid of the
membrane and do not induce lysis; other enzymes or combinations of enzymes
cause a partial hydrolysis of phospholipids without causing lysis; one enzyme
and several combinations cause lysis and simultaneously extensive or complete
breakdown of the membrane phospholipids. For the interpretation of the
results, the substrate specificity of the various phospholipases and the proposed
asymmetric phospholipid distribution has to be taken into account. Some suggestions were made to account for the distinction between lytic and non-lytic
action of pure phospholipases in erythrocytes. However, more work will be
necessary in order to arrive at more solid conclusions. Another question concerns the observation that a number of phospholipases are not capable of attacking their substrates from the exterior of the membrane, whereas others can find
their target. A shielding of phospholipids by proteins may be involved which
prevents the access of certain phospholipases to their substrates. On the other
hand, it was demonstrated that the lipid packing in a monolayer is critical for
the action of lipolytic enzymes (for a review cf. Dawson (13)). This leads us to
the following topic.

HYDROLYSIS OF PHOSPHOLIPID MONOLAYERS BY PHOSPHOLIPASES IN RELATION TO THE LIPID PACKING IN THE HUMAN ERYTHROCYTE MEMBRANE

Because some phospholipases can attack their substrates in the intact erythrocyte membrane, whereas other phospholipases remain inert, it was of interest
to study the action of these enzymes in a systematic manner on monomolecular
films of phospholipids (14). Both the chemical composition and the degree of

molecular packing of the phospholipids in monomolecular layers can be well controlled and the extent of enzymatic hydrolysis can be monitored by measurement of changes of surface pressure and surface radioactivity (see also Dawson (13)). As demonstrated in Fig. 4, pure phospholipase C from *B. cereus* injected underneath a film of a synthetic phosphatidylcholine containing a ^{14}C-methyl group in the choline moiety causes a decrease in surface radioactivity, which is due to release of labelled phosphorylcholine into the subphase. The conversion of phosphatidylcholine into diglyceride is accompanied by a simultaneous decrease in surface pressure. These results were obtained at an initial surface pressure of the phospholipid film of 30 dynes per cm. However, at an initial surface pressure of 35 dynes per cm, no enzymatic hydrolysis of phosphatidylcholine is detectable (Fig. 4). Further detailed studies demonstrated that at initial film pressures below 30 dynes per cm rapid hydrolysis of phosphatidylcholine by the phospholipase C occurs, but that at a pressure of 31 dynes per cm the enzymatic hydrolysis is completely inhibited (Fig. 5). Inasmuch as the outer layer of the human erythrocyte consists of phosphatidylcholine and sphingomyelin mainly, the effect of the insertion of the latter phospholipid into the monolayer was investigated. Using a ratio of phosphatidylcholine-sphingomyelin as present in the natural membrane, it was found that

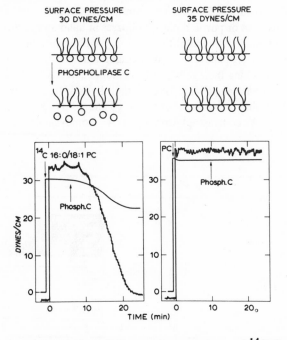

Fig. 4. Action of phospholipase C (B. cereus) on monolayers of ^{14}C-phosphatidylcholine.

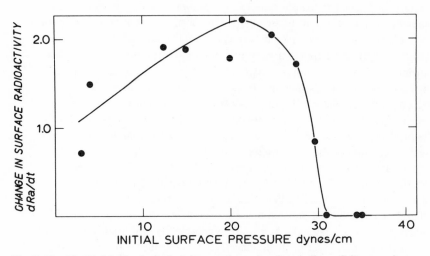

Fig. 5. Phosphatidylcholine hydrolysis in monolayers by phospholipase C (B. cereus) at various initial pressures.

the maximum surface pressure at which enzymatic degradation is possible was shifted only slightly to about 32 dynes/cm. Similar experiments were carried out with the various pure or purified phospholipases at hand and significant differences were observed in their ability to hydrolyse phospholipid monolayers. Phospholipase A_2 from pig pancreas and *Crotalus adamanteus*, which enzymes are not capable of attacking their substrates when added at the exterior of the human erythrocyte membrane (Table I), drop their hydrolytic activity on phosphatidylcholine films at pressures of about 16 and 23 dynes per cm, respectively (Fig. 6). By contrast, phospholipase A_2 from *Naja naja* and bee

PHOSPHOLIPASES		ERYTHROCYTES	MONOLAYERS
PHOSPHOLIPASE A_2	PIG PANCREAS	-	16.5 DYN/CM
PHOSPHOLIPASE D	CABBAGE	-	20.5
PHOSPHOLIPASE A_2	CROT. ADAMANTEUS	-	23
PHOSPHOLIPASE C	B. CEREUS	-	31
PHOSPHOLIPASE A_2	NAJA NAJA	+	34.8
↓ ↓	BEE VENOM	+	35.3
SPHINGOMYELINASE	S. AUREUS	+	>40
PHOSPHOLIPASE C	CL. WELCHII	+	>40
PHOSPHOLIPASE C	B. CEREUS +		
SPHINGOMYELINASE	S. AUREUS	+	>40

Fig. 6. Hydrolytic action of pure phospholipases on erythrocytes and monolayers.

venom, known to hydrolyse phosphatidylcholine in the intact erythrocyte membrane (Table I), were capable of acting on monolayers at pressures up to about 35 dynes per cm. Phospholipase D from cabbage is not active on the outside of intact erythrocytes, and revealed on monomolecular films of synthetic phosphatidylcholine hydrolytic activity at surface pressures below 20 dynes per cm only. Sphingomyelinase from *S. aureus*, which readily hydrolyses its substrate in the intact human erythrocyte membrane, appeared to be capable of degrading (radioactive) sphingomyelin at all film pressures attainable. This was true also in mixed films of sphingomyelin and phosphatidylcholine.

On the basis of these results, one is inclined to conclude that the ability of a phospholipase to penetrate and hydrolyse a phospholipid monolayer above a certain critical surface pressure is a good indication of its possible action on the exterior of the human erythrocyte. The results suggest that only those phospholipases which can act on phosphatidylcholine or mixed phosphatidylcholine-sphingomyelin monolayers above 30-32 dynes per cm have success in meeting and hydrolyzing their substrates in the outer monolayer of the human erythrocyte.

To a certain degree one can mimic the behaviour of combinations of phospholipases on erythrocytes in monomolecular films. Phospholipase C (*B. cereus*) does not act on intact erythrocytes and is not capable of hydrolysing phosphatidylcholine in a single or in mixed phosphatidylcholine films at pressures above 32 dynes per cm. However, when sphingomyelin in the film is degraded either first or simultaneously by sphingomyelinase from *S. aureus*, thereby producing ceramides, it appears that phospholipase C is at once capable of hydrolysing phosphatidylcholine even at a film pressure of 38 dynes per cm (Fig. 6). It was demonstrated above (Table III) that on the intact erythrocyte sphingomyelinase action, which is in itself not lytic, allows phospholipase C to exert its activity, and the combined effort of the enzymes leads to cell lysis and 100% hydrolysis of the major membrane phospholipids. A somewhat related behaviour was observed with purified phospholipase C from *Cl. welchii*, which enzyme alone is active on intact erythrocytes (Table II). In the model system, single phosphatidylcholine films were not hydrolysed at pressures above 28 dynes per cm, but sphingomyelin monolayers were found to be attacked at the highest surface pressures attainable. Mixtures of phosphatidylcholine and sphingomyelin appeared to facilitate the hydrolysis of the former phospholipid and allowed hydrolysis at a surface pressure of about 32 dynes per cm. Because of limits of space, it is not possible to present in this paper the effect of cholesterol or to discuss the action of phospholipases on monolayers of phosphatidylserine and phosphatidylethanolamine, which topic is important for the action of phospholipase at the inner side of the membrane (10,15).

Some conclusions

Although it is realized that several oversimplifications are made when comparing observations made on a complex natural membrane with those on phospholipid monolayers, the results obtained to date support the view that the ability of a phospholipase to act or not to act on the natural membrane depends on its capability to attack a rather tightly compressed lipid layer. We are tempted to conclude that the packing of the phospholipids in the outer layer of the human erythrocytes is comparable to a film with a lateral pressure between 31 and 35 dynes per cm. The results obtained with pure phospholipases support the view that the polar headgroups of phospholipids are at least in part readily available at the surface of the human erythrocyte membrane. The following section will confront us with a lipid-protein structure which is different in this respect.

LIPID–PROTEIN INTERACTION AND MODE OF ACTION OF A PHOSPHO-LIPID CARRIER PROTEIN

Several groups of investigators have detected and purified proteins of the cytosol from animal and plant cells which catalyse the transfer of phospholipids between membranes; for a recent review cf. Wirtz (16). Recently, a protein was isolated from beef liver (7) which specifically stimulates the exchange of phosphatidylcholine. The specificity of the protein was demonstrated in the exchange of phosphatidylcholine between microsomes and mitochondria, between these membrane fractions and liposomes, and between liposomes of different composition (cf. Fig. 7). The protein, which has no distinct pH optimum,

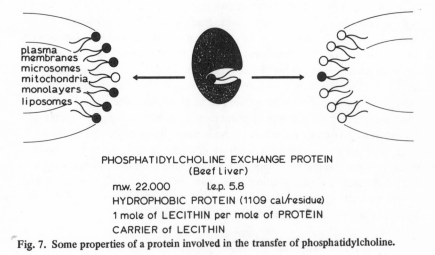

PHOSPHATIDYLCHOLINE EXCHANGE PROTEIN
(Beef Liver)
m.w. 22.000   l.e.p. 5.8
HYDROPHOBIC PROTEIN (1109 cal/residue)
1 mole of LECITHIN per mole of PROTEIN
CARRIER of LECITHIN

Fig. 7. Some properties of a protein involved in the transfer of phosphatidylcholine.

has an isoelectric point of 5.8. It consists of 190 amino acids, of which 38% contain charged residues and 38% contain non-polar side-chains; a disulphide bridge is present and the N-terminal amino acid was found to be glutamic acid. Lipid analyses on non-lyophilized protein demonstrated that the protein contains one equivalent of phosphatidylcholine. This phospholipid-protein ratio was found when the protein was injected underneath a monomolecular film of ^{14}C-labelled phosphatidylcholine (18). Under these conditions, the surface radioactivity decreased and the surface pressure remained constant, indicating that the phosphatidylcholine bound to the protein can exchange with the molecules present at the air-water interface. The ^{14}C-label could be detected in the protein collected from the subphase. Quantitative measurements of the decrease of surface radioactivity after equilibration with different amounts of injected protein confirmed the presence of 1 mol of phosphatidylcholine per 1 mol of protein. Having demonstrated that phosphatidylcholine on the protein can exchange with phosphatidylcholine in monolayers, this technique was used to demonstrate that the exchange protein acts as a true carrier (18). In two equivalent surface compartments, sharing a common subphase, a monolayer of ^{14}C-labelled substrate is separated from a second monolayer of unlabelled phosphatidylcholine (Fig. 8). Injection of the pure protein underneath the monolayers resulted in a decline of surface radioactivity in the labelled compartment and a concomitant increase of ^{14}C-label in the other compartment. This transfer of radioactive phosphatidylcholine by the carrier protein does not cause a change in surface pressure in the monolayers. The surface radioactivity is decreased quantitatively in agreement with an interchange with the known amount of unlabelled phosphatidylcholine originally bound to the protein. The

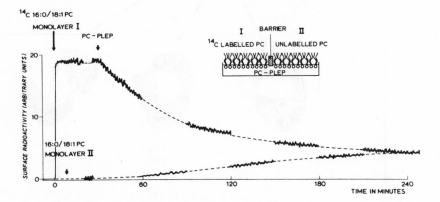

Fig. 8. Transport of phosphatidylcholine from one (labelled) monolayer to another monolayer and vice versa by phosphatidylcholine carrier protein (also denoted phosphatidylcholine exchange protein; PC-PLEP).

carrier function of the protein could also be demonstrated between monolayers and liposomes.

The exact mechanism by which the protein exerts its action of transferring a phospholipid molecule from one membrane to another provokes many questions. In this respect, it is of interest to study the nature of the binding of phosphatidylcholine to the carrier protein. Delipidation studies of the carrier protein with detergents showed that the associations between the phospholipid and protein molecule are partly of a hydrophobic nature (19). An extremely high degree of specificity of the protein with respect to the polar headgroup of phospholipids already suggests that the hydrophilic moiety of phosphatidylcholine may also play an intimate part in the lipid-protein association. Studies on a series of synthetic phosphatidylcholine analogues showed that the phosphorylcholine moiety and the carboxyl ester group are crucial for this specific interaction.

Incubation with phospholipases may give some clues about the disposition of the lipid on the protein molecule. The rate of hydrolysis of ^{14}C-labelled phosphatidylcholine in the protein by phospholipase A_2 from pig pancreas, *Naja naja* and bee venom turned out to be extremely slow (Fig. 9). The same

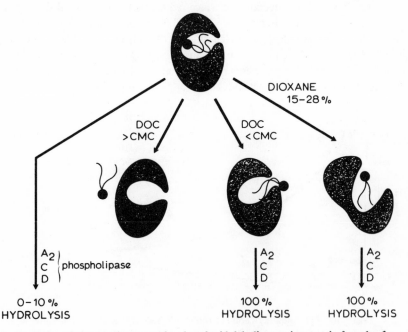

Fig. 9. Action of phospholipases on the phosphatidylcholine carrier protein from beef liver.

was true when incubations were carried out with phospholipase C from *B. cereus* and *Cl. welchii*, as well as with phospholipase D from cabbage, even when relatively high amounts of phospholipases were utilized (20). Previous studies from this laboratory have demonstrated that low concentrations of detergents may facilitate the action of phospholipases without a complete disruption of membrane structure. A systematic study was carried out on the action of a number of detergents on the binding of ^{14}C-phosphatidylcholine to the carrier protein (19). Using spectrophotometric and radioactivity analyses on poly-acrylamide gels of the phosphatidylcholine-protein complex treated in this manner, it was possible to determine the relationship between delipidation and critical micellar concentration of the detergents. It was subsequently established that under conditions where delipidation of the exchange protein did not occur, the hydrolytic action of all phospholipases was greatly enhanced (Fig. 9). The effect of deoxycholate (DOC) below the CMC is considered to involve a competition with the endogeneous phosphatidylcholine molecule, and to cause a translocation of the latter within the protein so as to exhibit its hydrophilic groups to the attacking phospholipase. Dioxane at concentrations which did not release any ^{14}C-phosphatidylcholine from the carrier protein also allowed various phospholipases to act on the phosphatidylcholine molecule. By contrast to deoxycholate, dioxane causes an irreversible inactivation of the carrier properties of the protein.

Concluding comments

The protein from beef liver which exchanges phosphatidylcholine between both artificial and natural membranes is a rather unique lipoprotein because it contains per molecule of protein one mole (non-covalently bound) of phosphatidylcholine. Apparently, the lipid is well hidden in the protein, which has a rather hydrophobic character. Further studies are directed to unravel the localization of phosphatidylcholine on the protein, and to learn which segments of the protein are involved in the hydrophobic and electrostatic association between phospholipid and protein. It will be of great interest to elucidate the mechanism by which the carrier protein exchanges its phosphatidylcholine molecule with the phospholipid present in the membrane.

REFERENCES

1. Gorter, E. and Grendel, F. (1925) *J. Exptl. Med. 41*, 439-443.
2. van Deenen, L.L.M. (1965) In *Progress in the Chemistry of Fats and other Lipids* (Holman, R.T., ed.) Vol. VI, pp. 1-127, Pergamon Press, London.
3. Bar, R.S., Deamer, D.W. and Cornwell, D.G. (1966) *Science 153*, 1010-1012.

4. van Deenen, L.L.M. and de Gier, J. (1974) In *The Red Blood Cell* (Surgenor, D.M., ed.) pp. 147-211, Academic Press, New York.

5. Bretscher, M. (1972) *J. Mol. Biol. 71*, 523-528.

6. Gordesky, S.E. and Marinetti, G.V. (1973) *Biochem. Biophys. Res. Commun. 50*, 1027-1031.

7. Colley, C.M., Zwaal, R.F.A., Roelofsen, B. and van Deenen, L.L.M. (1973) *Biochim. Biophys. Acta 307*, 74-82.

8. Verkleij, A.J., Zwaal, R.F.A., Roelofsen, B., Comfurius, P., Kastelijn, D. and van Deenen, L.L.M. (1973) *Biochim. Biophys. Acta 323*, 178-193.

9. Zwaal, R.F.A., Roelofsen, B. and Colley, C.M. (1973) *Biochim. Biophys. Acta 300*, 159-182.

10. Zwaal, R.F.A., Roelofsen, B., Comfurius, P. and van Deenen, L.L.M. Submitted.

11. Hoffman, J.F. (1962) *J. Gen. Physiol. 45*, 837-859.

12. Bodemann, H. and Passow, H. (1972) *J. Membrane Biol. 8*, 1-26.

13. Dawson, R.M.C. (1968) In *Biological Membranes* (Chapman, D., ed.) pp. 203-232, Academic Press, New York.

14. Demel, R.A., Geurts van Kessel, W.S.M., Zwaal, R.F.A., Roelofsen, B. and van Deenen, L.L.M. Submitted.

15. Low, M.G., Limbrick, A.R. and Finean, J.B. (1973) *FEBS Letters 34*, 1-4.

16. Wirtz, K.W.A. (1974) *Biochim. Biophys. Acta 344*, 95-117.

17. Kamp, H.H., Wirtz, K.W.A. and van Deenen, L.L.M. (1973) *Biochim. Biophys. Acta 318*, 313-325.

18. Demel, R.A., Wirtz, K.W.A., Kamp, H.H., Geurts van Kessel, W.S.M. and van Deenen, L.L.M. (1973) *Nature New Biol. 246*, 102-105.

19. Kamp, H.H., Wirtz, K.W.A. and van Deenen, L.L.M. Submitted.

20. Kamp, H.H., Sprengers, E.D., Westerman, J., Wirtz, K.W.A. and van Deenen, L.L.M. Submitted.

THE STRUCTURE OF AN ION PUMP

Walther Stoeckenius

Cardiovascular Research Institute
and
Department of Biochemistry and Biophysics
University of California
San Francisco, California 94143

AND

Ames Research Center
NASA
Moffett Field, California 94035

The general structure of biological membranes is thought to consist of a bi-layer of lipids with proteins embedded in the bilayer and also present on its surface. Some proteins apparently span the membrane and are accessible from both surfaces. Large areas on the surface of typical membrane proteins appar-ently expose mainly hydrophobic amino acid residues and these interact with the hydrocarbon chains of the lipids (1). This, however, is an insufficient de-scription for any given membrane, because a variety of vital functions are localized in membranes, and the many different functions will require different structures. The functions are mainly determined by the proteins and some closely associated lipid which has properties different from the bulk of the lipid forming the bilayer. Most functions are localized in circumscribed areas of the membrane and we call these areas specific functional sites. The sites may be small and occupy only a minor fraction of the total membrane area. Most averaging techniques such as chemical analysis, X-ray diffraction, etc., will therefore not give much information on the structure and composition of the functional sites; these are, however, the most important parts of the mem-brane for an understanding of membrane structure and function. Conventional electron microscopic techniques have insufficient resolution to study the func-tional sites at the level where structure and function merge. Most of the postu-lated functional sites cannot even be recognized by electron microscopy. Some, however, occur in high enough concentration so that they can be isolated from the membrane relatively easily and some are large enough to be readily recognized in electron micrographs. In addition, some have or can be induced to form a regular repeating structure which allows a structural analysis at high resolution. The purple membrane of halobacteria is such a functional site which shows a fortunate combination of these features.

When *Halobacterium halobium* is grown under low oxygen tension and in the light, its cell membrane contains a rhodopsin-like protein, bacteriorhodopsin, in high concentration. The chromoprotein molecules form crystalline arrays in the membrane which are only one molecule thick and exclude all other membrane proteins. The lipid content of these areas is reduced as compared to the rest of the surface membrane (2-4). The protein apparently spans the membrane and is all oriented across it in the same direction. It has a high α-helix content and the α-helices are oriented with their long axis at right angle to the plane of the membrane (5,6). Several such crystalline patches of bacteriorhodopsin may be present in one cell; they typically have a diameter of ~0.5 μm, a thickness of ~4.8 nm and may occupy more than 50% of the total cell membrane. They are readily recognized in electron micrographs of freeze-fractured whole cells. These patches are called the purple membrane because they appear deep purple when isolated. The color is due to the bacteriorhodopsin. Its chromophore retinal is bound as a protonated Schiff base to a lysine residue of the protein and further complexed with aromatic amino acids which form a hydrophobic pocket around it (7,8). This complexation and protonation shifts the 370 nm absorption maximum of the retinylidene protein to 570 nm.

The purple membrane functions as a light-driven proton pump (9,10). When the bacteriorhodopsin in purple membrane containing cells absorbs light, protons are ejected from the cells and an electrochemical gradient is established across the cell membrane, which then can drive ATP synthesis (11). One proton is ejected for every photon absorbed (12).

This function of the purple membrane implies that a photoreaction cycle exists in bacteriorhodopsin and such a cycle has been observed with low temperature and flash spectroscopy (13,14). When bacteriorhodopsin with an absorption maximum at 570 nm (bR_{570}) is cooled to 77° K, its spectrum is slightly red shifted and some fine structure becomes visible, but no other dramatic changes occur. When it is then illuminated with light absorbed by the 570 nm band, its absorption maximum is further red shifted to 610 nm and this red shift is fully reversible with 650 nm irradiation at 77° K. It is apparently due to the formation of a new complex with an absorption maximum at longer wavelength. Photosteady states with different ratios of the complexes can be obtained; they show an isosbestic point at 590 nm. Because the spectra overlap too much and because of the photoreversibility, the new complex has not been obtained in pure form. Its spectrum can, however, be calculated as will be shown below and it has been designated bK_{590}. When a mixture of bR_{570} and bK_{590} is warmed from $77°$ K to room temperature, bK_{590} is also converted back to bR_{570}. This, however, in contrast to the light-driven process, is clearly not a direct back reaction. It occurs through a series of intermediates, two of which have been identified spectroscopically and called bL_{550} and bM_{415}. They show rather large blue shifts of the absorption maxima, which are given by the subscripts. bL_{550} begins to appear at $\sim -140°$ and bM_{412} at $\sim -60°$ C.

Apparently the same intermediates can also be observed at physiological temperatures with flash spectroscopy of sufficient time resolution. The sequence of their appearance and decay is the same as that observed with low temperature spectroscopy, and at least one or possibly two more intermediates can be observed between bM_{412} and bR_{570}. They are designated bN_{520} and bO_{640}. The existence of bN_{520} is still somewhat doubtful. The rise and decay times of these intermediates are temperature and pH dependent. At room temperature and pH 4 to 8 the half times for the formation beginning with bK_{590} are ≤ 10 psec (15), 1.2 μsec (bL_{550}), 40 μsec (bM_{412}), 1 msec (bN_{520}), 3 msec (bO_{640}) and 5 msec (bR_{570}). A cycle is therefore completed in ~ 9 msec. The absorption spectra of the intermediates have been computed using the assumption that the absorbance of bM_{412} around 570 nm is negligible and that because of the large difference between its rise and decay time practically all of the cycling pigment is in the bM_{412} state at the time of its maximum concentration. The decrease in absorbance at 570 nm at this time is therefore a measure for the amount of pigment cycling. Only the first step in this reaction cycle requires light. Light energy must be stored in the first intermediate bK_{590} to complete the cycle and transport a proton against an electrochemical gradient.

While the spectral changes observed in the photoreaction cycle of bacteriorhodopsin are similar to those seen in the visual pigments, there is no evidence so far that any isomerization of the pigment occurs in the cycle described here. Both bR_{570} and bM_{412} apparently contain all-trans retinal. An isomerization from all-trans to 13-cis retinal has only been observed in the slow dark adaptation of the pigment (16,17). The physiological significance of the dark adaptation is unknown. The dark adapted pigment undergoes a photoreaction cycle similar to that of the light adapted pigment, and a small part of the total pigment is converted to the light-adapted complex in every cycle (14). Because the dark adaptation is slow, preillumination of the purple membrane before the experiment is used so that only reactions of the light-adapted pigment are observed.

The cycling of the pigment in the purple membrane is accompanied by a release and uptake of protons. This can be demonstrated either spectroscopically when a pH indicator is added to the membrane preparation or with a glass electrode in a modified membrane preparation where the second half of the photoreaction cycle is slowed down about 1000 times (18). We find that one proton is first released, then one proton taken up per bacteriorhodopsin molecule per reaction cycle (14). Under the conditions of our experiments, the proton is released slightly slower than bM_{412} appears and a proton is taken up during the decay of bO_{640}. The release and uptake of protons are apparently not directly and tightly coupled with the spectral changes, and the time constants for the appearance and disappearance of protons in the outside medium under different conditions may not vary in exactly the same way as the time constants for the spectral changes. A scheme of the cycle is shown in Fig. 1.

The pump function requires that the release and uptake of protons occurs on opposite sides of the membrane and that protons are transferred across the membrane from the uptake to the release site. The most likely structural feature in the protein to effect such a transfer of protons would be a chain of proton exchanging groups extending across the membrane either through or on the surface of the bacteriorhodopsin molecule. Each of the groups should be

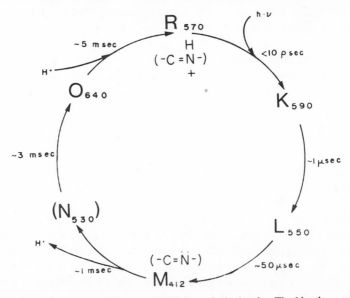

Fig. 1. Photoreaction cycle of light-adapted bacteriorhodopsin. The kinetic constants are given for a suspension of isolated purple membrane in water at room temperature. They show some variation with pH, salt concentration and temperature. The same reaction cycle occurs in intact cells with only small differences in the kinetic constants.

protonated and deprotonated once during the photoreaction cycle. One group which meets this requirement has been identified so far. Resonance Raman spectroscopy has shown that the Schiff base in bR_{570} is protonated and in bM_{412} is unprotonated (8). The Schiff base is rather inaccessible even to small water soluble reactants such as borohydride or hydroxylamine (3). When the purple membrane is transferred to D_2O, the exchange of the proton on the Schiff base for a deuteron is relatively slow in the dark. However, in the light it is fast (8). These observations are consistent with the assumption that the Schiff base is one link in the postulated chain of proton-exchanging groups. Because it is also part of the chromophore, where the energy driving the pump is absorbed, it is reasonable to assume that the energy input into the chain occurs at this link. A transient shift in the pK of the Schiff base would allow it to transfer its proton to the next group in the chain which is closer to the outer surface of the membrane and should have a pK lower than the pH of the medium. A conformational change occurring before the Schiff base return to its original high pK could prevent the back reaction and allow reprotonation only from the preceding group in the chain. Several other groups with suitable pK's on both sides of the groups directly exchanging protons with the Schiff base would complete the chain. Such a mechanism would require only small conformational changes in the protein and is compatible with the rigid crystal-line structure of the membrane.

This is, of course, a highly speculative model, not very detailed and probably oversimplified. It does, however, suggest experiments that would critically test it and should lead to either rejection or confirmation with the necessary modifications and refinements.

ACKNOWLEDGMENT

This work was supported by National Aeronautics and Space Administration Life Scientist Grant NGL 05-025-014 and National Heart and Lung Institute Program Project Grant HL-06285.

REFERENCES

1. Singer, S.J. and Nicolson, G.L. (1972) *Science 175*, 720-731.
2. Stoeckenius, W. and Kunau, W.H. (1968) *J. Cell Biol. 38*, 337-357.
3. Oesterhelt, D. and Stoeckenius, W. (1971) *Nature New Biol. 233*, 149-152.
4. Blaurock, A.E. and Stoeckenius, W. (1971) *Nature New Biol. 233*, 152-155.
5. Henderson, R. (1975) *J. Mol. Biol. 93*, 123-138.
6. Blaurock, A.E. (1975) *J. Mol. Biol. 93*, 139-158.
7. Oesterhelt, D. (1971) *Abstr. Comm. 7th Mtg. Eur. Biochem. Soc.*, p. 205, abstract 517.

8. Lewis, A., Spoonhower, J., Bogomolni, R.A., Lozier, R.H. and Stoeckenius, W. (1974) *Proc. Nat. Acad. Sci. (U.S.A.) 71*, 4462-4466.

9. Oesterhelt, D. and Stoeckenius, W. (1973) *Proc. Nat. Acad. Sci. (U.S.A.) 70*, 2853-2857.

10. Racker, E. and Stoeckenius, W. (1974) *J. Biol. Chem. 249*, 662-663.

11. Danon, A. and Stoeckenius, W. (1974) *Proc. Nat. Acad. Sci. (U.S.A.) 71*, 1234-1238.

12. Bogomolni, R.A., Baker, R.A., Lozier, R.H. and Stoeckenius, W. In preparation.

13. Stoeckenius, W. and Lozier, R.H. (1974) *J. Supramolec. Struc. 2*, 769-774.

14. Lozier, R.H., Bogomolni,, R.A. and Stoeckenius, W. (in press) *Biophys. J.*

15. Kaufmann, K.J., Rentzepis, P.M., Stoeckenius, W. and Lewis, A. In preparation.

16. Oesterhelt, D., Meentzen, M. and Schuhmann, L. (1973) *Eur. J. Biochem. 40*, 453-463.

17. Jan, L.K.-C.Y. (1974) Thesis, California Institute of Technology, Pasadena (May, 1974).

18. Oesterhelt, D. and Hess, B. (1973) *Eur. J. Biochem. 37*, 316-326.

STRUCTURE AND BIOSYNTHESIS OF
MITOCHONDRIAL CYTOCHROMES IN BAKER'S YEAST

Gottfried Schatz

Biocenter
University of Basel
CH-4056 Basel, Switzerland

The mitochondrial inner membrane is one of the most complex of all biological membranes. It contains at least several dozen polypeptides whose synthesis and assembly with each other is controlled by mitochondrial and cytoplasmic protein synthesis (1) and by environmental factors such as catabolite repression and availability of oxygen (2). In this brief review, I hope to show that the biogenesis of the mitochondrial inner membrane can be approached by studying the tightly membrane-bound cytochromes of the yeast respiratory chain. These cytochromes are not only essential catalysts for electron transport, but also important structural elements of the mitochondrial inner membrane. They are simple enough to be amenable to detailed chemical, biochemical and genetic investigation and yet their biosynthesis mirrors in many ways that of the mitochondrial inner membrane as a whole.

Much of our recent work has dealt with cytochrome c oxidase (3-8) and cytochrome c_1 (8,9). Cytochrome c oxidase appears to be a useful model for studying the interplay between mitochondrial and nuclear genes whereas the simpler cytochrome c_1 allows one to explore the control of mitochondrial assembly by small molecules such as oxygen and heme.

STRUCTURE AND BIOSYNTHESIS OF CYTOCHROME C OXIDASE

Cytochrome c oxidase from Baker's yeast consists of seven different polypeptide subunits. The three large ones are synthesized on mitochondrial ribosomes and the four small ones are synthesized on cytoplasmic ribosomes (3,10). Some of the characteristic features of these subunits are summarized in Table I.

R.O. Poyton and W. Birchmeier have isolated each of these seven polypep-

TABLE I

Properties of the Subunits of Yeast Cytochrome *c* Oxidase

Property	Subunit Number						
	I	II	III	IV	V	VI	VII
Molecular weight (SDS-acrylamide gel electrophoresis)	40,000	33,000	23,500	14,500	13,000	12,500	4,700
(gel filtration in guanidine HCl)	41,600	28,300	21,700	14,700	13,100	13,100	4,500
(amino acid analysis)	36,930	33,800	---	14,000	12,100	12,100	4,600
N-terminal amino acid	---	---	---	Glx	Ala	Ser	Ala
C-terminal amino acid	---	---	---	Leu	Ala	Ala	Lys
Polarity index[a]	35	42	---	48	50	51	41
Site of translation	mitochondria				cytoplasm		

--- Not determined

[a] Defined as mole percent of hydrophilic amino acids Lys, His, Arg, Asp, Thr, Ser, Glu.

tides by exploiting differences in polypeptide charge, solubility, and size (5,10). A critical step in the procedure is the use of guanidine—HCl to partially dissociate the enzyme. If yeast cytochrome c oxidase is exposed to 6 M guanidine—HCl and reducing agents at room temperature, subunits IV and VI (cf. Table I) are removed from an aggregate of the other five subunits and can be further purified by conventional methods. The remaining, less soluble subunits can be separated from each other in the presence of formic acid or sodium dodecyl sulfate. The amino acid composition of six of the seven polypeptides has been determined. The two large subunits are rather hydrophobic, whereas the cytoplasmically-synthesized subunits IV, V and VI exhibit polarity indices of around 50%, which is typical of most water-soluble proteins. Subunit VII has a somewhat lower polarity index, but the significance of this parameter may be questioned because of the small size of this polypeptide. Although no chemical information is as yet available for subunit III, indirect evidence indicates that it, too, is very hydrophobic.

W. Birchmeier and A. Tsugita are currently studying the chemical properties of the four cytoplasmically-synthesized subunits in more detail. The amino and carboxy terminal amino acids have been identified and the complete amino acid sequences are being determined. Work on the amino acid sequence of subunit VI is almost completed. Significantly, one of the two histidine peptides obtained by tryptic digestion bears a striking resemblance to one of the histidine peptides of cytochrome b_5 from bovine liver. This preliminary result raises the possibility that the amino acid sequences of the individual subunits may tell us more about the function of each subunit, the location of heme a in the enzyme, and the evolution of cytochrome c oxidase from simpler heme proteins.

In order to study the function of each of these seven subunits, R.O. Poyton tested the effects of subunit-specific antisera on the enzymic activity of solubilized and membrane-bound cytochrome c oxidase. He found that each of the subunit-specific antisera could inhibit the enzyme (6). It was particularly interesting that an antiserum against the mitochondrially-made subunit II inhibited the enzyme almost completely. Although by no means conclusive, this result implied that subunit II was not simply a tightly-bound reproducible contaminant but a functional component of cytochrome c oxidase. These results suggest that both mitochondrially and cytoplasmically-made subunits *participate* in the enzymic function of cytochrome c oxidase. It is still open which subunit is *essential* for activity, particularly if activity is measured *in vitro* with the solubilized enzyme. This question can only be settled by reconstituting the enzyme from its purified subunits. This has not yet been achieved.

G. Eytan has studied the three-dimensional arrangement of the seven subunits in solubilized and membrane-bound cytochrome c oxidase (12). Most of these experiments employed surface probes such as diazo coupling with p-diazonium benzene sulfonate and iodination with lactoperoxidase. In addition, externally-located subunits in the solubilized enzyme were identified by linking them to bovine serum albumin carrying a covalently-bound isocyanate group. With this last-mentioned method, any accessible subunit will be covalently coupled to the large bovine serum albumin molecule and, upon sodium dodecyl sulfate-polyacrylamide gel electrophoresis, will be eliminated from the typical pattern of cytochrome c oxidase subunits. The results obtained by these different methods with the membrane-bound and solubilized enzyme indicate that subunit I is almost completely inaccessible to surface probes. Subunit II is also largely inaccessible. All other subunits (with the possible exception of subunit IV) are accessible. Thus, the two largest mitochondrially-made subunits appear to form a hydrophobic internal "core" whereas most or all of the other subunits are closer to the surface. We suspect, therefore, that the two largest subunits of the enzyme may be intrinsic membrane proteins (13) which link the other subunits to the mitochondrial inner membrane.

Genetic evidence by E. Ebner supports this contention. If synthesis of the mitochondrially-made subunits is blocked by the extrachromosomal petite mutation, the remaining cytoplasmically-made subunits are only loosely bound to the mitochondrial inner membrane and can be removed by brief sonication (7). This finding (together with other work on cytochrome c oxidase-less yeast mutants not reviewed here (cf. Ref. 14-16) clearly illustrates the importance of mitochondrially-made proteins for the proper assembly of the mitochondrial inner membrane.

STRUCTURE AND BIOSYNTHESIS OF CYTOCHROME C_1

Cytochrome c_1 has long been one of the least known of the four cytochromes of the mitochondrial respiratory chain. This is probably explained by the fact that its characteristic α-absorption peak is obscured by that of cytochrome c unless spectroscopy is carried out at liquid nitrogen temperature. Information on the biosynthesis of cytochrome c_1 has also been scarce and contradictory. Some reports suggested that cytochrome c_1 was synthesized by mitochondria whereas other studies implicated synthesis on cytoplasmic ribosomes (1).

Several years ago, E. Ross set out to study the structure and the biosynthesis of cytochrome c_1 by chemical, spectroscopic and immunological methods. In the first stage of his investigation (8,9,17), he purified cytochrome c_1 from *Saccharomyces cerevisiae* mitochondria by solubilization with cholate, ammonium sulfate fractionation, disruption of the cytochrome bc_1 complex with mercaptoethanol plus detergents and chromatography on DEAE cellulose. A summary of this purification procedure is given in Table II. The final product is spectrally pure (Fig. 1), contains 32 nmoles of covalently-bound heme per mg of protein and does not measurably react with oxygen or carbon monoxide.

TABLE II

Purification of Cytochrome c_1 from *S. cerevisiae*

Fraction	Total protein	Total heme c_1	Heme c_1/protein	Purification	Yield
	mg	nmoles	nmoles/mg	fold	%
Submito-chondrial particles	17,800	2190	0.16	(1)	(100)
Ammonium sulfate frac-tionation	1,300	950	0.52	3.3	43
Cytochrome bc_1 precip-itate	325	572	2.50	16.0	26
Crude cyto-chrome c_1	43	412	9.70	61.0	19
DEAE-cellu-lose chro-matography	13	341	27.00	169.0	16

Cytochrome c_1 was purified from submitochondrial particles as described in ref. 9. The fraction obtained in the final step represents the pooled fractions from DEAE chromatography which had a heme content of at least 26 nmoles heme $c_1 \cdot$mg protein[-1]. This represents about 85% of the cytochrome c_1 recovered from the column. The remainder was re-chromatographed. In the experiment documented here the heme content of the peak fraction from the column was 29 nmoles heme $c_1 \cdot$mg protein[-1].

Sodium dodecyl sulfate disaggregates the purified cytochrome into a single 31,000 dalton subunit carrying the covalently-attached heme group (Fig. 2). Many cytochrome c_1 preparations contain in addition an 18,500 dalton polypeptide which is devoid of covalently-bound heme. Since this polypeptide can be removed from the heme-carrying polypeptide by ion exchange chromatography or by immunoprecipitation (Fig. 3), it is probably not an essential subunit of cytochrome c_1.

Cytochrome c_1 is extremely sensitive to proteolysis. If it is purified in the absence of protease inhibitors, a family of heme polypeptides with molecular

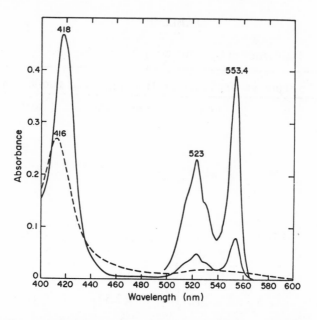

Fig. 1. Absolute room temperature spectra of yeast cytochrome c_1. Solid lines: reduced with $Na_2S_2O_4$. Dashed line: oxidized with $Na_2S_2O_8$. Protein concentration was 0.11 mg/ ml with the two lower curves and 0.54 mg/ml with the upper curve. Light path: 10 mm. The protein was dissolved in 0.1 M sodium phosphate–0.5 mM EDTA, pH 7.5.

weights of 29,000, 27,000 and 25,000 daltons is obtained. In the presence of the protease inhibitor phenylmethyl sulfonyl fluoride, the purification yields predominantly a 31,000 dalton heme protein with only little contamination by degradation products.

In order to obtain a more reliable estimate of the molecular weight of undegraded cytochrome c_1, the cytochrome c_1 heme protein was identified directly in a lysate of cells which had been grown in the presence of the heme-precursor $[^3H]$ -δ-aminolevulinic acid. Since cytochromes c and c_1 are the only major heme proteins with covalently-bound heme groups, cytochrome c_1 can be easily displayed by mixing intact labelled spheroplasts with protease inhibitors, adding an excess of boiling dodecyl sulfate solution, and subjecting the dissociated mixture to electrophoresis in dodecyl sulfate-acrylamide gels. This procedure, which would be expected to minimize proteolysis, again yields a molecular weight of 31,000.

We conclude that the molecular weight of the cytochrome c_1 heme protein *in vivo* is 31,000 daltons. It is, of course, possible that other polypeptides are also intrinsic subunits of the native cytochrome. This question awaits further study.

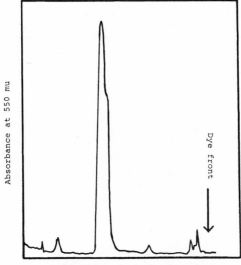

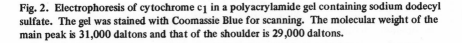

Fig. 2. Electrophoresis of cytochrome c_1 in a polyacrylamide gel containing sodium dodecyl sulfate. The gel was stained with Coomassie Blue for scanning. The molecular weight of the main peak is 31,000 daltons and that of the shoulder is 29,000 daltons.

In order to determine the site of translation of the cytochrome c_1 heme protein, yeast cells were labelled with [^{3}H] leucine under the following conditions: (a) in the absence of inhibitors; (b) in the presence of acriflavin, an inhibitor of mitochondrial translation; (c) in the presence of cycloheximide, an inhibitor of cytoplasmic translation. The incorporation of radioactivity into the heme protein was measured by immunoprecipitating it from mitochondrial extracts and analyzing the precipitate by dodecyl sulfate—polyacrylamide gel electrophoresis. Label was incorporated into the cytochrome c_1 apoprotein only in the presence of acriflavin or in the absence of inhibitor, but not in the presence of cycloheximide (Table III). Cytochrome c_1 is thus a cytoplasmic translation product.

This conclusion was further supported by the demonstration that an extra-chromosomal petite mutant lacking mitochondrial protein synthesis still contained holo-cytochrome c_1 that was indistinguishable from cytochrome c_1 of the wild-type with respect to molecular weight, the presence of a covalently-bound heme group (Fig. 4), absorption spectrum and antigenic properties (not shown). The concentration of cytochrome c_1 in the mutant mitochondria approached that of wild-type mitochondria, but the lability of the mitochondrially-bound cytochrome c_1 to proteolysis appeared to be increased. A mitochondrial translation product may thus be necessary for the correct conformation or ori-

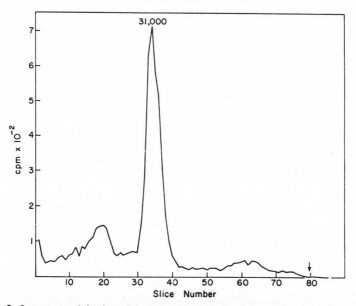

Fig. 3. Immunoprecipitation of the cytochrome c_1 hemeprotein from wild-type yeast mitochondria uniformly labeled with $[^3H]$ leucine. The cells were grown for about 10 generations in semi-synthetic medium containing $[^3H]$ leucine, and the mitochondria were isolated and extracted with detergent in the presence of phenylmethyl sulfonyl fluoride (9). The extract was mixed with anticytochrome c_1 antiserum and the resulting immunoprecipitate was analyzed by dodecyl sulfate-polyacrylamide gel electrophoresis.

entation of cytochrome c_1 in the mitochondrial inner membrane. Since heme is an essential component of cytochrome c_1, it was of interest to know whether the synthesis (or the accumulation) of the apo-cytochrome was regulated by the availability of heme within the cell. E. Ross approached this question with the yeast mutant, A2180-38, which lacks δ-aminolevulinic acid synthetase and, as a consequence, is auxotrophic for heme (18). When grown in the absence of δ-aminolevulinic acid, the mutant cannot synthesize porphyrins, lacks all mitochondrial cytochromes, and is respiration-deficient; it also requires ergosterol as well as oleic acid for growth since the synthesis of these essential lipids is mediated by heme proteins. However, when grown in the presence of δ-amino-levulinic acid, the mutant is phenotypically wild-type. The mutant A2180-38 was grown in a δ-aminolevulinic acid-deficient medium in the presence of $[^3H]$-leucine and converted to spheroplasts. The spheroplasts were then mixed with spheroplasts from the same mutant grown in the presence of δ-aminolevulinic acid and $[^{14}C]$ leucine. The mixed spheroplasts were lysed, the mitochondria were isolated and extracted, and cytochrome c_1 was immunoprecipitated from

TABLE III

Apo-Cytochrome c_1 is Synthesized on Cytoplasmic Ribosomes

Inhibitor	Specific [^{3}H] Activity (cpm/mg protein)		Inhibition (%)
	Mitochondria	Cytochrome c_1	Cytochrome c_1
None	8.5×10^6	5.0×10^6	(0)
Acriflavin	8.3×10^6	4.9×10^6	1
Cycloheximide	0.8×10^6	0.09×10^6	99

Wild-type yeast cells were grown in semi-synthetic medium containing 0.3% galactose, 0.2% yeast extract and 0.4 mCi of [U-^{14}C]leucine/1 (3). Cells were harvested, washed by semi-sterile techniques, suspended to 12.5 mg wet weight/ml in 200 mM galactose–40 mM sodium phosphate, pH 7.4, and, where indicated, 12.5 μg of acriflavin/ml or 100 μg of cycloheximide/ml. After 10 min at 28°, [4,5-^{3}H]leucine was added to a final concentration of either 25 μCi/ml (no inhibitor; plus acriflavin) or 60 μCi/ml (plus cycloheximide). After 75 min at 28°, labeling was stopped by adding unlabeled leucine to a final concentration of 1 mM and incubating for an additional 10 min. The cells were washed twice with 40 mM sodium phosphate, pH 7.4, and the mitochondria were isolated. They were subjected to immunoprecipitation with anti-cytochrome c_1 antiserum, and the immunoprecipitates were dissociated with dodecyl sulfate and analyzed by dodecyl sulfate-acrylamide gel electrophoresis. The specific radioactivities of ^{14}C and ^{3}H in the isolated mitochondria were determined by relating radioactivity to total protein. The amount of ^{14}C was then used as a convenient measure to assess cytochrome c_1 protein. The "Specific [^{3}H] Activity" of cytochrome c_1 was calculated by determining the ^{3}H/^{14}C ratio in the cytochrome c_1 peak of the polyacrylamide gels and converting the denominator to protein. Appropriate corrections were applied for differences in the concentration of [^{3}H]leucine between the incubation mixtures.

the combined extracts. Electrophoretic analysis of the immunoprecipitate revealed that the mutant grown in the presence of δ-aminolevulinic acid contained cytochrome c_1, as would be expected; in contrast, little, if any, cytochrome c_1 protein could be immunoprecipitated from the mutant grown in the absence of δ-aminolevulinic acid. These experiments indicate that the heme-carrying polypeptide of cytochrome c_1 is not significantly accumulated unless heme is present.

Analogous experiments showed that the cytochrome-deficient promitochondria of anaerobically-grown yeast cells (19) are likewise deficient in apocytochrome c_1. This is the first indication that the functional deficiencies of promitochondria are at least partially explained by the lack of specific membrane polypeptides.

We realize that this conclusion rests so far only on immunochemical evidence

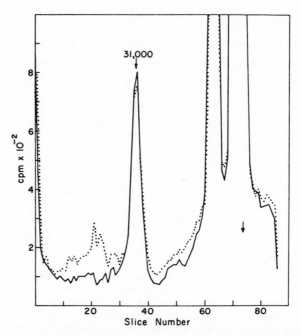

Fig. 4. Determination of the molecular weight and the presence of covalently-bound heme
for cytochrome c_1 in a cytoplasmic petite mutant. The wild-type D273-10B and the derived cytoplasmic petite mutant (7) were grown in semisynthetic medium containing 0.1%
yeast extract and 2 μCi/ml of [4-^{14}C]-δ-aminolevulinic acid or 8μCi/ml of [3,4-^{3}H]-δ-
aminolevulinic acid, respectively. Cells were converted to spheroplasts, the spheroplasts
were mixed and lysed in boiling 2.5% sodium dodecyl sulfate containing 1 mM phenyl
methyl sulfonyl fluoride and 57 μM diisopropylfluorophosphate. The lysate was analyzed
by dodecyl sulfate-polyacrylamide gel electrophoresis. Solid trace: ^{14}C; dotted trace:
^{3}H.

and that we have not shown that the antisera used by us would react with heme-
free cytochrome c_1 apoprotein. However, we consider this unlikely since these
experiments were conducted with antisera from two different rabbits; one of
these sera appeared to be rather non-selective since it immunoprecipitated cytochrome c_1 that had been acetylated with [^{3}H] acetic anhydride and denatured
with dodecyl sulfate.

There are at.least three possible explanations for the absence of cytochrome
c_1 apoprotein in the heme-free cells. According to the first explanation, heme
controls the translation of the apoprotein on cytoplasmic ribosomes. Such a
control mechanism has been established for globin synthesis where heme is required for the translation of globin mRNA (20). A second explanation would
be that apocytochrome c_1 is synthesized at a normal rate in the absence of heme

but is rapidly degraded by intracellular proteases since it cannot assume its normal configuration. E. Ross is currently undertaking short-term pulse-labeling experiments in the presence of protease inhibitors to test this possibility. Finally, it is also possible that apocytochrome c_1 simply cannot be integrated into the mitochondrial inner membrane unless it has combined with heme. Whatever the correct explanation, the situation in yeast mitochondria seems to be different from that in bacteria since heme-deficient bacterial mutants accumulate the protein moieties of most or all of the normal cytochromes (21).

CONCLUDING REMARKS

The studies summarized above reflect our belief that the function and the biogenesis of the mitochondrial inner membrane will only be understood once its major polypeptides have been purified and characterized in considerable detail. Only then will it be possible to devise unambiguous experiments on how these polypeptides are arranged within the membrane, how they are assembled, how they change their conformations during different functional states, and how they are controlled genetically. This may seem like an overly pessimistic viewpoint since it could be argued that the mitochondrial inner membrane contains too many polypeptides for this approach. Being prejudiced, we would counter that information of general significance can already be obtained by the detailed study of a few key components such as the two cytochromes discussed here.

ACKNOWLEDGEMENTS

This study was supported by the following grants: GM 16320 from the U.S. Public Health Service, GB-40541X from the U.S. National Science Foundation, NY (C) 181406 Hatch Project Grant from the U.S. Department of Agriculture, and 3.2350.74 from the Swiss National Foundation.

REFERENCES

1. Schatz, G. and Mason, T.L. (1974) *Ann. Rev. Biochem. 43*, 51-87.
2. Linnane, A.W., Haslam, J.M., Lukins, H.B. and Nagley, P. (1972) *Ann. Rev. Microbiol. 26*, 163-198.
3. Mason, T.L., Poyton, R.O., Wharton, D.C. and Schatz, G. (1973) *J. Biol. Chem. 248*, 1346-1354.
4. Mason, T.L. and Schatz, G. (1973) *J. Biol. Chem. 248*, 1355-1360.
5. Poyton, R.O. and Schatz, G. (1975) *J. Biol. Chem. 250*, 752-761.
6. Poyton, R.O. and Schatz, G. (1975) *J. Biol. Chem. 250*, 762-766.
7. Ebner, E., Mason, T.L. and Schatz, G. (1973) *J. Biol. Chem. 248*, 5369-5378.

8. Ross, E., Ebner, E., Poyton, R.O., Mason, T.L., Ono, B. and Schatz, G. (1974) In *The Biogenesis of Mitochondria* (Kroon, A.M. and Saccone, C., eds.) pp. 477-489, Academic Press, New York.

9. Ross, E. and Schatz, G. Submitted for publication.

10. Rubin, M.S. and Tzagoloff, A. (1973) *J. Biol. Chem. 248*, 4275-4279.

11. Birchmeier, W. and Schatz, G. In preparation.

12. Eytan, G. and Schatz, G. (1975) *J. Biol. Chem. 250*, 767-774.

13. Singer, S.J. and Nicolson, G.L. (1972) *Science 175*, 720-731.

14. Ebner, E., Mennucci, L. and Schatz, G. (1973) *J. Biol. Chem. 248*, 5360-5368.

15. Ebner, E. and Schatz, G. (1973) *J. Biol. Chem. 248*, 5379-5384.

16. Ono, B., Fink, G.R. and Schatz, G. (1975) *J. Biol. Chem. 250*, 775-782.

17. Ross, E. (1975) *Fed. Proc. 34*, 1563.

18. Golub, E.G. and Sprinson, D.B. In preparation.

19. Criddle, R.S. and Schatz, G. (1969) *Biochemistry 9*, 322-334.

20. Beuzard, Y. and London, I. (1974) *Proc. Nat. Acad. Sci. (U.S.A.) 71*, 2863-2866.

21. Haddock, B.A. and Schairer, H.U. (1973) *Eur. J. Biochem. 35*, 34-45.

THE LATERAL ORGANISATION OF LIPIDS AROUND A CALCIUM TRANSPORT PROTEIN: EVIDENCE FOR A PHOSPHOLIPID ANNULUS THAT MODULATES FUNCTION

J.C. Metcalfe, J.P. Bennett, T.R. Hesketh,
M.D. Houslay, G.A. Smith and G.B. Warren

Department of Biochemistry
University of Cambridge
Cambridge CB2 1QW, England

INTRODUCTION

We have developed a technique of lipid substitution which enables the phospholipids in cell membranes to be almost completely exchanged with pure, defined phospholipids (1-3). The substitution can be achieved without the loss of function of a range of membrane proteins, provided that lipids of appropriate structure are substituted for the endogenous lipids. The technique can therefore be used to define how lipids interact with membrane proteins and modulate their functions.

The (Ca^{2+}, Mg^{2+})-dependent ATPase of sarcoplasmic reticulum (SR), which is responsible for the stoichiometric uptake of calcium into purified SR vesicles, provides a convenient system for the analysis of lipid-protein interactions because the ATPase comprises 80% of the membrane protein and is easily separated from the minor membrane proteins (1). The interaction of substituted lipids with the ATPase provides evidence for a phospholipid annulus around the protein which has a dominant effect on its function as an ATPase.

The ATPase requires about 30 phospholipid molecules to maintain maximal ATPase activity (2), and spin label experiments indicate that about the same number of lipids interact directly with the ATPase. We have therefore proposed a model for the ATPase-lipid complex in which about 30 phospholipids form the first shell of lipid bilayer on the hydrophobic surface of the ATPase which spans the bilayer. We term this first shell of lipid the annulus, and it is distinguished by spin label experiments as being relatively immobilised on the protein surface compared with lipid molecules in the bilayer outside the annulus.

The lipid composition of the annulus is apparently determined by the tertiary structure of the protein. Cholesterol is rigorously excluded from the annulus (4), and the phospholipids are selected on the basis of their polar headgroups. In the complexes which have been examined so far, this lateral segregation of lipid classes around the protein serves to maintain the activity of the ATPase. For example, if cholesterol is forced to replace phospholipid molecules in the annulus, it causes a complete and reversible inactivation of the ATPase at equimolar proportions with phospholipid. Complexes of the ATPase with dipalmitoyl lecithin (DPL) provide evidence that the saturated lecithin molecules in the annulus are unable to undergo a normal phase transition, and that it is the annular DPL molecules which are mainly responsible for determining the temperature-activity profile of this complex. These experiments suggest that although the SR membrane is highly fluid, there is a short-range dynamic organisation of lipids around the ATPase that is not random, and that both the chemical and physical properties of the phospholipids have a critical effect on the function of the protein.

BIOCHEMICAL EVIDENCE FOR A PHOSPHOLIPID ANNULUS

Purified sarcoplasmic reticulum contains about 100 phospholipid molecules per ATPase. The ATPase can be purified, with some depletion of this proportion of endogenous lipid, by dispersing SR in cholate and centrifuging into a sucrose gradient (1,2). The detergent and minor membrane problems are left in the supernatant together with the excess lipid, and the ATPase with its remaining endogenous lipid is recovered as a particulate band from the gradient. The proportion of lipid associated with the ATPase decreases with increasing cholate concentration, but the ATPase retains maximal activity until about 30 phospholipids per ATPase remain (2). The removal of further phospholipid molecules results in a rapid and irreversible loss of activity, and below about 15 phospholipids per ATPase there is not significant activity (Fig. 1). This suggests that about 30 phospholipid molecules form the minimal lipid environment for each ATPase molecule needed to maintain full enzyme activity. Similar amounts of phospholipid were also found to be required for maximal activity when different lecithins were substituted for the endogenous phospholipid by the technique of lipid substitution, suggesting that there is an approximate stoichiometry in the phospholipid ATPase complexes, which is not substantially affected by the structure of the phospholipid. For example, when DPL is substituted for endogenous phospholipid, about 38 molecules were needed to support maximal ATPase activity (Fig. 1).

Further evidence from spin label experiments suggests that about 30 phospholipid molecules form the first shell of bilayer surrounding the penetrant part

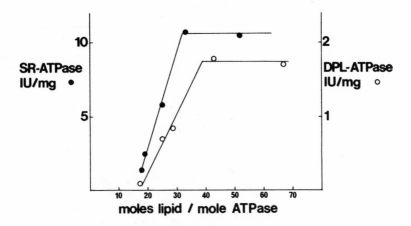

Fig. 1. The effect of lipid depletion by cholate on the ATPase activity of the calcium transport protein associated with SR lipid or DPL at 37° C.

of the ATPase, and this represents the minimal amount of phospholipid needed to cover the hydrophobic surface of the protein. In this model, about 15 molecules of phospholipid surround the ATPase in each half of the bilayer. Results that have been obtained using Triton X-100 to delipidate SR are consistent with this idea. The and Hasselbach (5) have shown that 70-106 fatty acid molecules per ATPase are needed to fully reactivate the enzyme (equivalent to 35-53 phospholipid molecules), while it has recently been found that about 35 phospholipid molecules are required for reactivation (N.M. Green, personal communication).

THE PHOSPHOLIPID ANNULUS OF DPL—ATP-ASE DOES NOT UNDERGO A NORMAL PHASE TRANSITION

The ability to prepare ATPase-lipid complexes with defined lipids in different proportions has enabled us to determine the relative importance for the function of the protein of the annular phospholipids compared with those lipids in the bilayer not interacting directly with the protein. For example, we find that while the ATPase is able to sense the phase transition which occurs in extra-annular DPL molecules, it is the annular DPL molecules which have a predominant influence on the activity of the enzyme, and these molecules do not undergo a normal phase transition.

ATPase complexes with lecithins show a reversible inactivation of the enzyme either at or below the phase transition temperature of the lipid (Fig. 2). For example, dimyristoyl lecithin (DML)-ATPase shows a 150-fold activation be-

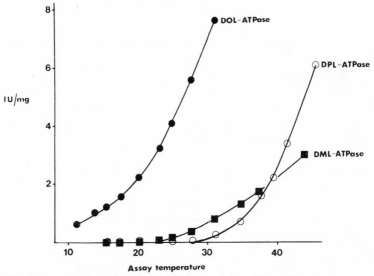

Fig. 2. The temperature dependence of the ATPase activity of the calcium transport protein associated with DOL, DML or DPL.

tween 25° C and 35° C, and DPL-ATPase shows a similar activation between 29° C and 39° C. The complex with DPL is of particular interest because it retains significant ATPase activity down to about 30° C, well below the phase transition temperature at 41° C. We find that in DPL-ATPase complexes, the lipid to protein ratio has a relatively small effect on the temperature-activity profile of the enzyme, provided that the protein has a complete phospholipid annulus (Fig. 3). This indicates that either the ATPase activity is determined

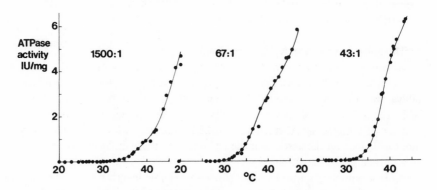

Fig. 3. The effect of the DPL:ATPase ratio on the temperature-activity profiles of DPL-ATPase complexes.

almost entirely by its annular lipids, or that the annular lipids are indistinguishable from those in the bilayer away from the protein. This latter possibility seems unlikely since a complex with about 30 phospholipids per ATPase in which all of the phospholipid interacts directly with the protein is unlikely to undergo a normal phase transition, whereas the extra-annular lipid at high lipid to protein ratios should be relatively unaffected by the protein. This is clearly demonstrated by spin label experiments, using 5-doxyl stearic acid (6). Pure DPL and complexes with high DPL:ATPase ratios show a well-defined phase transition at about 41° C in Arrhenius plots of 2 $T_{\parallel}$ but this transition becomes much less pronounced at lower ratios, and cannot be defined in complexes with 25-30 DPL molecules per ATPase (Fig. 4). This clearly indicates that a normal phase transition does not occur in the lipid annulus, but does occur in the DPL bilayer not interacting directly with the protein. At the same time, the 2 $T_{\parallel}$ plots indicate that there may be a fairly sharp increase in the fluidity of the annulus at around 27-30° C for the complexes with low DPL: ATPase ratios, coinciding with the development of significant ATPase activity (see Fig. 5). Taken together these data are most simply interpreted by assuming that the annulus is the primary determinant of ATPase activity and that the extra-annular lipid has relatively little effect. However, a close examination of the Arrhenius plots of ATPase activity indicates that a second-order effect of the phase transition in the DPL bilayer at high DPL:ATPase ratios can be detect-

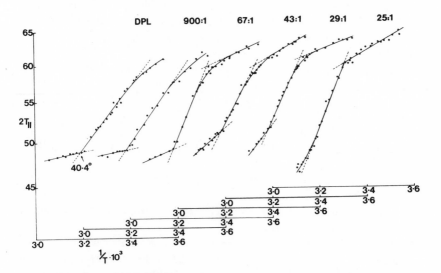

Fig. 4. The temperature dependence of 2 $T_{\parallel}$ (gauss) for 5-doxyl stearic acid in DPL and DPL-ATPase complexes at different DPL:ATPase ratios.

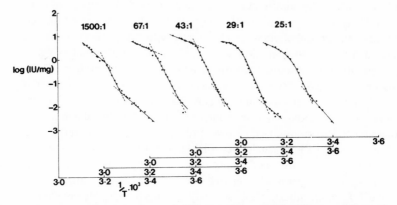

Fig. 5. The temperature dependence of ATPase activity of DPL-ATPase complexes at different DPL:ATPase ratios.

ed. A fairly well-defined break occurs at 38-40° C for these complexes, whereas only a continual change in slope can be defined at low DPL:ATPase ratios. There is a second break at 28-31° C for all complexes, which we tentatively assign to a change in the interaction of the annular DPL molecules with the ATPase as the complex develops activity. In this view of the activity of the complexes, it is the *rigidity* of the interaction of annular DPL molecules with the ATPase which dominates the temperature profile, rather than the phase transition in the extra-annular lipids, which is sensed in the ATPase only indirectly by transmission through the annulus.

We can now ask whether the chemical composition of the annulus is determined by the protein. We have approached this question by determining whether the ATPase can maintain its function in a fluid bilayer which contains substantial proportions of lipid which will not support its function.

CHOLESTEROL IS EXCLUDED FROM THE PHOSPHOLIPID ANNULUS

The lipid substitution and titration techniques were adapted to prepare ATPase-lipid complexes which contain cholesterol. Comparison of the properties of the various complexes provides evidence for the exclusion of cholesterol from the phospholipid annulus (4).

When phospholipids are added to complexes of pure ATPase with a complete annulus of about 30 SR phospholipids (termed SR-ATPase), there is a rapid and complete equilibration of the phospholipid pools in the presence of appropriate concentrations of cholate. The activity of the ATPase is then determined by the composition of the phospholipids in the annulus: for example, DML and

DPL inhibit the activity of pure SR-ATPase complexes to about the same extent, and the inhibition is a linear function of the lipid composition (Fig. 6). In contrast, if SR-ATPase is titrated in the presence of cholate with an excess of DOL, DML or DPL containing increasing proportions of cholesterol, the activity of the equilibrated mixtures is virtually unaffected by the presence of cholesterol at up to equimolar proportions with the added lecithins (Fig. 7). Control experi-

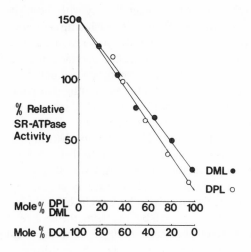

Fig. 6. Titrations of ATPase activity with mixtures of DOL/DML or DOL/DPL.

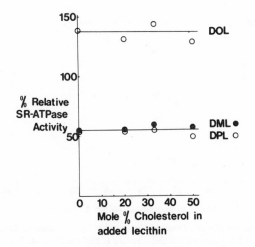

Fig. 7. Titrations of ATPase activity with lecithin-cholesterol mixtures at 37° C.

ments showed that the presence of cholesterol does not prevent the normal equilibration of the phospholipids by cholate. These results are most simply interpreted by assuming that cholesterol does not significantly affect the inter-action of the phospholipids with the ATPase, and that the phospholipids exert the same effect on the activity of the enzyme, irrespective of the presence of cholesterol. We conclude that cholesterol is excluded from the phospholipid annulus.

In order to force cholesterol to interact directly with the ATPase, it is nec-essary to use high concentrations of cholate which strip phospholipids from the annulus. If this stripping is performed in the presence of cholesterol, it sub-stitutes for the displaced phospholipids and the enzyme is maintained in a stable but inactive conformation. These complexes can only be reactivated fully by displacing the cholesterol from the annulus by phospholipid.

Figure 8 shows the relative ATPase activities of 20 separate preparations of ATPase-lipid complexes prepared by treating SR with cholesterol in the presence of cholate at various concentrations. The activity of the complexes is maximal and unaffected by the presence of cholesterol until there are less than about 30 phospholipid molecules per ATPase in the complexes. Below this proportion of phospholipid, there is a sharp drop in ATPase activity until below about 15 phospholipids per ATPase the complexes have negligible activity. This activity profile is very similar to that observed by increasing the cholate concentration in the absence of cholesterol, also shown in Fig. 8. The important difference is that the complexes inactivated by cholesterol can be reactivated by adding back phospholipid (SR or DOL) in the presence of cholate. Several reactivation ex-

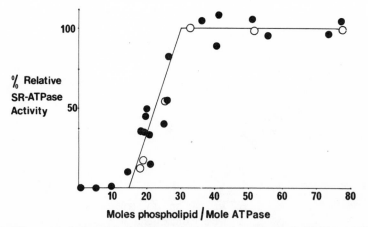

Fig. 8. Effect of replacing phospholipids in the annulus by cholesterol. Lipid was removed from SR by cholate in the presence (●) and absence (○) of cholesterol. Activities were measured at 37° C.

periments are shown in Fig. 9, in which the ATPase-lipid complexes have been titrated with either the precise amount of phospholipid required to complete an annulus of 30 phospholipid molecules, or a much larger excess of phospholipid to achieve the maximal reactivation. The partially inactivated complexes can be fully reactivated provided they contain a total of at least 25-30 phospholipid plus cholesterol molecules per ATPase. Complexes with a lower lipid complement prepared by using very high cholate concentrations can only be partially reactivated. For example, one very inactive complex containing only 0.7 mole of SR phospholipid and 12 moles of cholesterol was reactivated 380-fold by a large excess of phospholipid to 10% of the maximal ATPase activity. It can also be seen from Fig. 9 that a stoichiometric addition of phospholipid calculated to give a total of 30 phospholipids per ATPase restores 75-80% of the maximal activity of the complex.

These experiments show that cholesterol causes a potent inhibition of activity when it replaces phospholipids in the annulus, and that equimolar cholesterol and phospholipid in the annulus completely and reversibly inhibits ATPase. Most of the inhibition due to cholesterol is abolished by sufficient phospholipid to reform a complete annulus.

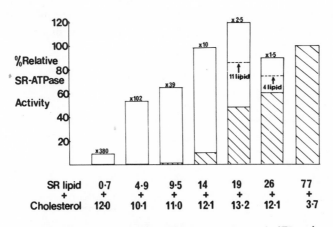

Fig. 9. Reactivation of ATPase by displacing cholesterol bound to protein by SR lipids. Shaded area defines activity of complexes before addition of lipid; open area, reactivation produced by added lipid at 37° C. Reactivation factor is shown above each open area and the dashed lines represent experiments in which sufficient phospholipid was added back to complete an annulus of 30 SR phospholipid molecules.

THE COMPOSITION OF THE PHOSPHOLIPID ANNULUS IS DETERMINED BY THE ATP-ASE

In Fig. 6 there is a linear change in ATPase activity as the composition of the lecithin mixtures is varied. By isolating the complexes by centrifugation and analysing the composition of the annulus, it was found that the lecithin species are present in the proportions expected for complete mixing of the lipids. The simplest interpretation of these observations is that the activity is directly determined by the composition of the annulus. Making this assumption, we can attribute a non-linear response in ATPase activity to a preferential selection of lipids into the annulus. A markedly non-linear titration curve (Fig. 10) is obtained with mixtures of DOL and dioleoyl phosphatidic acid (DOPA), and when these complexes were isolated, there was a much higher proportion of DOL in the annulus than would be expected from the composition of the DOL/DOPA mixture which had been added. However, if the titrations were performed at much higher cholate concentrations, the DOL/DOPA titration curve became approximately linear, indicating that the equilibrating effect of cholate eventually overrides the lipid specificity expressed by the protein. These results suggest that the tertiary structure of the protein specifies its immediate lipid environment. In this example, the effect of the specificity is to maintain maximal ATPase activity in the presence of substantial proportions of a phospholipid which strongly inhibits the enzyme. Similar results have been obtained for β-hydroxybutyrate dehydrogenase, which specifically requires lecithin for activity.

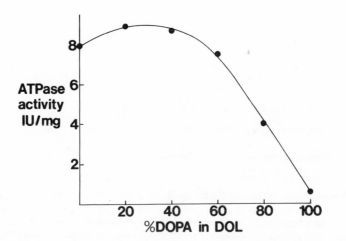

Fig. 10. Titrations of ATPase activity with mixtures of DOL and DOPA at 37° C.

REFERENCES

1. Warren, G.B., Toon, P.A., Birdsall, N.J.M., Lee, A.G. and Metcalfe, J.C. (1974) *FEBS Letters 41*, 122-124.
2. Warren, G.B., Toon, P.A., Birdsall, N.J.M., Lee, A.G. and Metcalfe, J.C. (1974) *Biochemistry 13*, 5501-5507.
3. Houslay, M.D., Warren, G.B., Birdsall, N.J.M. and Metcalfe, J.C. (1975) *FEBS Letters 51*, 146-151.
4. Warren, G.B., Houslay, M.D., Metcalfe, J.C. and Birdsall, N.J.M. (1975) *Nature 255*, 684-687.
5. The, R. and Hasselbach, W. (1973) *Eur. J. Biochem. 39*, 63-68.
6. Hesketh, T.R., Houslay, M.D., Metcalfe, J.C., Smith, G.A., McGill, K.A. and Warren, G.B. (in press) *Biochemistry*.

ROLE OF ADENINE NUCLEOTIDES IN THE STRUCTURE
AND FUNCTION OF THE ATP-ASE
OF THE INNER MITOCHONDRIAL MEMBRANE

Harvey S. Penefsky, Neil E. Garrett and Ta-min Chang

Department of Biochemistry
The Public Health Research Institute of the City of New York, Inc.
New York, New York 10016

Physical and chemical studies of beef heart mitochondrial ATPase (F_1) have revealed a molecular architecture of considerable complexity. The molecular weight of the enzyme has been reported to be 347,000 (1) and 360,000 (2) and it contains five chemically distinct subunits of molecular weights 54,000, 50,000, 33,000, 17,000 and 5700 (1). Although the subunit stoichiometry is as yet uncertain, it appears likely that there may be as many as ten per oligomer. Studies on the interactions of adenine nucleotides with F_1 indicate a degree of complexity fully commensurate with that of the sub-unit structure. Whereas initial investigations produced evidence for one high affinity binding site per molecule for ATP or ADP (3,4), it was later found from equilibrium dialysis-type binding studies that two binding sites for ADP were present (5,6) characterized by considerably different affinities, namely 0.3 μM and 47 μM (6). It was, moreover, recently reported from Dr. Slater's laboratory that isolated, homogeneous F_1 contained 5 moles of adenine nucleotide tightly bound per mole of enzyme; that is, the nucleotides resisted removal by repeated precipitation with ammonium sulfate or treatment with charcoal, but were released following protein denaturation (7,8). The relationship, if any, between the two binding sites which engaged in rapid, reversible binding of added nucleotides (6) and the five sites which contained tightly bound adenine nucleotides (7,8) thus becomes of considerable interest, especially in view of the limited slow exchange which occurred between added nucleotides and those tightly bound to the enzyme (7).

1. *Adenine Nucleotide Binding Sites on F_1.* Studies of F_1 prepared in this

laboratory (9) have shown that the isolated enzyme contained 3 moles of tightly bound adenine nucleotide per mole of protein (10). "Tightly bound" nucleotides are here defined as those which resisted removal by repeated precipitation of the enzyme with ammonium sulfate or chromatography on columns of Sephadex or Dowex-chloride. On the average, F_1 contained 2 moles of ADP and 1 mole of ATP, Table I.

TABLE I

Adenine Nucleotide Content of F_1

ATP	ADP	AMP
0.9 ± 0.4 (12)	2.3 ± 0.4 (11)	0.3 ± 0.2 (7)

Samples of the enzyme were chromatographed on Sephadex columns equilibrated with dilute buffers in order to remove unbound or loosely bound adenine nucleotides. Nucleotides were measured on perchloric acid extracts of the enzyme (10).

Such enzyme preparations also exhibited two binding sites which engaged in readily reversible binding of the ATP analog adenylylimidodiphosphate (AMP-PNP). The sites appeared to be equivalent with a K_D of 1.3 μM (10). The results suggested, in agreement with the implications of the findings of Rosing et al. (7), that each molecule of F_1 indeed contained five adenine nucleotide binding sites. The present results differ, however, in that of the five sites, three are here characterized as "tight" binding sites and two as sites which engaged in readily reversible binding of added nucleotides. Direct evidence for the existence of at least five sites per molecule is presented below.

2. *Adenine Nucleotide-Dependent Properties of F_1*. The experiments described in this section were carried out with samples of F_1 which were subjected to Sephadex or Dowex-chloride chromatography (10,11). The enzyme was thus free of unbound nucleotides as well as those nucleotides bound to the readily reversible binding sites discussed above. The enzyme did, however, contain 3 moles of tightly-bound nucleotide per mole of protein. Addition of adenine nucleotides to such preparations induced changes in the properties of the molecule which could be detected by aurovertin, ultraviolet difference spectroscopy and covalently linked pyrenebutyrate.

Aurovertin. Aurovertin is an antibiotic of considerable interest to the study of F_1 because it forms a fluorescent complex with F_1 but with no other component of the mitochondrial membrane (11,12,13). This fact facilitates considerably the interpretation of experiments with aurovertin and submito-chondrial particles. If as a result of a specific manipulation, a change in fluorescence intensity occurs, the change must reflect an altered interaction between aurovertin and membrane-bound F_1. These interactions can influence the fluorescence quantum yield of the bound fluorophore, the binding stoichiometry, or both (11,13). Studies of the interaction between aurovertin and soluble F_1 indicated that, in the absence of added adenine nucleotide, each molecule of enzyme exhibited one binding site for the probe with a K_D of 0.013 μM (Mg^{++} present) or 0.07 μM (ADP present). However, addition of ATP, which presumably was bound at the readily reversible site(s) on the enzyme, caused a change in the conformation of the enzyme which exposed a second binding site for aurovertin and reduced the affinity of both sites for the antibiotic (11). The observed quenching of fluorescence of the aurovertin-F_1 complex by ATP was due predominantly to a change in fluorescence quantum yield, although under appropriate conditions partial dissociation of the complex occurred. An enhancement of fluorescence intensity by ADP was assigned entirely to an increase in quantum yield.

The influence of adenine nucleotides on the fluorescent complex formed between aurovertin and submitochondrial particles, that is, with membrane-bound F_1, was closely similar, both qualitatively and quantitatively, to that in the soluble system. Further, energization of the mitochondrial membrane by addition of an oxidizable substrate such as succinate to mitochondria or sub-mitochondrial particles led to a rapid, energy-dependent enhancement of fluores-cence of aurovertin complexed with membrane-bound F_1 (13,14). The half-time of the enhancement, 36 msec (13) or 20 msec (14), was sufficiently rapid that it might well reflect responses to initial energy conservation steps in the respiratory chain. It thus appears that aurovertin functions as a highly sensitive, rapidly responding fluorescent reporter molecule of conformational changes in-duced in soluble F_1 by ATP and in membrane-bound F_1 by either ATP or by energization of the membrane.

A variety of observations suggest that the ATP-induced conformational changes in F_1 which are reported by aurovertin are localized or limited in extent. For example, neither the shape nor the intensity of the bands in the CD spectrum of F_1 (Fig. 1) was altered by the addition of adenine nucleo-tides, aurovertin or both. Similarly, careful measurements of the sedimenta-tion velocity of F_1 in dilute buffers and in the presence of ATP or ADP failed to reveal departures from the $s_{20,w}^0$ of 12.9S reported earlier (15).

Ultraviolet difference spectroscopy. Examination of the ultraviolet

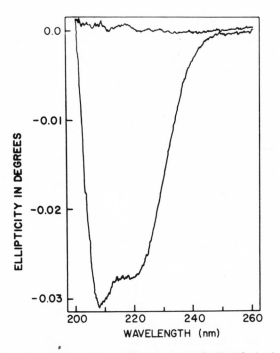

Fig. 1. Circular dichroism of F_1. The enzyme was dissolved in 1 ml of a buffer which contained 0.1 M sucrose, 1 mM EDTA, 0.5 mM ATP and 50 mM Pi, pH 7.4, and was dialyzed in a collodion bag (1) against 100 ml of the same buffer for 4 hr. A sample of the dialyzed enzyme (2.1 mg/ml) was placed in a thermostatted cell (25 ± 0.1°) with a path length of 0.14 mm. The spectrum was obtained in a Cary model 61 recording instrument. The average molar ellipticity (at 200 nm) per peptide bond of a number of different samples of enzyme was 10,400 degcm^2/mole. A mean residue weight of 110 mg/mmole peptide bond was used in the calculation.

difference spectrum of F_1 provides further evidence for interaction of the protein with adenine nucleotides. It may be seen in Fig. 2A that a typical tyrosine spectrum became apparent when F_1 was dissolved in 6 M guanidine hydrochloride. There is no tryptophan in the molecule (1). Addition of ATP, or ADP, to F_1 freed of nucleotides at the readily reversible binding sites gave rise to a difference spectrum with a maximum at 280 nm but which lacked the full tyrosine spectrum characteristic of tyrosine (Fig. 2B). The hypochromicity at 260 nm, suggestive of some ordering of nucleotide with respect to protein, may also be noted. The optical changes were observed with ATP or ADP but not with AMP. The latter nucleotide did not bind to F_1 prepared from rat liver mitochondria (16) or to the beef heart enzyme.

Pyrenebutyryl-F$_1$. A second approach to a fluorescent probe of the structure of F_1 made use of the pyrenebutyric anhydride reagent of Knopp

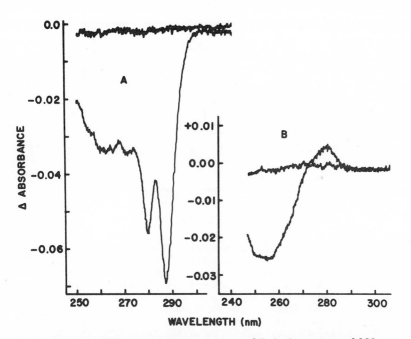

Fig. 2. Ultraviolet difference absorbance spectrum of F_1 in the presence of 6 M guanidinium hydrochloride or ADP. The enzyme was equilibrated on Sephadex columns with a buffer which contained 0.1 M sucrose, 2 mM EDTA and 50 mM Tris-SO_4, pH 7.4. The compartments of a pair of tandem cells (pyrocell, Westwood, N.J., U.S.A.) were filled and matched by weighing. The final volume of each solution was 3.2 ml. Spectra were recorded with a Cary model 14 spectrophotometer using the 0-0.1 absorbance slidewire. The sample compartment contained 1.68 μM F_1 in the above buffer. Guanidinium hydrochloride (6 M) was present in the enzyme compartment of the sample beam and in the buffer compartment of the reference beam. B. The concentration of F_1 was 6.71 μM and that of ADP was 30 μM. The straight line in each panel is the spectrum recorded prior to addition of guanidine hydrochloride or ADP, as shown.

and Weber (17). The reaction between F_1 and the mixed anhydride occurred under mild conditions of pH and temperature and samples of F_1 with 1 or 2 moles of pyrene fluorophore per mole of protein were readily obtained. The fluorescence lifetime of the pyrenebutyryl conjugates, measured by quantum counting, was 105 msec, a value in good agreement with that reported for pyrenebutyryl-bovine serum albumin (18). Fluorescence polarization studies of the labeled enzyme were carried out at constant temperature but varying viscosity by altering the concentration of sucrose, up to 1.2 M. Typical Perrin plots constructed from the data are shown in Fig. 3. It may be seen that the mean harmonic rotational relaxation time of pyrenebutyryl-F_1 was 740 nsec in the presence of ATP and approximately 2000 nsec in the absence of nucleotide. The difference in axial ratio implied

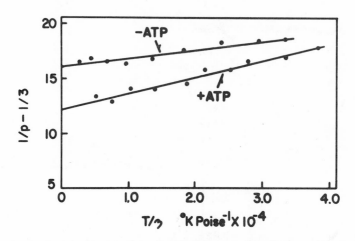

Fig. 3. Rotational relaxation time of pyrenebutyryl-F_1. Pyrenebutyryl-F_1 was prepared by reacting the enzyme with pyrenebutyric anhydride. The labeled enzyme was separated from other products by precipitation two times with ammonium sulfate (50%) followed by chromatography, first on Sephadex G-50 and then on a column of Dowex-chloride. The enzyme was dissolved in a buffer containing 50 mM Tris-SO_4, pH 8, 2 mM EDTA and, where indicated, 4 mM ATP. T/η was varied at constant temperature by addition of sucrose, up to 1.2 M. Polarization of fluorescence was measured at $21°$, lower curve, or $26.3°$, upper curve, in an instrument described earlier (11). Lower curve: the cuvette contained 4 mM ATP and 2.7 μM pyrenebutyryl-F_1 with 0.57 mole of pyrene fluorophore per mole protein. Upper curve: the cuvette contained 3.9 μM pyrenebutyryl-F_1 with 0.83 mole of pyrene fluorophore per mole protein. p is the polarization of fluorescence, T is the absolute temperature in $°K$ and η is the viscosity.

by these numbers was confirmed by other measurements such as ultracentrifugation only when the enzyme was dissolved in concentrated solutions of polyols as discussed below.

3. *Adenine Nucleotide-Depleted F_1.* In order to examine all of the five adenine nucleotide binding sites on F_1 directly, it was necessary to remove nucleotides from the three tight sites without denaturing the enzyme. Chromatography of F_1 on columns of Sephadex equilibrated with 50% glycerol indeed resulted in enzyme preparations essentially depleted of adenine nucleotides. When stored in 50% glycerol, these preparations retained indefinitely the ATPase and coupling factor activity characteristic of native F_1. Glycerol, at concentrations of 50%, was more effective in removing nucleotides than sucrose. Whereas the oligomeric structure of the protein was preserved in the presence of either polyol, the shape of the molecule in such solvents underwent considerable change (Fig. 4, inset). It may be seen that in the presence of 1.5 M sucrose the sedimentation coefficient of F_1 decreased, 8.4 S compared to a value of 11.9 S at this protein concentration in dilute buffers. Upon removal of the polyol, the sedimentation coefficient of native

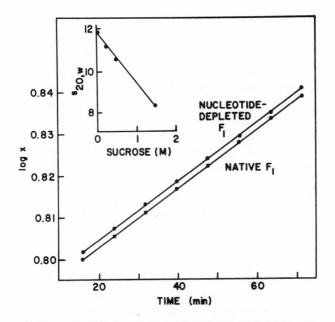

Fig. 4. Sedimentation velocity of native and nucleotide-depleted F_1. F_1 was depleted of adenine nucleotides as described earlier (10). Both the nucleotide-depleted enzyme and the native enzyme were equilibrated with a buffer containing 0.25 M sucrose, 50 mM Tris-sulfate, pH 8.0, and 2 mM EDTA by chromatography on columns of Sephadex G-50. The protein concentration was 4.1 mg/ml. The sedimentation coefficient for the native and nucleotide-depleted enzyme was 11.2. Inset: $s_{20,w}$ for native F_1. The buffer was the same as that given above except for the concentration of sucrose. Movement of the sedimenting boundaries in 0.5 and 1.5 M sucrose was monitored at 280 nm with the ultraviolet scanner attachment to the ultracentrifuge. All of the observed values were corrected for the effects of solvent viscosity and density (20).

F_1, as well as that of nucleotide-depleted F_1, returned to the value characteristic of the native enzyme (Fig. 4). The non-destructive removal of adenine nucleotides from F_1 is thus associated with a reversible, polyol-induced change in shape of the protein which may be accompanied by a decrease in the affinity of the binding sites. Since slow chromatography was more effective in removing nucleotides than rapid flow rates, it may be inferred that the "off" rate of nucleotides dissociating from F_1 is slow and that an important aspect of the efficacy of glycerol columns relates to stabilization of the enzyme during the procedure. It is also clear from Fig. 4 that the shape of F_1 in solution is not directly dependent on the presence of adenine nucleotides at any of the five binding sites on the molecule. The longer term stability of either nucleotide-depleted or native F_1 in dilute buffer does, however, require the presence of millimolar concentration of ATP.

These polyol-induced changes in the shape of F_1 are to be distinguished from the localized, adenine nucleotide-induced conformational changes in the enzyme which are reported by aurovertin. Moreover, the apparent adenine nucleotide-dependent differences in axial ratio indicated in the rotational relaxation time measurements with pyrenebutyryl-F_1 described above are more readily explained as solvent effects against which adenine nucleotides offer some protection. These observations on the effects of high concentrations of polyols on the structure of F_1 are thus consistent with the observations of Myers and Jakoby (18) who have reported polyol-induced conformational changes in a variety of enzymes.

4. *Role of Adenine Nucleotide Binding Sites in F_1.* It was originally suggested by Hillborn and Hammes that of the two ADP binding sites on F_1, the site with K_D of 47 μM was the hydrolytic site. The second site was not accessible to nucleotides in the membrane-bound enzyme (6). The presence, however, of at least five adenine nucleotide binding sites on F_1 (8,10) raises anew the question of the role of such sites in ATP hydrolysis and in ATP synthesis during oxidative phosphorylation.

One or more sites would appear to participate in adenine nucleotide-induced conformational changes in the enzyme. Since only the two sites which engaged in readily reversible binding were unoccupied prior to the addition of nucleotides in these experiments, it is possible that the conformational changes reported by aurovertin reflected occupation of these sites alone. However, an unambiguous assignment of sites is not yet possible since nucleotides bound at the three tight sites can be at least partially displaced by externally added nucleotides (8,10) and the observed conformational changes may well reflect displacement of, for example, ADP in the tight sites by added ATP.

A possible role of some importance for the tight sites is maintenance of the oligomeric structure of the molecule via either intra- or intersubunit links. As pointed out above, no immediate changes in shape occurred when the enzyme was depleted of nucleotides. Thus the 3-dimensional configuration of the molecule is not immediately dependent on the presence of nucleotides at any of the binding sites. Bound nucleotides may, however, make important contributions to the long-term stability of the protein.

It is not clear at present whether all of the five sites must be occupied by nucleotides during catalysis, since the nucleotide-depleted enzyme exhibited undiminished ATPase and coupling factor activities. While there is no apparent difference in initial rates of reactions catalyzed by native or nucleotide-depleted F_1, it may be that rapid binding on a millisecond time scale precedes initiation of catalytic function.

5. *Reconstitution of Nucleotide-Depleted F_1*. Adenine nucleotides are restored to the binding sites on nucleotide-depleted F_1 following incubation of the enzyme with appropriate nucleotides and Mg^{++}. If ADP is the added ligand, up to 4 moles of ADP are bound per mole of F_1. Two or three of the sites occupied by ADP may be characterized as tight sites, because bound ligand is not removed from these sites by subsequent chromatography on columns of Sephadex. Upon incubation of nucleotide-depleted F_1 with ATP, both ADP and ATP, in a molar ratio of 2 to 1, are found on the tight sites. It will be recalled from Table I that this ratio is characteristic of the native enzyme.

While occupancy of five sites was observed infrequently when ADP or ATP was the added ligand, stoichiometric binding of AMP-PNP was readily achieved. The bound analog appeared to be tightly bound to all five sites since the enzyme emerged from Sephadex columns with approximately five sites occupied (4.7 to 4.9 moles AMP-PNP/mole F_1).

The catalytic properties of samples of F_1 containing varying amounts of AMP-PNP bound were examined. After incubation of nucleotide-depleted F_1 with AMP-PNP for 26 min, 2 moles of analog were bound per mole of protein (Table II). ATPase activity was inhibited 80%. A slow increase in the amount of analog bound upon further incubation was not accompanied by appreciable change in enzyme activity. Incubation for about 15 hours was required to occupy the five binding sites on the molecule. In a separate experiment, B, a sample of F_1 which had been incubated overnight and contained 4.5 moles of AMP-PNP per mole protein was incubated with 4 mM ATP in order to displace the bound analog. Experiments A and B are not directly comparable since equivalent sites may not have been uncovered. Nevertheless, it may be seen in B that ATPase was inhibited 70% when 3 moles of analog remained on the protein. Several points of interest emerge from these studies. First, F_1 contains, as a lower limit, five binding sites for adenine nucleotide per molecule. Second, binding of AMP-PNP to the protein causes reversible inhibition of enzyme activity. Third, it is unlikely that the inhibition represented direct binding of AMP-PNP to the hydrolytic site in competition with ATP. The maximum concentration of AMP-PNP which may have been present in the reaction mixture (0.05 μM) would have been rapidly displaced from the hydrolytic site on F_1 by the 5 mM ATP used in the ATPase assay (19). The data indicate, therefore, that binding of AMP-PNP to a site or sites other than catalytic ones can strongly influence catalytic activity. It seems reasonable to suggest that these sites are control sites which modulate enzyme activity.

TABLE II

Inhibition of ATPase Activity by AMP-PNP
Tightly Bound to F_1

Incubation Time	AMP-PNP/F_1	ATPase	Inhibition
min	moles/mole	units/mg	%
(A) 0	0.0	100	0
26	2.0	22	78
120	2.3	19	81
200	2.5	17	82
(B) 0	4.5	10	90
11	3.0	30	70
60	2.5	54	46
180	2.0	70	30

Nucleotide-depleted F_1 was prepared as described (10). In A, 5.2 μM F_1 was incubated with 25 μM ^{3}H-AMP-PNP. Aliquots of the reaction mixture were withdrawn at the times shown and chromatographed on Sephadex columns to remove unbound nucleotides. ATPase activity was measured as described (9). A unit of ATPase activity is defined as the amount of protein which catalyzes the hydrolysis of 1 μmole of ATP per min. In B, 2 μM F_1 was incubated overnight with 50 μM ^{3}H-AMP-PNP. The ratio of analog to F_1 shown represents that amount of analog which remained bound to F_1 following chromatography on Sephadex columns. Displacement of ^{3}H-AMP-PNP from F_1 was monitored by adding 4 mM ATP at zero time and measuring analog bound to the enzyme after column chromatography (20).

REFERENCES

1. Knowles, A.F. and Penefsky, H.S. (1972) *J. Biol. Chem. 247*, 6624-6630.
2. Lambeth, D., Lardy, H.A., Senior, A.E. and Brooks, J.C. (1971) *FEBS Letters 17*, 330-332.
3. Zalkin, H., Pullman, M.E. and Racker, E. (1965) *J. Biol. Chem. 240*, 4011-4016.
4. Penefsky, H.S. (1967) *J. Biol. Chem. 242*, 5789-5795.
5. Sanadi, D.R., Sani, B.P., Fisher, R.J., Li, O. and Taggart, W.V. (1971) In *Energy Transduction in Respiration and Photosynthesis*

(Quagliariello, E., Papa, S. and Rossi, C.S., eds.) p. 89, Adriatica, Bari.

6. Hilborn, D.A. and Hammes, G.G. (1973) *Biochemistry 12*, 983-990.
7. Harris, D.A., Rosing, J., Van De Stadt, R.J. and Slater, E.C. (1973) *Biochim. Biophys. Acta 314*, 149-153.
8. Rosing, J., Harris, D.A., Kemp. A., Jr. and Slater, E.C. (1975) *Biochim. Biophys. Acta 376*, 13-26.
9. Knowles, A.F. and Penefsky, H.S. (1972) *J. Biol. Chem. 247*, 6617-6623.
10. Garrett, N.E. and Penefsky, H.S. (in press) *J. Biol. Chem.*
11. Chang, Ta-min and Penefsky, H.S. (1973) *J. Biol. Chem. 248*, 2746-2754.
12. Lardy, H.A. and Lin, G.-H.C. (1969) In *Inhibitors, Tools for Cell Research* (Buecher, T., ed.) pp. 279-281, Springer Verlag, New York.
13. Chang, Ta-min and Penefsky, H.S. (1974) *J. Biol. Chem. 249*, 1090-1098.
14. Van De Stadt, R.J., Van Dam, K. and Slater, E.C. (1974) *Biochim. Biophys. Acta 347*, 224-236.
15. Penefsky, H.S. and Warner, R.C. (1967) *J. Biol. Chem. 240*, 4694-4702.
16. Catterall, W.A. and Pedersen, P.L. (1972) *J. Biol. Chem. 247*, 7969-7976.
17. Knopp, J.A. and Weber, G. (1969) *J. Biol. Chem. 244*, 6309-6315.
18. Myers, J.S. and Jakoby, W.S. (1975) *J. Biol. Chem. 250*, 3785-3789.
19. Penefsky, H.S. (1974) *J. Biol. Chem. 249*, 3579-3585.
20. Garrett, N.E. and Penefsky, H.S. (in press) *J. Supramolecular Struct.*

III
Transport in Membranes

PATHWAYS AND STOICHIOMETRY OF H^+ AND Ca^{2+} TRANSPORT COUPLED TO ELECTRON TRANSPORT

Albert L. Lehninger and Martin D. Brand

Department of Physiological Chemistry
The Johns Hopkins University School of Medicine
Baltimore, Maryland 21205

SUMMARY

Quantitative measurements of the uptake of ^{14}C-labeled weak acids such as β-hydroxybutyrate, with $^{45}Ca^{2+}$ by rat liver mitochondria respiring on succinate in the presence of rotenone showed that 2.0 weak acid anions are accumulated per Ca^{2+} ion. Moreover, further measurements showed that 2 Ca^{2+} ions and 4 weak acid anions are accumulated per 2 e^- per energy-conserving site. These findings are consistent with an $\rightarrow H^+/\sim$ ratio of 4.0; the 4 protons so ejected are carried back into the alkaline mitochondrial matrix in the form of 4 molecules of the protonated weak acid. The relevance of these findings to other energy-dependent ion transport processes and to chemiosmotic coupling is briefly discussed.

INTRODUCTION

This paper reports new data on the stoichiometry of H^+ ejection during electron transport stimulated by Ca^{2+}. Early reports from this laboratory (1,2) established the stoichiometric relationship between the accumulation of Ca^{2+} and the flow of electrons along the respiratory chain of mitochondria, as well as the relationship between the accumulation of Ca^{2+} and of the counteranion phosphate. Moreover, the relationship between Ca^{2+} concentration and respiratory stimulation has been established (3). There appears to be universal agreement (cf. 4) that 2 Ca^{2+} ions are accumulated per 2 e^- per site (hereafter called the $Ca^{2+}/\sim$ ratio). Under the usual conditions (i.e. phosphate as anion, pH about 7) approximately one H^+ is ejected per Ca^{2+} accumulated, thus yielding a net $H^+/\sim$ ratio of about 2.0 (4,5,6). However, the H^+/Ca^{2+} ratio is a function of the anions present. In the presence of acetate, the $H^+/\sim$ ratio is 0.2 or less (6,7); under other conditions H^+/Ca^{2+} ratios approaching 2.0 have been reported (8,9), equivalent to a $H^+/\sim$ ratio of 4.0. These stoichiometric ratios may be compared with the $H^+/\sim$ ratio of 2.0 postulated by Mitchell in his chemiosmotic hypothesis for the development of protonmotive force during electron transport.

Closely relevant to the stoichiometry of H^+ transport is the question of the pathway and mechanism of Ca^{2+} transport. The first concrete evidence on the nature of the Ca^{2+} transport process came from experiments of Selwyn *et al.* (10) and also Lehninger (11,12) on passive Ca^{2+} movements into non-respiring mitochondria, i.e. down a concentration gradient. It was found, as expected, that mitochondria do not swell in 83 mM (isotonic) $CaCl_2$ or Ca isethionate, since chloride and isethionate do not pass through the mitochondrial membrane. Nor do they swell in 83 mM Ca acetate unless a proton-conducting uncoupler is added, to allow the exit of the proton carried into the mitochondria by undissociated acetic acid. Mitochondria do, however, swell readily in 83 mM solutions of Ca^{2+} salts of NO_3^-, SCN^-, ClO_3^-, and ClO_4^-, anions which permeate freely into mitochondria (11). Since a proton-conducting uncoupler was not required and did not affect swelling with these anions, they were obviously passing through the membrane as such, and not as undissociated acids, together with the Ca^{2+}. The osmotic swelling induced by Ca^{2+} salts of these anions was inhibited by both La^{3+} and ruthenium red, inhibitors of respiration-dependent Ca^{2+} transport by mitochondria. From these observations it was concluded that Ca^{2+} passes through the membrane in respiration-inhibited mitochondria by a passive process of electrogenic uniport (Fig. 1), without exchange or co-transport with H^+. Ca^{2+}/H^+ antiport processes, whether electrogenic (Ca^{2+}/H^+) or neutral $(Ca^{2+}/2\ H^+)$, were excluded, since they cannot occur except in the presence of proton-conducting uncouplers.

We may now consider the mechanism by which Ca^{2+} enters the matrix of *respiring* mitochondria, *up* a gradient. It has long been thought that any permeant anion will enter respiring mitochondria with Ca^{2+}. However, this is not the case. We have shown (11) that permeant anions may be divided into two classes – those that support and accompany Ca^{2+} uptake into respiring mitochondria and those that do not. The former group includes phosphate, acetate, butyrate, β-hydroxybutyrate, and bicarbonate (HCO_3^-– CO_2); in general, salts of lipid-soluble weak acids will support and accompany Ca^{2+} uptake. On the other hand, the permeant anions NO_3^-, SCN^-, ClO_3^-, and ClO_4^- do not accompany Ca^{2+} into respiring mitochondria; in fact, no

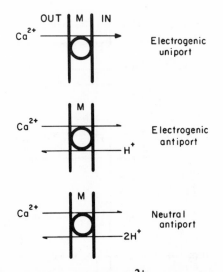

Fig. 1. **Vectorial diagrams of various models for Ca^{2+} transport through the mitochondrial membrane. The electrogenic uniport model is the only one consistent with the results of osmotic swelling experiments (see text).**

Ca^{2+} uptake takes place if any one of this latter group is present as the major anion (11). Very similar findings were made independently by Brierley *et al.* (13) in the case of respiration-dependent accumulation of K^+. We concluded from these observations that the effective anions listed are active because they pass through the membrane in a form that is potentially capable of giving up a proton to the respiration-alkalinized matrix. For example, phosphate enters as the protonated form $H_2PO_4^-$ in exchange for OH^-; acetate and other anions of weak lipophilic acids enter as undissociated acids, as is well known; and bicarbonate actually passes through the membrane as dissolved CO_2, to be rehydrated by mitochondrial carbonic anhydrase to H_2CO_3, which in turn gives up H^+ to the alkaline matrix (14). On the other hand, the permeant anions NO_3^-, SCN^-, ClO_3^-, and ClO_4^- pass through the matrix as such and thus cannot give up a proton to the alkaline matrix.

Entry of the proton-yielding acids into respiring mitochondria converts the electrochemical H^+ gradient generated by electron transport into a neutral (in the acid-base sense) negative-inside anion gradient (Fig. 2). Such gradients were concluded to be responsible for "pulling" Ca^{2+} into the matrix electrophoretically via the electrogenic uniport process, to yield an electroneutral energy-dependent accumulation of the Ca^{2+} salt. Actually, the entry of Ca^{2+} may either precede or follow the entry of proton-yielding acid, as shown in Fig. 3; however, available observations cannot distinguish

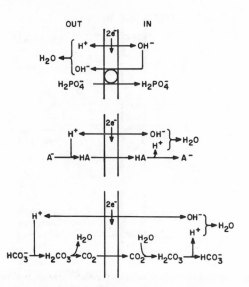

Fig. 2. Transport of proton-yielding acids through the membrane in response to the protonmotive force generated by electron transport.

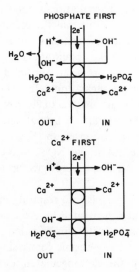

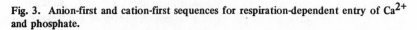

Fig. 3. Anion-first and cation-first sequences for respiration-dependent entry of Ca^{2+} and phosphate.

between the two possible sequences. In the intact liver cell, in which mitochondria are respiring between states 4 and 3, and in which both phosphate and bicarbonate buffers are present in the cytosol in high concentrations, it may be presumed there is a standing respiration-dependent anion gradient across the membrane at all times. For this reason we prefer the anion-first formulation of Fig. 3.

The vectorial equations in Figs. 2 and 3 are not balanced for charge or mass; they merely show the qualitative nature of the respiration-generated pulling force bringing Ca^{2+} and the anion into the matrix. It is, in fact, difficult to establish the stoichiometry precisely and to write such equations in balanced form when phosphate is the anion, because it may occur in three different ionic species and because the precise matrix pH is unknown. Somewhat easier to interpret is the stoichiometry of entry of CO_2 with Ca^{2+}. Here we found, with ^{14}C-labeled $HCO_3^-–CO_2$ mixtures, that the accumulated salt has a $^{45}Ca/^{14}C$ ratio of exactly 1.0 (14), corresponding to the salt $CaCO_3$. In this case, therefore, 2.0 Ca^{2+} and 2.0 CO_2 were found to enter per $\sim$. As is shown in Fig. 4, this stoichiometry is best accounted for by assuming that the $H^+/\sim$ ratio is 4.0, rather than 2.0.

In order to obtain more accurate and totally unambiguous measurements of Ca^{2+} uptake, anion uptake, and H^+ ejection, without the complications of multiple ionization stages in the anion, we have carried out a quantitative study (15) of the movements of Ca^{2+} and the anions of various weak

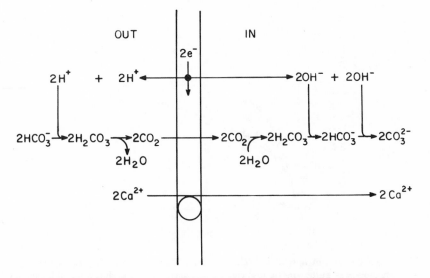

Fig. 4. **Vectorial diagram accounting for the accumulation of CaCO₃ by respiring mitochondria.**

monocarboxylic acids such as acetic, propionic, butyric and 3-hydroxy-
butyric acids, accompanying the transport of Ca^{2+} into respiring rat liver
mitochondria. These weak lipophilic acids pass the inner membrane as such.
In the matrix they dissociate to yield their anions and protons; the latter
are absorbed by the alkalinity developed in the mitochondrial matrix by
electron transport. Mitochondria were suspended in an isotonic medium
weakly buffered with Tris or HEPES buffer supplemented with the potassium
salt of the ^{14}C-labeled weak acid. The substrate was succinate and the
system was supplemented with rotenone to abolish respiration from endogenous site 1 substrates. After a period of state 4 respiration, a pulse of
$^{45}Ca^{2+}$ was then added and the oxygen consumption and proton movements
recorded during the ensuing respiratory stimulation, followed by centrifugation of the mitochondria. The ^{14}C-anion and $^{45}Ca^{2+}$ in the mitochondrial
pellet as well as in the remaining medium were determined. Corrections
were applied for the extra-mitochondrial water space of the pellets with
^{14}C-sucrose and ^3H_2O.

Fig. 5 shows the time course of Ca^{2+} and β-hydroxybutyrate (BOH⁻)
uptakes, which were corrected for non-energized distribution of BOH⁻ across
the membrane. These data already show a BOH⁻/Ca^{2+} accumulation ratio of
~ 2.0. Fig. 6 shows data collected from several experiments with different
mitochondrial preparations and different time intervals. The amounts of

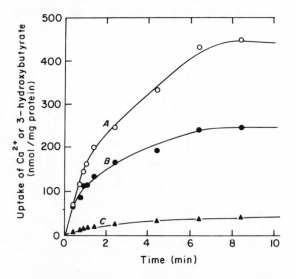

Fig. 5. Time course of uptake of $^{45}Ca^{2+}$ (B) and ^{14}C-labeled β-hydroxybutyrate (A) in
rotenone-poisoned mitochondria respiring on succinate. Curve C shows the non-energized uptake of β-hydroxybutyrate.

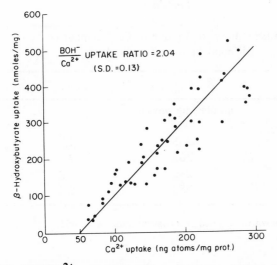

Fig. 6. The BOH⁻/Ca^{2+} ratio, obtained from the slopes of plots of Ca^{2+} vs. BOH⁻ uptake in a series of experiments.

Ca^{2+} taken up in such experiments were quite large, thus minimizing the contribution of super-stoichiometric Ca^{2+} uptake, which is relatively small (16) and can in any case be abolished with oligomycin (17). A best line was drawn through the points by the method of orthogonal regression; its slope is equal to the BOH⁻/Ca^{2+} ratio. By this procedure no corrections are necessary for respiration-independent Ca^{2+} binding. The slope was found to be 2.04 in this series of measurements; thus, for every Ca^{2+} accumulated 2 BOH⁻ also enter to yield the soluble $Ca(BOH)_2$ salt. Table I shows the results of similar experiments with acetate, propionate, and butyrate. In all cases the anion⁻/Ca^{2+} ratio approached 2.0.

The precise value of the Ca^{2+}/O and anion/O ratios were then determined, from which the Ca^{2+}/∼ and anion⁻/∼ ratios were obtained. Such experiments were carried out with a wide range of Ca^{2+} and anion concentrations. Typical results are given in Fig. 7, in which the lines were fitted by the least squares method. From the slopes the Ca^{2+}/∼ ratio was found to be 1.9 and the anion⁻/∼ was 4.1. These data, taken together with the stoichiometry of $CaCO_3$ loading reported earlier (14), clearly show that four positive charges in the form of two Ca^{2+} ions and four negative charges in the form of four monocarboxylate anions are transported per 2 e⁻ per site.

We may now construct vectorial equations to account for the observed stoichiometry, assuming that Ca^{2+} enters by electrophoretic uniport, the only pathway of Ca^{2+} entry consistent with the anaerobic osmotic swelling

TABLE I

The Stoichiometry of Uptake of Ca^{2+} Supported by
Different Monocarboxylic Acids

Acid	Ca^{2+} uptake	Acid/Ca^{2+}	$\pm$S.D.
	nmoles/mg protein		
Acetic	265	1.7	$\pm$0.1
Propionic	265	2.2	$\pm$0.2
Butyric	285	2.0	$\pm$0.2
3-Hydroxybutyric	250	2.0	$\pm$0.1

The various acids shown were added at a concentration of 20 mM; Ca^{2+} was present
at 400 nmoles/mg protein.

experiments, indeed, the only pathway with any experimental support. The
observed stoichiometry is most simply accounted for by assuming electron
transport causes ejection of 4 H^+ (or 4 positive charges) per 2 e^- per site
(Fig. 8). Ejection of 4 H^+ can pull in four monocarboxylate anions (in the
form of HA), to generate a negative inside gradient of A^-, which is neutralized
by electrophoretic uptake of 2 Ca^{2+}. Ejection of 4 H^+ per 2 e^- per site also

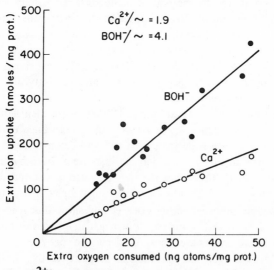

Fig. 7. The $Ca^{2+}/\sim$ and $BOH^-/\sim$ ratios obtained from a series of measurements of the
Ca^{2+}/O and BOH^-/O ratios.

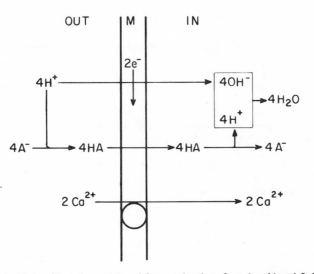

Fig. 8. Vectorial equations for uptake of four molecules of weak acid and 2 Ca²⁺ ions per 2 e⁻ per site.

accounts for accumulation of two molecules of $CaCO_3$ per 2 e⁻ per site observed earlier (14).

It will be noted that if 4 H⁺ are ejected and used to bring in 4 BOH⁻ anions there should be no *net* respiration-dependent H⁺ ejection during Ca²⁺ uptake in the presence of BOH⁻. This was verified by appropriate measurements.

The findings reported here do not necessarily call the chemiosmotic hypothesis into question; they only suggest that some consideration must be given to a basic H⁺/~ ratio of 4 rather than 2. A H⁺/~ ratio of 4 is thermodynamically feasible; indeed, it is consistent with other types of energized ion transport mechanisms. Table II shows that the Na⁺K⁺-stimulated ATPase of the plasma membrane and the Ca²⁺-stimulated ATPase of sarcoplasmic reticulum both move 3 or 4 positive charges per ~ . Moreover, laser-flash single-turnover experiments on photosynthetic electron transport give H⁺/ATP ratios of at least 3 and more likely 4.0 (18). A report by Avron elsewhere in this volume also indicates strongly that the H⁺/~ ratio in photosynthetic electron transport is greater than 2.0 and may be as high as 4.0. Moreover, ATP-dependent accumulation of Ca²⁺ and phosphate in respiration-inhibited mitochondria gives a Ca²⁺/ATP ratio of 2, also equivalent to transport of 4 positive charges per ~P hydrolyzed (19).

We have confirmed Mitchell and Moyle's finding of a H⁺/~ ratio of about 2.0 (20) in oxygen pulse experiments under their conditions. However, as

TABLE II

Stoichiometry of Some Energy-Coupled Ion Transport Systems

		Positive charges transported per $\sim$
Na^+K^+-ATPase	$Na^+/\sim = 3$	3
Ca^{2+}-ATPase (s.r.)	$Ca^{2+}/\sim = 2.0$	4.0
Ca^{2+}-ATPase (mito.)	$Ca^{2+}/\sim = 2.0$	4.0
Ca^{2+}-mitochondrial electron transport	$Ca^{2+}/\sim = 2.0$	4
H^+-chloroplasts, single turnover	$H^+/\sim = 3\text{-}4$	3-4

Mitchell has noted (21) H^+ ejection in such experiments is accompanied by re-uptake of endogenous Ca^{2+} leaking from the mitochondria anaerobically (22). Under these conditions the H^+/Ca^{2+} ratio can be expected to be about 1.0, i.e. the ratio observed in the presence of phosphate at pH near 7.0, thus resulting in observed $H^+/\sim$ ratios of about 2.0. Clearly, the $H^+/\sim$ ratios of electron transport in mitochondria need to be re-evaluated with more complete accounting of all associated ion movements.

REFERENCES

1. Rossi, C. and Lehninger, A.L. (1963) *Biochem. Z. 338*, 698-713.
2. Rossi, C. and Lehninger, A.L. (1964) *J. Biol. Chem. 239*, 3971-3980.
3. Chance, B. (1965) *J. Biol. Chem. 240*, 2729-2748.
4. Lehninger, A.L., Carafoli, E. and Rossi, C.S. (1967) *Adv. Enzymol 29*, 259-320.
5. Carafoli, E., Gamble, R.L., Rossi, C.S. and Lehninger, A.L. (1967) *J. Biol. Chem. 242*, 1199-1204.
6. Rasmussen, H., Chance, B. and Ogata, E. (1965) *Proc. Nat. Acad. Sci. (U.S.A.) 53*, 1069-1076.
7. Gear, A.R.L., Rossi, C.S., Reynafarje, B. and Lehninger, A.L. (1967) *J. Biol. Chem. 242*, 3403-3413.
8. Rossi, C., Azzone, G.F. and Azzi, A. (1967) *Eur. J. Biochem. 1*, 141-146.
9. Wenner, C.E. and Hackney, J.H. (1967) *J. Biol. Chem. 242*, 5053-5058.

10. Selwyn, M.J., Dawson, A.P. and Dunnett, S.J. (1970) *FEBS Letters* *10*, 1-5.
11. Lehninger, A.L. (1974) *Proc. Nat. Acad. Sci. (U.S.A.)* *71*, 1520-1524.
12. Lehninger, A.L. (1972) In *The Molecular Basis of Electron Transport,* Miami Winter Symposia (Schultz, J. and Cameron, B.F., eds.) Vol. 4, pp. 133-146, Academic Press, New York.
13. Brierley, G.P., Jurkowitz, M., Scott, K.M. and Merola, A.J. (1971) *Arch. Biochem. Biophys. 147*, 545-556.
14. Elder, J.A. and Lehninger, A.L. (1973) *Biochemistry 12*, 976-982.
15. Brand, M.D., Chen, C.-H. and Lehninger, A.L. , submitted to *Biochemistry.*
16. Reynafarje, B. and Lehninger, A.L. (1974) *J. Biol. Chem. 249*, 6067-6073.
17. Brand, M.D. and Lehninger, A.L., submitted to *J. Biol. Chem.*
18. Witt, H.T. (1971) *Q. Rev. Biophys. 4*, 365-477.
19. Bielawski, J. and Lehninger, A.L. (1966) *J. Biol. Chem. 241*, 4316-4322.
20. Mitchell, P. and Moyle, J. (1968) *Biochem. J. 105*, 1147-1162.
21. Mitchell, P. (1969) In *The Molecular Basis of Membrane Functions* (Tosteson, D.C., ed.) pp. 493-498, Prentice-Hall, Englewood Cliffs.
22. Rossi, C.S., Siliprandi, N., Carafoli, E., Bielawski, J. and Lehninger, A.L. (1967) *Eur. J. Biochem. 2*, 332-340.

GLUTATHIONE, γ-GLUTAMYL TRANSPEPTIDASE, AND THE γ-GLUTAMYL CYCLE IN AMINO ACID TRANSPORT

Alton Meister

Department of Biochemistry
Cornell University Medical College
New York, New York 10021

The γ-glutamyl cycle, a series of six enzyme-catalyzed reactions which account for the synthesis and amino acid-dependent degradation of glutathione (Fig. 1 (1)), was formulated in the course of studies in our laboratory on the enzymology of glutamate and γ-glutamyl derivatives. This work, which began with investigations on the synthesis of such γ-glutamyl com-

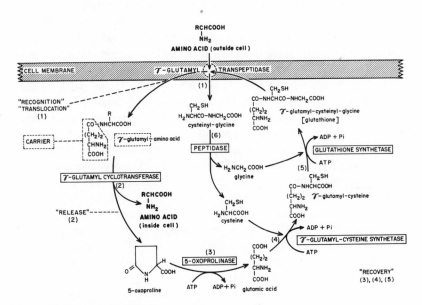

Fig. 1. The γ-glutamyl cycle (1).

95

pounds as glutamine (2-5) and glutathione (5-10), led to isolation of highly purified preparations of the enzymes involved and also provided information about the mechanisms of these reactions, which involve participation of enzyme-bound acyl phosphate intermediates. The γ-glutamyl cycle emphasizes the γ-glutamyl moiety of glutathione rather than the sulfhydryl group of this molecule, but there seem to be significant relationships between these two interesting structural features of this tripeptide, which is found in virtually all mammalian tissues in concentrations ranging from about 0.5 to about 10 mM. Glutathione is not only the most abundant intracellular thiol compound, but it is also, along with glutamine, a quantitatively significant intracellular γ-glutamyl compound (11).

The initial reaction involved in the amino acid-dependent breakdown of glutathione is catalyzed by γ-glutamyl transpeptidase, which transfers the γ-glutamyl moiety of glutathione and of other γ-glutamyl compounds to an amino acid (or peptide) acceptor. The reaction catalyzed by γ-glutamyl

$$\text{Glutathione} + \text{L-amino acid} \rightarrow \text{L-}\gamma\text{-glutamyl-L-amino acid} + \text{L-cysteinylglycine} \tag{1}$$

transpeptidase was elucidated by Hanes *et al.* (12), who suggested that it might play a role in protein synthesis; at about the same time, several workers (13-16) suggested that γ-glutamyl transpeptidase might function in amino acid transport, and this idea has been mentioned again by others. γ-Glutamyl transpeptidase is a membrane-bound enzyme which is associated with the epithelial structures of many mammalian tissues including those of the proximal renal tubule (17-19), choroid plexus (20,21), ciliary body (22), intestinal villi (22,23), seminal vesicles (24,25) and others (26). Earlier work on the purification of the enzyme led to isolation of relatively high molecular weight complexes ($> 250{,}000$)(27,28), which now appear to contain other proteins as well as γ-glutamyl transpeptidase (29). In recent work, enzyme preparations exhibiting much higher specific activities and a much lower molecular weight ($\sim 60{,}000$) have been obtained by procedures involving enzymatic fractionation of such complexes (29). γ-Glutamyl transpeptidase catalyzes transfer of the γ-glutamyl group to a variety of amino acids. Cystine is one of the most active (30); glutamine, methionine, serine, alanine, and glycine are also very active acceptors of the γ-glutamyl moiety (31). It is notable that cystine is a good acceptor of the γ-glutamyl group even at low concentrations (0.06-0.2 mM)(30); the concentrations of cystine in body fluids and tissues are in this low range. Glutamine is also a good acceptor of the γ-glutamyl group; this amino acid is present in relatively high concentrations in blood plasma (about 0.6 mM) and in most tissues.

γ-Glutamyl transpeptidase also exhibits a broad specificity with respect to the γ-glutamyl donor; thus, it reacts effectively with many γ-glutamyl amino acids and with S-substituted glutathione derivatives. However, it interacts only at very low rates with glutathione disulfide and with glutamine (although this amino acid is a good acceptor of the γ-glutamyl moiety). The enzyme acts rapidly on γ-glutamyl-p-nitroanilide to yield p-nitroaniline, thus providing a simple colorimetric procedure for the determination of enzymatic activity (32). It also acts on γ-glutamyl naphthylamide, a substrate which has been widely used in histochemical studies of the enzyme (17-26). In this approach, the liberated amine is coupled with a diazonium salt to yield an insoluble azo dye. Application of this method has shown that the enzyme has a membraneous localization in many mammalian tissues. Localization to the brush border of the proximal renal tubule has also been established in studies in which cell fractionation procedures were employed (33,34). The cysteinyl-glycine formed in the transpeptidation reaction with glutathione is cleaved by dipeptidase (35). The γ-glutamyl amino acids formed in transpeptidation are substrates of γ-glutamyl cyclotransferase (36-41), which catalyzes the following reaction:

$$\text{L-}\gamma\text{-Glutamyl-L-amino acid} \rightarrow \text{L-amino acid} + \text{5-oxo-L-proline*} \qquad (2)$$

The best substrates for this enzyme include the γ-glutamyl derivatives of glutamine, cystine, cysteine, methionine and alanine. Other γ-glutamyl amino acids can be converted to 5-oxoproline and the corresponding free amino acids by coupled reactions involving γ-glutamyl transpeptidase and γ-glutamyl cyclotransferase, as previously discussed (1); glutamine may play an important role in such reactions.

Two observations in our laboratory were of particular importance in leading to the formulation of the γ-glutamyl cycle. The first of these was the finding that rat kidney exhibits high activities of γ-glutamylcysteine and glutathione synthetases (42). The other was the observation that 5-oxo-proline is rapidly metabolized by intact animals (43), and also by tissue slices (44); the major initial product formed was found to be glutamate (44,45). Kidney contains high activities of γ-glutamyl transpeptidase and γ-glutamyl cyclotransferase, and since the steady state concentration of glutathione in rat kidney is about 2-3mM, it seemed that kidney must have the enzymes that catalyze the synthesis of glutathione. It was found that the potential enzymatic activity available for glutathione synthesis in the

---

*   **Synonyms: pyrrolidone carboxylate; pyroglutamate.**

kidney was similar to that available for its degradation. However, the breakdown of glutathione leads to the formation of 5-oxoproline, a compound that had not previously been thought to be a metabolite. Indeed, this compound has long been known as a non-enzymatic cyclization product of γ-glutamyl derivatives such as glutamine. 5-Oxoproline does not accumulate in kidney, nor is it excreted in the urine. We therefore looked for an enzyme that might act on 5-oxoproline, and this led to the discovery of 5-oxoprolinase, which catalyzes the ATP-dependent decyclization of 5-oxoproline according to the following reaction (45):

$$\text{5-Oxo-L-proline} + \text{ATP} + 2 \text{ H}_2\text{O} \;\rightarrow\; \text{L-glutamate} + \text{ADP} + \text{Pi} \qquad (3)$$

The findings indicate that 5-oxoproline is a metabolite of glutathione and that it is not formed as an artifact. The reaction catalyzed by 5-oxoprolinase is of significance in that it links the reactions involved in the amino acid-dependent degradation of glutathione with those which lead to its synthesis, and thus makes it possible to visualize a cycle (Fig. 1).

To turn, the cycle requires ATP, glutathione and an amino acid. Each turn of the cycle requires a molecule of amino acid and forms and uses one molecule each of 5-oxoproline and glutathione. The cycle appears to fulfill the requirements of a transport system. Thus, γ-glutamyl transpeptidase, a membrane bound enzyme, might function in the translocation step of transport. This enzyme is postulated to interact with extracellular amino acid and intracellular glutathione. A hypothetical scheme has been proposed (1) according to which the extracellular amino acid binds non-covalently to a site on the membrane surface. A group on the membrane-bound transpeptidase interacts with the γ-glutamyl moiety of glutathione (intracellular) to yield a γ-glutamyl-enzyme. Attack of the amino acid nitrogen atom on the γ-carbon atom of the γ-glutamyl-enzyme leads to formation of a γ-glutamyl-amino acid, thus removing the amino acid from its initial binding site. The amino acid then enters the cell as a γ-glutamyl derivative. Such movement might be accompanied by conformational changes in the membrane associated with the binding of substrates to the membrane. The γ-glutamyl carrier is cleaved from the γ-glutamyl amino acid within the cell by the action of intracellular γ-glutamyl cyclotransferase.

It has often been suggested that a mobile carrier is involved in amino acid transport (see, for example, (46)). Such a carrier would bind the amino acid and carry it across the membrane; the carrier would be degraded to an inactive form after delivering its amino acid inside the cell, and it would then need to be reactivated by processes requiring energy. The γ-glutamyl cycle might function as such a transport system; with γ-glutamyl transpeptidase functioning in recognition and translocation. γ-Glutamyl cyclotransferase would release the amino acid from its γ-glutamyl carrier. The cycle includes several energy-requiring "recovery" steps (Fig. 1).

There are now many data which support the view that the γ-glutamyl cycle functions in a number of mammalian tissues. The enzymes of the cycle are widely distributed, as is glutathione, and there is evidence that 5-oxoproline is a significant metabolite of glutathione. These considerations indicate that the γ-glutamyl cycle exists in many tissues as a metabolic pathway. It is clearly more difficult to prove the hypothesis that the cycle functions in amino acid transport. Nevertheless, a number of findings are consistent with this idea. It should be emphasized that we are not proposing that the γ-glutamyl cycle is the only amino acid transport system; there are undoubtedly others. The broad amino acid specificity of the membrane-bound transpeptidase suggests that the γ-glutamyl cycle might mediate transport of the "high capacity-low specificity" type (47). The most active amino acid acceptors include a number of neutral amino acids such as cystine, glutamine, methionine, serine, alanine and glycine (30,31). L-Cystine is an active acceptor of the γ-glutamyl group at very low concentrations which are not far from the concentrations of L-cystine found in mammalian blood plasma (30). At low concentrations, L-cystine is somewhat more active than L-glutamine. γ-Glutamylcystine is a good substrate of γ-glutamyl cyclotransferase, but it is possible that this substrate is also reduced to yield cysteine and γ-glutamylcysteine, thus yielding substrates for both glutathione and γ-glutamylcysteine synthetases (30). It may be significant that glutathione, which is the most abundant intracellular form, is a much better γ-glutamyl donor than glutathione disulfide, while the predominant extracellular form, cystine, is a better γ-glutamyl acceptor substrate than cysteine.

Evidence that the cycle functions *in vivo* has come from studies in which mice were given the 5-oxoprolinase inhibitor, L-2-imidazolidone-4-carboxylate. When the inhibitor is injected into mice the utilization of 5-oxoproline is markedly decreased and therefore 5-oxoproline accumulates in the tissues, e.g., brain, kidney, liver, eye; in addition, unusually large amounts of 5-oxoproline are excreted in the urine (48,49). In experiments in which the inhibitor was administered together with one of several L-amino acids, the accumulation of 5-oxoproline in the tissues was significantly increased, as was the amount excreted in the urine. These observations indicate that the cycle functions *in vivo*, and they also support the conclusion that 5-oxoproline is a quantitatively significant metabolite of glutathione.

In other experiments, rats were injected with tracer amounts of L-[^{14}C]-glutamate; this was followed by rapid labeling of the 5-oxoproline of liver and kidney (50). In another approach, the overall turnover of glutathione in kidney and liver was measured in experiments in which L-[^{14}C] glutamate was administered to mice (51). Turnover was much more rapid in kidney than in liver; the first-order rate constants for glutathione synthesis from glutamate in liver and kidney were not far from those found for the synthesis of glutathione from 5-oxoproline in these tissues. Thus, 5-oxoproline and glutamate are used in an essentially equivalent manner for glutathione synthesis. It is well known that the kidney is very active in amino acid transport and the finding of high activities of the γ-glutamyl cycle enzymes in this organ, as well as a rapid overall turnover of glutathione, are consistent with the view that the cycle plays a role in renal amino acid transport. It may be relevant to cite the finding of small amounts of several γ-glutamyl amino acids in normal human urine (52).

High activities of the γ-glutamyl cycle enzymes have also been found in the choroid plexus (21); there is good evidence that the choroid plexus functions in transport phenomena at the blood cerebro-spinal fluid barrier. There would seem to be an analogy between the formation of urine by the nephron and the secretion of cerebrospinal fluid by the choroid plexuses which are probably the sites at which transport occurs. A number of γ-glutamyl amino acids have been found in brain (53-56).

The ciliary body of the eye also contains high activities of the γ-glutamyl cycle enzymes; the transpeptidase is localized in the basal portions of the epithelial cells of the ciliary processes (22). These findings suggest that the γ-glutamyl cycle may function in the transport of amino acids across the blood-aqueous humor barrier. The epithelial cells of the intestine, especially the jejunum, contain high activities of the γ-glutamyl cycle enzymes and the transpeptidase is localized in the epithelial cells located at the tips of the villi (22,23,57).

Thus far, two kinds of inborn errors of metabolism in man have been found in which there are specific enzymatic blocks of the γ-glutamyl cycle (58). In one of these (5-oxoprolinuria) the patients excrete large amounts of 5-oxoproline in the urine; they exhibit metabolic acidosis which requires continuous therapy with sodium bicarbonate (59,60). Studies in our laboratory on erythrocytes, fibroblasts, and placenta from these patients showed that they have a marked deficiency of glutathione synthetase and that their 5-oxoprolinuria is secondary to this enzyme defect (61). The excessive excretion of 5-oxoproline seems to result from overproduction of γ-glutamylcysteine which, in contrast to glutathione, is a good substrate for γ-glutamyl cyclotransferase. The evidence indicates that the patients with 5-oxoprolinuria have a modified form of the γ-glutamyl cycle, which uses only four enzymes (γ-glutamyl transpeptidase, γ-glutamyl cyclotransferase, 5-oxoprolinase and γ-glutamylcysteine synthetase). γ-Glutamylcysteine is a substrate of γ-glutamyl transpeptidase and also of γ-glutamyl cyclotransferase. Thus, in this disease there is a greater than normal formation of 5-oxoproline due to increased formation of γ-glutamylcysteine. The formation of 5-oxoproline exceeds the capacity of 5-oxoprolinase to convert this substrate to glutamate and, therefore, some of the 5-oxoproline formed (about 25% of the total (60)) is excreted in the urine. There is thus a futile cycle of γ-glutamylcysteine synthesis followed by its conversion to 5-oxoproline and cysteine. Since γ-glutamylcysteine functions in transpeptidation, the modified γ-glutamyl cycle in this disease can mediate amino acid transport and, therefore, a defect in amino acid transport would not be expected, nor has such been observed in these patients. In related work in our laboratory, it was shown that glutathione exerts feedback inhibition on γ-glutamylcysteine synthetase (62). Inhibition is evidently due to binding of glutathione to the glutamate and cysteine sites of γ-glutamylcysteine synthetase; such inhibition is therefore non-allosteric in type. Inhibition by glutathione probably plays an important role in the regulation of glutathione biosynthesis under normal conditions, but in the disease 5-oxoprolinuria, the relatively low levels of glutathione lead to increased synthesis of γ-glutamylcysteine.

In another inborn error of metabolism, there is a block at the γ-glutamylcysteine synthetase step of the cycle (63); such a block should stop the cycle and produce a defect in amino acid transport, e.g., aminoaciduria, if indeed the hypothesis that the cycle plays a role in renal handling of amino acids is correct. Two such patients who exhibit γ-glutamylcysteine synthetase deficiency of an apparently generalized nature have been studied; these patients have hemolytic anemia associated with decreased erythrocyte glutathione concentrations and they also have serious central nervous system disease. In contrast to the patients with 5-oxoprolinuria, these patients are deficient in both glutathione and in γ-glutamylcysteine and are therefore markedly deficient in substrate for γ-glutamyltranspeptidase. It is of interest in relation to our hypothesis that these patients exhibit aminoaciduria (1,64).

It is evident that a block at the γ-glutamyl transpeptidase step should also stop the cycle and produce defects in amino acid transport. Recent studies in our laboratory are consistent with the view that the γ-glutamyl transpeptidase reaction involves at least two steps (31): (a) formation of a γ-glutamyl-enzyme; and (b) reaction of the γ-glutamyl-enzyme with an acceptor. Maleate was found to inhibit the second step (65). Thus, when the γ-glutamyl transpeptidase reaction was carried out in the presence of maleate, there was an increase in hydrolysis of the γ-glutamyl donor and a marked decrease in the transpeptidase reaction. Presumably maleate acts so as to facilitate reaction of the γ-glutamyl-enzyme with water and thus impairs the ability of the enzyme to transfer the γ-glutamyl moiety to an amino acid acceptor. If such an effect were to occur *in vivo*, one would expect aminoaciduria. Such experiments have been carried out in several laboratories and it has been established that administration of maleate to experimental animals produces aminoaciduria (66-70). In studies in which amino acid transport in the kidney was studied *in vivo* by use of a micro-injection technique, it was concluded that maleate produces aminoaciduria by increasing efflux of amino acid, leading to, or secondary to, a loss of cellular amino acid accumulation (70). These *in vivo* experiments when considered in relation to the *in vitro* experiments on the effect of maleate of γ-glutamyl transpeptidase are consistent with the proposed function of the transpeptidase in amino acid transport. However, it is known that administration of maleate also produces excessive excretion of phosphate and glucose (66,71), and there is evidence that maleate can interfere with enzymes other than transpeptidase and that it can react with sulfhydryl compounds (72,73). Thus, although the findings are consistent with our hypothesis, other interpretations of the findings are possible. It would be of considerable interest to carry out studies with a specific transpeptidase inhibitor, and we hope to do such experiments.

CONCLUSIONS

The evidence indicates that the γ-glutamyl cycle functions *in vivo* in a number of mammalian tissues.

The data show that 5-oxoproline is a significant metabolite of glutathione and that it is converted by the action of 5-oxo-L-prolinase to L-glutamate in a reaction coupled with the cleavage of ATP to ADP and inorganic phosphate.

Glutathione inhibits its own synthesis by inhibiting the activity of γ-glutamylcysteine synthetase; this seems to be a physiologically significant feedback control mechanism.

Studies in which blocks or partial blocks of the γ-glutamyl cycle were observed or achieved are consistent with its proposed function in amino acid transport: (a) Patients with an inborn error of metabolism associated with marked deficiency of γ-glutamylcysteine synthetase exhibit aminoaciduria. (b) Aminoaciduria occurs after administration of maleate to animals; this result can be ascribed to a block by maleate of the γ-glutamyl transpeptidase step of the cycle. (c) Administration to animals of an inhibitor of 5-oxoprolinase does not produce aminoaciduria, but 5-oxoproline accumulates in the tissues. This block would not be expected to stop the cycle because there are other pathways of glutamate formation (e.g., reductive amination of α-ketoglutarate and hydrolysis of glutamine to glutamate). (d) A block at the glutathione synthetase step of the cycle does not produce aminoaciduria in patients with the disease 5-oxoprolinuria; such individuals are able to produce γ-glutamylcysteine, and this γ-glutamyl donor can be used in transpeptidation reactions by means of a modified form of the γ-glutamyl cycle.

The findings support the view that the γ-glutamyl cycle is one of the systems that functions in the active transport of amino acids across cell membranes in mammalian tissues. It is not proposed that this is the only amino acid transport system; the quantitative significance of the γ-glutamyl cycle may vary in different tissues and at different stages of development. While the data support the idea that the cycle functions in certain mammalian tissues, they do not conclusively demonstrate its proposed role in amino acid transport, which nevertheless remains as an attractive working hypothesis which hopefully may stimulate further investigations.

REFERENCES

1. Meister, A. (1973) *Science 180*, 33-39; Mini-review (1974) *Life Sciences 15*, 177-190.
2. Levintow, L. and Meister, A. (1953) *J. Am. Chem. Soc. 75*, 3039; (1954) *J. Biol. Chem. 209*, 265-280.
3. Krishnaswamy, P.R., Pamiljans, V. and Meister, A. (1960) *J. Biol. Chem. 236*, PC 39-40; (1962) *ibid. 237*, 2932-2940.
4. Meister, A. (1962) *The Enzymes 6* (second edition), 443-448; (1974) *ibid. 10* (third edition), 699-754.
5. Meister, A. (1968) *Adv. Enzymol. 31*, 183-218; (1969) *Harvey Lectures 63*, 139-178.
6. Nishimura, J.S., Dodd, E.A. and Meister, A. (1963) *J. Biol. Chem. 238*, PC 1179-1180; (1964) *ibid. 239*, 2553-2558.
7. Mooz, E., Dodd, E.A. and Meister, A. (1967) *Biochemistry 6*, 1722-1734.
8. Orlowski, M. and Meister, A. (1971) *Biochemistry 10*, 372-380; (1969) *J. Biol. Chem. 244*, 7095-7105.
9. Richman, P.G., Orlowski, M. and Meister, A. (1973) *J. Biol. Chem. 248*, 6684-6690.
10. Meister, A. (1974) *The Enzymes 10* (third edition), 671-697.
11. Meister, A. (1975) In *Biochemistry of Glutathione, Metabolism of Sulfur Compounds* (Greenberg, D.M., ed.) pp. 101-188, Academic Press, New York.
12. Hanes, C.S., Hird, F.J.R. and Isherwood, F.A. (1950) *Nature 166*, 288-292; (1952) *Biochem. J. 51*, 25-35.
13. Hird, F.J.R. (1950) *The γ-Glutamyl Transpeptidation Reaction.* Doctoral dissertation, Cambridge University, England.
14. Binkley, F. (1951) *Nature 167*, 888-889; (1954) In *Symposium on*

Glutathione (Colowick *et al.*, eds.) p. 160, Academic Press, New York.

15. Springell, P.H. (1953) *Amino Acid Metabolism with Special Reference to Peptide Bond Transfer.* Doctoral thesis, Melbourne University, Australia.

16. Ball, E.G., Cooper, O. and Clarke, E.C. (1953) *Biol. Bull. 105*, 369-370.

17. Albert, Z., Orlowski, M. and Szewczuk, A. (1961) *Nature 191*, 767-768.

18. Glenner, G.G. and Folk, J.E. (1961) *Nature 192*, 338-339.

19. Glenner, G.G., Folk, J.E. and McMillan, P.J. (1962) *J. Histochem. Cytochem. 10*, 481-489.

20. Albert, Z., Orlowski, M., Rzucidlo, Z. and Orlowska, J. (1966) *Acta Histochem. 25*, 312-320.

21. Tate, S.S., Ross, L.L. and Meister, A. (1973) *Proc. Nat. Acad. Sci. (U.S.A.) 70*, 1447-1449.

22. Ross, L.L., Barber, L., Tate, S.S. and Meister, A. (1973) *Proc. Nat. Acad. Sci. (U.S.A.) 70*, 2211-2214.

23. Greenberg, E., Wollaeger, E.E., Fleisher, G.A. and Engstrom, G.W. (1967) *Clin. Chim. Acta 16*, 79-89.

24. Goldbarg, J.A., Friedman, O.M., Pineda, P., Smith, E.E., Chatterji, R., Stein, E.H. and Rutenberg, A.M. (1960) *Arch. Biochem. Biophys. 91*, 61-70.

25. DeLap, L.W., Tate, S.S. and Meister, A. (1975) *Life Sciences 16*, 691-704.

26. Meister, A., Tate, S.S. and Ross, L.L. (in press) In *Membrane Bound Enzymes* (Martonosi, A., ed.) Plenum Press, New York.

27. Szewczuk, A. and Barnowski, T. (1963) *Biochem. Zeit. 338*, 317-329.

28. Orlowski, M. and Meister, A. (1965) *J. Biol. Chem. 240*, 338-347.

29. Tate, S.S. and Meister, A. (in press) *J. Biol. Chem.*

30. Thompson, G.A. and Meister, A. (in press) *Proc. Nat. Acad. Sci. (U.S.A.)*

31. Tate, S.S. and Meister, A. (1974) *J. Biol. Chem. 249*, 7593-7602.

32. Orlowski, M. and Meister, A. (1963) *Biochim. Biophys. Acta 73*, 679-681.

33. Glossman, H. and Neville, D.M. (1972) *FEBS Letters 19*, 340-344.

34. George, S.G. and Kenny, A.J. (1973) *Biochem. J. 134*, 43-57.

35. Semenza, G. (1957) *Biochim. Biophys. Acta 24*, 401-413.

36. Connell, G.E. and Hanes, C.S. (1956) *Nature 177*, 377-378.

37. Cliffe, E.E. and Waley, S.G. (1961) *Biochem. J. 79*, 118-128.

38. Connell, G.E. and Szewczuk, A. (1967) *Clin. Chim. Acta 17*, 423-430.

39. Kakimoto, Y., Kanazawa, A. and Sano, I. (1967) *Biochim. Biophys. Acta 132*, 472-480.

40. Orlowski, M., Richman, P.G. and Meister, A. (1969) *Biochemistry 8*, 1048-1055.

41. Adamson, E.D., Szewczuk, A. and Connell, G.E. (1971) *J. Biochem.* *49*, 218-226.
42. Orlowski, M. and Meister, A. (1970) *Proc. Nat. Acad. Sci. (U.S.A.) 67*, 1248-1255.
43. Richman, P.G., data obtained in this laboratory in 1969-1970; (1974) Doctoral dissertation, Cornell University Graduate School of Medical Sciences, New York.
44. Van Der Werf, P., data obtained in this laboratory in 1970; (1974) Doctoral dissertation, Cornell University Graduate School of Medical Sciences, New York.
45. Van Der Werf, P., Orlowski, M. and Meister, A. (1971) *Proc. Nat. Acad. Sci. (U.S.A.) 68*, 2982-2985.
46. Heinz, E. (1972) *Metabolic Pathways 6* (Transport of Amino Acids by Animal Cells), 455-501.
47. Young, J.A. and Freedman, B.S. (1971) *Clin. Chem. 17*, 245-266.
48. Van Der Werf, P., Stephani, R.A., Orlowski, M. and Meister, A. (1973) *Proc. Nat. Acad. Sci. (U.S.A.) 70*, 759-761.
49. Van Der Werf, P., Stephani, R.A. and Meister, A. (1974) *Proc. Nat. Acad. Sci. (U.S.A.) 71*, 1026-1029.
50. Van Der Werf, P. and Meister, A. (in press) *Adv. Enzymol.*
51. Sekura, R. and Meister, A. (1974) *Proc. Nat. Acad. Sci. (U.S.A.) 71*, 2969-2972.
52. Buchanan, D.L., Haley, E.E. and Markiw, R.T. (1962) *Biochemistry 1*, 612-620.
53. Kanazawa, A., Kakimoto, Y., Nakajima, T. and Sano, J. (1965) *Biochim. Biophys. Acta 111*, 90-95.
54. Kakimoto, Y., Nakajima, T., Kanazawa, M., Takesada, M. and Sano, J. (1964) *Biochim. Biophys. Acta 93*, 333-338.
55. Reichelt, K.L. (1970) *J. Neurochem. 17*, 19-25.
56. Versteeg, D.H.G. and Witter, A. (1970) *J. Neurochem. 17*, 41-52.
57. Cornell, J., unpublished studies in this laboratory.
58. Meister, A. (1974) *Ann. Int. Med. 81*, 247-253.
59. Jellum, E., Kluge, T., Borresen, H.C., Stokke, O. and Eldjarn, L. (1970) *Scand. J. Clin. Lab. Invest. 26*, 327-335.
60. Hagenfeldt, L., Larsson, A. and Zetterstrom, R. (1973) *Acta Paediat. Scand. 62*, 1-8.
61. Wellner, V.P., Sekura, R., Meister, A. and Larsson, A. (1974) *Proc. Nat. Acad. Sci. (U.S.A.) 71*, 2505-2509.
62. Richman, P. and Meister, A. (1975) *J. Biol. Chem. 250*, 1422-1426.
63. Konrad, P.N., Richards, F., Valentine, W.N., and Paglia, D.E. (1972) *New Eng. J. Med. 286*, 557.

64. Richards, F., Cooper, M.R., Pearce, L.A., Cowan, R.J. and Spurr, C.L. (1974) *Arch. Intern. Med. 134*, 534-537.
65. Tate, S.S. and Meister, A. (1974) *Proc. Nat. Acad. Sci. (U.S.A.) 71*, 3329-3333.
66. Harrison, H.E. and Harrison, H.C. (1954) *Science 120*, 606-608.
67. Worthen, H.G. (1963) *Lab. Investig. 12*, 791-801.
68. Rosenberg, L.E. and Segal, S. (1964) *Biochem. J. 92*, 345-352.
69. Angielski, S., Niemiro, R., Makarewicz, W. and Rogulski, J. (1958) *Acta Biochim. Polonica 5*, 431-436.
70. Bergeron, M. and Vadeboncoeur, M. (1971) *Nephron. 8*, 367-374.
71. Berliner, R.W., Kennedy, T.J. and Hilton, J.G. (1950) *Proc. Soc. Exp. Biol. Med. 75*, 791-794.
72. Webb, J.L. (1966) In *Enzyme and Metabolic Inhibitors*, Vol. 3, pp. 285-335, Academic Press, New York.
73. Morgan, E.J. and Friedmann, E. (1938) *Biochem. J. 32*, 733-742.

MOLECULAR ASPECTS OF ACTIVE TRANSPORT

H.R. Kaback, G. Rudnick, S. Schuldiner, S.A. Short and P. Stroobant

The Roche Institute of Molecular Biology
Nutley, New Jersey 07110

Bacterial cytoplasmic membrane vesicles provide a useful model system for the study of certain active transport systems (1-4). These vesicles are devoid of the cytoplasmic constituents of the intact cell and their metabolic activities are restricted to those provided by the enzymes of the membrane itself, constituting a considerable advantage over intact cells. Since transport by membrane vesicles *per se* is practically nil, the energy source for transport of a particular substrate can be determined by studying which compounds stimulate solute accumulation. In addition, metabolic conversion of the transport substrate and the energy source is minimal, allowing clear definition of the reactions involved.

Vesicles isolated from a variety of bacterial cells catalyze active transport of many different solutes by a respiration-dependent mechanism that does not involve the generation or utilization of ATP or other high-energy phosphate compounds (1-7). In *Escherichia coli and Salmonella typhimurium* vesicles, most of these transport systems are coupled primarily to the oxidation of D-lactate or reduced phenazine methosulfate [PMS] (or pyocyaine) via a membrane-bound respiratory chain with oxygen or, under appropriate conditions (8-10), fumarate or nitrate as terminal electron acceptors. On the other hand, in vesicles prepared from *Staphylococcus aureus*, active transport of amino acids is coupled exclusively to either L-lactate dehydrogenase or α-glycerol-P dehydrogenase depending upon the growth conditions of the parent cells (11). In many cases, the vesicles transport metabolites at rates which are comparable to those of the parent intact cells (12,13).

This discussion is not intended as a general review of the current work in the field of transport, but is concerned with studies of respiration-linked active transport in membrane vesicles isolated from *E. coli*, and emphasis is placed upon recent observations. Thus, methods for the preparation of bacterial membrane vesicles, their morphology and other general properties, and experimental observations related to the role of the P-enolpyruvate-P-transferase system in the vectorial phosphorylation of certain sugars will not be discussed; the reader is referred to other publications (14-17).

GENERAL PROPERTIES OF RESPIRATION–LINKED ACTIVE TRANSPORT IN BACTERIAL MEMBRANE VESICLES

The experiment presented in Fig. 1 represents a typical transport study carried out with vesicles prepared from *E. coli* ML 308-225, an organism which is constitutive for the β-galactoside transport system but lacks β-galactosidase, the enzyme responsible for the first step in the metabolism of lactose in *E. coli*. Many of the experiments carried out with *E. coli* vesicles utilize this transport system because of the wealth of genetic information available for the *lac* operon (18), because it has been demonstrated that the *lac y* gene codes for a single protein (the *lac* carrier protein or M protein)(19), and final

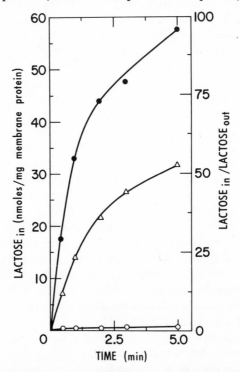

Fig. 1. Lactose transport by E. coli **ML 308-225** membrane vesicles. Experiment was carried out as described in Ref. **15.** ●—●, Ascorbate-phenazine methosulfate; △—△, D-lactate; ○—○, no additions.

ly because this transport system is non-specific enough to accommodate β-galactoside analogues which can be modified chemically in various ways (see below). As shown in Fig. 1, ML 308-225 vesicles exhibit very little lactose uptake in the absence of exogenous electron donors. When D-lactate is added to the reaction mixture, however, there is a dramatic stimulation of both the initial rate and steady-state level of lactose accumulation, and the effect is even more marked when an artificial electron donor system consisting of ascorbate plus PMS is added. If, at the steady-state, the reaction mixtures are gassed with argon or certain electron transport inhibitors or uncoupling agents are added, the lactose accumulated by the vesicles is lost virtually quantitatively within a short time (20). This observation is consistent with the observation that essentially all of the lactose taken up by the vesicles can be recovered in an unmodified form. It is also noteworthy that membrane vesicles prepared from cells which are devoid of the β-galactoside transport system do not exhibit significant transport of a variety of β-galactosides.

Kinetic studies demonstrate that the apparent K_m for D-lactate-dependent lactose influx in the vesicles is considerably lower than the apparent K_m for efflux, although the apparent V_{max}'s for influx and efflux are equal (20). These findings suggest that one of the primary effects of energy-coupling is to decrease the affinity of the carrier for its substrate on the inner surface of the membrane. Moreover, the ability of the vesicles to concentrate substrate against a given concentration gradient is related to the quotient between the apparent K_m's for efflux and influx. For instance, the apparent K_m for lactose efflux is approximately 20 mM, while the analogous value for influx is approximately 0.2 mM, yielding a quotient of 100. Since the vesicles contain about 2.2 μl of intravesicular fluid per mg of membrane protein, it can be calculated readily from the data presented in Fig. 1 that the vesicles establish a steady-state concentration gradient of approximately 100 in the presence of ascorbate-PMS, assuming that all of the intravesicular lactose is in free solution. Similar relationships have been observed qualitatively with many of the other transport systems studied in the vesicles (11,12).

SIDEDNESS OF MEMBRANE VESICLES AND SPECIFICITY OF D-LACTATE AS AN ELECTRON DONOR FOR ACTIVE TRANSPORT

One of the most striking and therefore controversial aspects of the vesicle system is the degree of specificity of the physiologic electron donors which drive active transport. Thus, in *E. coli* vesicles, of a large number of potential energy sources tested, only L-lactate, succinate, α-glycerol-P, and NADH replace D-lactate or reduced PMS to any extent whatsoever, and none of these compounds is as effective although they are oxidized at least as rapidly as D-lactate (12,20-23). It should be emphasized that incubation of the vesicles with radioactive D-lactate, L-lactate, succinate, or α-glycerol-P results in stoichiometric conversion to pyruvate, fumarate, or dihydroxyacetone-P, respectively (7,21,22). Thus, in each case, the ability of these compounds to drive active transport is related to a clearly defined enzymatic reaction.

Since each electron donor reduces the same membrane-bound cytochromes, both qualitatively and quantitatively (7,23), it was suggested that the energy-coupling site for active transport is located in a relatively specific segment of the respiratory chain between D-lactate dehydrogenase (D-LDH) and cytochrome b_1 (1-4,20,23), the first cytochrome in the common portion of the *E. coli* respiratory chain. It has been argued, however, that a significant number of membrane vesicles become inverted during preparation, and that these inverted vesicles oxidize NADH and other electron donors but do not catalyze active transport (24-26). There is a considerable body of evidence that this cannot be the case, and that all of the vesicles in the preparation have the same orientation as the membrane in the intact cell. Some of this evidence is as follows:

(1) Initial rates of transport in the vesicles are, in many cases, similar to those observed in whole cells (12,18). Moreover, in most instances, the steady-state level of accumulation of transport substrates is comparable to that observed in the intact cell.

(2) Freeze fracture studies of membrane vesicles in at least three different laboratories (1-4,14,27,28) demonstrate that the "texture" of the convex surface of the vesicles is distinctly different than that of the concave surface, and that the vesicles are homogeneous in this respect. Moreover, the texture observed on the respective surfaces is exactly the same as that observed in the intact cell.

(3) As mentioned previously, all of the electron donors which are oxidized by the vesicles reduce the same cytochromes both qualitatively and quantitatively (7,23). If a percentage of the vesicles were inverted, and only these inverted vesicles oxidized NADH, it is difficult to understand how NADH could reduce all of the cytochrome in the preparations.

(4) Although NADH is generally a poor electron donor for transport in *E. coli* vesicles, it is the best physiological electron donor for transport in *B. subtilis* vesicles which are prepared in a similar manner (6). Moreover, recent experiments carried out in this laboratory (29) demonstrate that addition of ubiquinone (CoQ_1) to *E. coli* ML 308-225 vesicles in the presence of NADH results in rates and extents of lactose and amino acid transport which are comparable to those observed with D-lactate. Since this effect of CoQ_1 is not observed in the presence of NADPH nor in vesicles from which NADH dehydrogenase has been removed, it seems apparent that CoQ_1 is able to shunt electrons from NADH dehydrogenase to an energy-coupling site which is not located in that portion of the respiratory chain between NADH dehydro-

genase and the cytochromes. As such, these observations provide direct evidence for specific localization of the energy-coupling site.

(5) Studies by Reeves *et al.* (30) demonstrate that fluorescence of 1-anilino-8-naphthalene sulfonate (ANS) is dramatically quenched upon addition of D-lactate to *E. coli* ML 308-225 membrane vesicles, an observation similar to that observed in energized mitochondria and ethylenediaminotetraacetic acid-treated intact *E. coli*. In chloroplasts and submitochondrial particles, in which the polarity of the membrane is opposite to that of intact mitochondria, ANS fluorescence is enhanced upon energization. It follows that any inverted membrane vesicles in the preparations would exhibit ANS fluorescence in the presence of D-lactate. Thus, if 50% of the vesicles were inverted, no net change in ANS fluorescence should have been observed by Reeves *et al.* because half of the vesicles would exhibit quenching and half would exhibit enhancement.

Similarly, Rosen and McClees (31) have demonstrated that inverted membrane preparations catalyze calcium accumulation but do not catalyze D-lactate-dependent proline transport. In contrast, vesicles prepared by osmotic lysis (14) do not exhibit calcium transport but accumulate proline effectively in the presence of D-lactate.

(6) Although D-LDH mutants exhibit normal transport and vesicles prepared from these mutants do not exhibit D-lactate-dependent transport, addition of succinate to these vesicles drives transport to the same extent as D-lactate in wild-type vesicles (32). Since succinate oxidation by both wild-type and mutant vesicles is similar, it seems apparent that the coupling between succinate dehydrogenase and transport is increased in the mutant vesicles. In vesicles prepared from double mutants defective in both D-LDH and succinate dehydrogenase, the coupling between L-lactate dehydrogenase and transport is increased, and L-lactate is the best physiological electron donor for transport (32a). Moreover, in vesicles prepared from a triple mutant defective in D-LDH, succinate dehydrogenase, and L-lactate dehydrogenase, the coupling between NADH dehydrogenase and transport is markedly increased, and NADH drives transport as well as D-lactate in wild-type vesicles (32a). In addition, it is noteworthy that vesicles prepared from galactose-grown *E. coli* exhibit high rates and extents of lactose transport in the presence of NADH. These observations indicate that the coupling between a particular dehydrogenase and the energy-coupling site for transport is subject to regulation, and that it may be difficult, if not impossible, to demonstrate specificity of energy-coupling in the intact cell. In some bacteria, however, evidence in favor of this hypothesis has been presented with intact cells. In *Arthrobacter pyridinolis* (33), hexose transport in both intact cells and membrane vesicles is coupled to malate dehydrogenase; and in a marine pseudomonas (34), it has been shown that amino acid transport in whole cells and membrane vesi-

cles is coupled to alcohol dehydrogenase.

(7) As will be discussed below (see section on Reconstitution), studies with antibodies against D-LDH and calcium, magnesium-stimulated ATPase demonstrate that both of these membrane-bound enzymes are present on the inner surface of the vesicle membrane (35,36). These observations provide direct evidence in support of the contention that none of the vesicle can be inverted or sufficiently damaged to allow access of antibody to the interior surface of the vesicle membrane.

(8) Recent experiments (37) demonstrate that each vesicle in these preparations catalyzes active transport. The experiments supporting this conclusion are described in the next few paragraphs.

2-Hydroxy-3-butynoic acid (38) irreversibly inactivates D- and L-LDH's and D-lactate-dependent active transport in membrane vesicles isolated from *E. coli* (39,40). The compound is a substrate for the membrane-bound, flavin-linked D-LDH which undergoes 15 to 30 turnovers prior to inactivation. Inactivation is due to covalent attachment of a reactive intermediate to FAD at the active site of the enzyme. The proposed reaction sequence is shown below (Fig. 2, Reaction II). Both the hydroxy function and the alkyne linkage in 2-hydroxy-3-butynoate are critical for inactivation. Thus, 3-butynoate has no effect on the enzyme; and vinylglycolate (2-hydroxy-3-butenoic acid) serves as a substrate for D-LDH, and is an effective electron donor for transport.

Inactivation of D- and L-LDH and D-lactate-dependent transport by

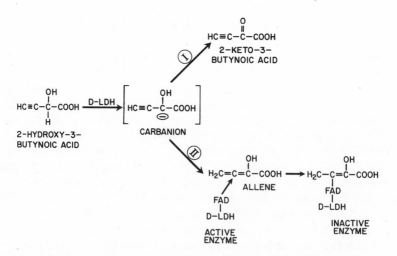

Fig. 2. **Inactivation of D-lactate dehydrogenase by 2-hydroxy-3-butynoic acid. From Kaback, H.R. (4).**

2-hydroxy-3-butynoate is highly specific. Other membrane-bound dehydrogenases are not inhibited, transport in the presence of ascorbate-PMS is not affected, and α-glycerol-P-dependent transport in *Staphylococcus aureus* vesicles is not inactivated by the acetylenic hydroxy acid. Moreover, inactivation of D-lactate-dependent transport is blocked by D-lactate but not by succinate and NADH.

In view of this high degree of specificity, it was surprising when subsequent experiments demonstrated that hydroxy butynoate inactivates vectorial phosphorylation catalyzed by the P-enolpyruvate-P-transferase system, and that vinylglycolate (2-hydroxy-3-butenoate), a substrate for D-LDH, is 50 to 100 times more potent. The key to the puzzle came with the realization that Reaction I in Fig. 2, although inconsequential for D-LDH, leads to the formation of a highly reactive electrophile. Shortly thereafter, it was demonstrated that vinylglycolate inactivates Enzyme I of the P-transferase system, and by this means, blocks vectorial phosphorylation in whole cells and membrane vesicles of *E. coli* (40,41). The relative lack of potency of hydroxybutynoate is due to inactivation of D- and L-LDH's by this compound. Generation of 2-keto-3-butynoate (Fig. 2) is limited therefore by inactivation of the enzymes which catalyze its formation. Vinylglycolate, on the other hand, is a noninactivating substrate, and the putative electrophile (2-keto-3-butenoic acid) is generated at a rapid rate and for extended periods of time. Synthesis of isotopically labeled vinylglycolate has allowed a detailed study of the biochemical properties of this compound (37,42).

Prior to inactivation of the P-transferase system, vinylglycolate is transported by the lactate transport system. Subsequently, it is oxidized by membrane-bound D- and L-LDH's to yield a reactive electrophile (presumably 2-keto-3-butenoate) which then reacts with Enzyme I and many other sulfhydryl-containing proteins on the membrane (Fig. 3).

There is considerable evidence supporting these conclusions (37,40,41,42); however, only two points are critical for this discussion: (i) vinylglycolate transport is the limiting step for labeling the membrane proteins; and (ii) almost all of the vinylglycolate taken up is covalently bound to the vesicles. In experimental terms, the rate of covalent binding of vinylglycolate is stimulated at least 10-fold by ascorbate-PMS; and stimulation is completely abolished by uncoupling agents or phospholipase treatment, neither of which affect vinylglycolate oxidation.

With this background, vinylglycolate can be utilized to estimate the transport activity of individual membrane vesicles. Using extremely high specific activity [^{3}H] vinylglycolate, vesicles have been labeled for an appropriate time in the presence of ascorbate-PMS, and examined by radioautography in the electron microscope (37). Each vesicle that takes up vinylglycolate is overlaid with exposed silver grains. Examination of the preparations reveals that

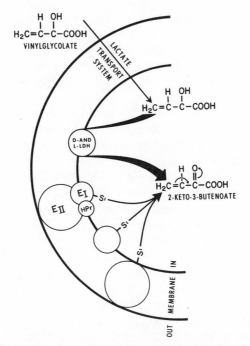

Fig. 3. Proposed sequence of events in uptake and covalent binding of vinylglycolate (2-hydroxy-3-butenoate) by E. coli membrane vesicles. D- and L-LDH, D- and L-lactate dehydrogenase; E II, enzyme II of the phosphoenolpyruvate-P-transferase system; E I, enzyme I of the phosphotransferase system; HPr, histidine-containing protein; S, sulfhydryl groups contained in membrane proteins. Reaction of 2-keto-3-butenoate with sulfhydryl groups exclusively on the inner surface of the membrane is a presumption at the present time. From Kaback, H.R. (4).

85-90% of the vesicles are labeled. It should be emphasized that this is a minimal estimation. Virtually all of the large vesicles are labeled, while the size of the smaller vesicles is such that their proximity to individual silver grains in the emulsion may be limiting. Moreover, essentially identical radioautographic results are obtained with [^{3}H] acetic anhydride, a reagent which reacts non-specifically with the vesicles. These studies provide strong evidence that most, if not all, of the vesicles in the preparation catalyze active transport. This type of study is not possible with the usual transport substrates because they are not covalently bound by the vesicles, and are readily lost with even the most gentle manipulations.

ENERGY-DEPENDENT BINDING OF β-GALACTOSIDES TO THE *LAC* CARRIER PROTEIN

Fluorescent β-galactosides [1-(N-dansyl)amino-β-D-galactopyranoside (DG$_0$),

2-(N-dansyl)aminoethyl-β-D-thiogalactopyranoside (DG_2), 2-(N-dansyl)amino-
ethyl-β-D-galactopyranoside (oxy-DG_2), and 6-(N-dansyl)aminohexyl-β-D-
thiogalactopyranoside (DG_6)] competitively inhibit lactose transport by
membrane vesicles from *Escherichia coli* ML 308-225, but are not actively
transported (43-45). An increase in the fluorescence of these dansylgalactosides is observed upon addition of D-lactate, imposition of a membrane diffusion potential (exterior positive), or dilution-induced, carrier-mediated
lactose efflux. The increase is not observed with 2-(N-dansyl)aminoethyl-β-
D-thioglucopyranoside nor with membrane vesicles lacking the β-galactoside
transport system. Moreover, the D-lactate-induced fluorescence increase is
blocked and/or rapidly reversed by addition of β-galactosides, sulfhydryl
reagents, certain inhibitors of D-lactate oxidation, or uncoupling agents. The
fluorescence increase exhibits an emission maximum at 500 nm and excitation maxima at 345 nm and 292 nm. The latter excitation maximum is
absent unless D-lactate is added, indicating that the bound dansyl-galactoside
molecules are excited by energy transfer from the membrane proteins. Titration of vesicles with dansylgalactosides in the presence of D-lactate demonstrates that the *lac* carrier protein constitutes 3 to 6% of the total membrane
protein, and that the affinity of the carrier for substrate is directly related
to the length of the alkyl chain between the galactosidic and the dansyl
moieties of the dansylgalactosides. In addition, there is excellent agreement
between the affinity constants of the various dansylgalactosides as determined
by fluorimetric titration and their apparent K_i's for lactose transport (K_D's
and/or apparent K_i's are approximately 550 μM, 30 μM, 40 μM and 5 μM
for DG_0, DG_2, oxy-DG_2, and DG_6, respectively).

Anisotropy of fluorescence measurements with 2-(N-dansyl)aminoethyl-β-
D-thiogalactopyranoside (DG_2) and 6-(N-dansyl)-aminohexyl-β-D-thiogalacto-
pyranoside (DG_6) demonstrates a dramatic increase in polarization on addition of D-lactate which is reversed by anoxia or addition of lactose (44).
These findings provide strong evidence for the contention that the fluorescence changes observed on "energization" of the membrane are due to binding of the dansylgalactosides *per se*, rather than binding followed by transfer
into the hydrophobic interior of the membrane.

Fluorescence lifetime studies corroborate the results discussed above (46).
In the absence of D-lactate, DG_2 fluorescence exhibits a half-life of 3 nanoseconds. On addition of D-lactate to membrane vesicles containing the *lac*
transport system, a percentage of the DG_2 molecules exhibits a half-life of
18 nanoseconds. Using the experimental values for anisotropy and excited
state lifetime derived from the above experiments, it can be calculated that
the rotational relaxation time (ρ) of DG_2 is increased by approximately a
factor of 200 when it is bound to the *lac* carrier protein.

Finally, it should be emphasized that Schuldiner *et al.* (47) have recently corroborated much of the fluorescence data both qualitatively and quantitatively by direct measurement of binding using flow dialysis with high specific activity $[^3H]DG_6$.

The results are consistent with the suggestion that the *lac* carrier protein is inaccessible to the external medium unless energy is provided, and that energy is coupled to one of the initial steps in transport. It is proposed, furthermore, that at least one aspect of "energization" may be the generation of a membrane potential (exterior positive) resulting in increased accessibility of the *lac* carrier protein to the external solvent.

The ultimate strength of many of the conclusions discussed above rests heavily on observations which indicate that although the dansylgalactosides bind specifically to the *lac* carrier protein, they are not actively transported. Recent experimental evidence (45) provides definitive evidence that these compounds are not transported to any significant extent by either whole cells or isolated membrane vesicles: (i) Although DG_2 and oxy-DG_2 are competitive inhibitors of lactose transport in intact cells of *E. coli* and induce the synthesis of β-galactosidase *in vitro*, the compounds do not induce β-galactosidase *in vivo*. (ii) *p*-Chloromercuribenzenesulfonate (*p*-CMBS) does not cause efflux of lactose from the intravesicular pool, but causes rapid reversal of D-lactate-induced dansylgalactoside fluorescence. This difference between the effects of *p*-CMBS on β-galactoside transport and on the fluorescence behavior of the dansylgalactosides indicates that the fluorescence changes do not reflect events that occur subsequent to transport of the fluorescent galactosides into the vesicles. (iii) Dansylgalactosides inhibit dilution-induced, carrier-mediated efflux of lactose from the intravesicular pool. The results are consistent with the interpretation that the dansylgalactosides inhibit lactose efflux by preventing the unloaded carrier from returning to the inner surface of the membrane.

Studies with photoreactive β-galactosides, 2-nitro-4-azidophenyl-β-D-thiogalactopyranoside (APG_0)(48) and 2-(2-nitro-4-azidophenyl)aminoethyl-β-D-thiogalactopyranoside (APG_2)(49) provide independent support for the conclusions derived from the dansylgalactoside experiments. The rationale behind the use of this class of compounds is that irradiation with visible light causes photolysis of the azido group to form a highly reactive nitrene which then reacts covalently with the macromolecule to which the azido-containing ligand is bound.

APG_0 is a competitive inhibitor of lactose transport in membrane vesicles isolated from *E. coli* ML 308-225, exhibiting an apparent K_i of 75 μM. The initial rate and steady-state level of $[^3H]APG$ accumulation are markedly stimulated by the addition of D-lactate to vesicles containing the *lac* transport system, and kinetic studies reveal an apparent K_m of 75 μM. Membrane vesicles devoid of the *lac* transport system do not take up significant amounts of APG_0 in the presence or absence of D-lactate. When exposed to visible light in the presence of D-lactate, APG irreversibly inactivates the *lac* transport system. Strikingly, photolytic inactivation is not observed in the absence of D-lactate. Kinetic studies of the inactivation process yield a K_D of 77 μM. Since lactose protects against photolytic inactivation and APG_0 does not inactivate amino acid transport, it is apparent that these effects are specific for the *lac* transport system. Analogous studies carried out with APG_2 demonstrate that this compound behaves similarly with respect to photoinactivation, but is not transported by the vesicles (like DG_2), and has a higher affinity for the *lac* carrier protein (i.e., the K_i for competitive inhibition of lactose transport and the K_D for photolytic inactivation in the presence of D-lactate are 35 μM). Moreover, it has been demonstrated that in addition to D-lactate oxidation, APG_2-dependent photolytic inactivation can also be induced by an artificially imposed membrane potential (exterior positive).

It seems clear from these studies with both dansyl- and azidophenylgalactosides that the *lac* carrier protein is not accessible from the outside surface of the vesicle membrane in the absence of energy-coupling. Moreover, all of the data are consistent with the notion that a membrane potential (exterior positive) causes the carrier to "move" to the outside of the membrane where it is able to bind ligand. Evidence in support of the hypothesis that D-lactate oxidation also generates such a membrane potential will be discussed below (Mechanism of Active Transport). Clearly, in view of these findings, it is appropriate to postulate that the *lac* carrier protein itself, or a part of it, might carry a net negative charge.

Recently, it has been demonstrated (50,51) that *p*-CMBS, a potent inhibitor of lactose transport which inhibits all carrier-mediated aspects of this transport system (20), does not block D-lactate-induced APG_2-dependent photoinactivation. Even more striking is the observation that *p*-CMBS induces APG_2 photoinactivation in the absence of D-lactate. The increase in susceptibility to APG_2 is reflected in the dissociation constant of APG_2 for *p*-CMBS treated membranes which is about 20 μM. This value is very similar to that determined for D-lactate-induced APG_2-dependent photoinactivation. A reasonable interpretation of these results is that *p*-CMBS reacts with a sulfhydryl group which is exposed in the high affinity form of the *lac* carrier, trapping the protein in that configuration and preventing substrate translocation. The covalent nature of this modification is demonstrated by the fact that the uncoupler carbonylcyanide *m*-chlorophenylhydrazone (CCCP) does not inhibit *p*-CMBS-induced APG_2 photoinactivation. Thus, a membrane potential is not required for the *p*CMBS effect. It is also clear from the findings that p-CMBS does not block the binding site on the carrier since not only does APG_2 bind with high affinity, but the p-CMBS treated *lac* carrier protein is completely protected from APG_2 by lactose, thiodigalactoside, and melibiose (but not by DG_2). Additional evidence that sulfhydryl groups are exposed upon conversion of the *lac* carrier protein to its high affinity form is demonstrated by the observation that inactivation of lactose transport by N-ethylmaleimide is increased 2- to 4-fold in the presence of reduced phenazine methosulfate (50,51). The results as a whole indicate that in the presence of a membrane potential, exterior positive, a sulfhydryl group in the *lac* carrier protein which is not in the binding site becomes more accessible, and simultaneously, the carrier is converted into a form which has a high affinity for external substrate. Possibly, the sulfhydryl group in the *lac* carrier protein may exist in an ionized form in the hydrophobic milieu of the membrane and that it may be this functional group in the protein that responds to the membrane potential.

RECONSTITUTION OF D-LACTATE-DEPENDENT ACTIVE TRANSPORT IN D-LACTATE DEHYDROGENASE MUTANTS

The membrane-bound D-LDH of *E. coli* has been solubilized and purified to homogeneity (52,53). The enzyme has a molecular weight of 75,000 ± 7%, contains approximately 1 mole of flavin adenine dinucleotide per mole of enzyme, and exhibits low activity towards L-lactate. Oxidized diphosphopyridine nucleotide (NAD) has no effect on the catalytic conversion of D-lactate to pyruvate.

While this work was in progress Reeves *et al.* (54) demonstrated that guanidine·HCl extracts from wild-type membrane vesicles containing D-LDH activity are able to reconstitute D-lactate-dependent oxygen consumption and active transport in membrane vesicles from *E. coli* and *S. typhimurium* mutants defective in D-LDH (*dld⁻*). These studies have been confirmed and extended by Short *et al.* (55) using the homogeneous preparation of D-LDH described above, and Futai (56) has independently confirmed many of the observations.

Reconstituted *dld⁻* membranes carry out D-lactate oxidation and catalyze the transport of a number of substrates when supplied with D-lactate. D-lactate is not oxidized, and will not support transport of any of these substrates in unreconstituted *dld⁻* vesicles. Binding of enzyme to wild-type membranes produces an increase in D-lactate oxidation but has little or no effect on the ability of the membrane to catalyze active transport. Reconstitution of *dld⁻* membranes with increasing amounts of D-LDH produces a corresponding increase in D-lactate oxidation, and transport approaches an upper limit which is similar to the specific transport activity of wild-type membrane vesicles. However, the quantity of enzyme required to achieve maximum initial rates of transport varies somewhat with different transport systems.

Binding of 2-(N-dansyl)aminoethyl-β-D-thiogalactoside (DG$_2$) to membrane vesicles containing the *lac* transport system is dependent upon D-lactate oxidation, and this fluorescent probe can be utilized to quantitate the number of *lac* carrier proteins in the membrane vesicles (see previous discussion). When *dld-3* membrane vesicles are reconstituted with increasing amounts of D-LDH, there is a corresponding increase in the binding of DG$_2$. Assuming that each *lac* carrier protein molecule binds one molecule of DG$_2$, it can be estimated that there is at least a 7- to 8-fold excess of *lac* carrier protein relative to functional D-LDH in reconstituted *dld*$^-$ vesicles. A similar determination can be made for wild-type vesicles. These vesicles contain approximately 0.07 nmole of D-LDH per mg membrane protein (based on the specific activity of the homogeneous enzyme preparation), about 1.1 nmoles of *lac* carrier protein per mg membrane protein, yielding a ratio of about 15 for *lac* carrier protein relative to D-LDH.

Although the rate and extent of transport increases dramatically with reconstitution, the rate and extent of labeling of *dld*$^-$ vesicles with radioactive vinylglycolate remains constant. As discussed above, this compound is transported via the lactate transport system, and oxidized to a reactive product by D- and L-LDH's on the inner surface of the vesicle membrane. The observation that reconstituted *dld*$^-$ membranes do not exhibit enhanced labeling by vinylglycolate suggests that bound D-LDH is present on the outer surface of the vesicles. In this case, the reactive product released from D-LDH would be diluted into the external medium, whereas if the enzyme were on the inner surface of the vesicle membrane, the rate of labeling would be expected to increase with reconstitution, since the reactive product should accumulate within the vesicles to higher effective concentrations.

The suggestion that D-LDH is localized on the outer surface of reconstituted *dld*$^-$ membrane vesicles, as opposed to the inner surface of native ML 308-225 vesicles has received strong support from recent experiments with antibody against D-LDH (35,36). Incubation of ML 308-225 membrane vesicles with anti-D-LDH does not inhibit D-LDH activity (assayed by tetrazolium dye reduction, oxygen uptake, and/or D-lactate-dependent transport) unless the vesicles are disrupted physically or spheroplasts are lysed in the presence of antibody. In contrast, treatment of reconstituted *dld*$^-$ vesicles with anti-D-LDH results in marked inhibition of D-LDH activity. The titration curves obtained with reconstituted *dld-3* membrane vesicles are almost identical quantitatively to that obtained with the homogeneous preparation of D-LDH. In addition to providing information about the localization of D-LDH in native and reconstituted vesicles, the results with the native vesicles are consistent with other experiments which demonstrate that essentially all of the vesicles catalyze active transport (cf. above) and therefore cannot be inverted or sufficiently damaged to allow access of anti-D-LDH to D-LDH.

The conclusion that D-LDH is able to drive transport from the outer surface of the vesicle membrane is also consistent with recent experiments of Konings (57) and Short *et al.* (36) demonstrating that reduced 5-N-methylphenazonium-3-sulfonate, an impermeable electron carrier, drives transport as well as reduced PMS, its lipophilic analogue.

In view of the findings with antibody against D-LDH, it is pertinent to discuss recent experiments carried out with antibody against calcium, magnesium-stimulated ATPase (36). It has been reported (58,59) that incubation of membrane vesicles with antiserum prepared against this enzyme results in 30-50% of membrane-bound ATPase activity, leading to the conclusion that a significant number of vesicles are inverted or damaged (26,59,60), or that ATPase becomes dislocated to the outer surface of the vesicle membrane during lysis (27,58). It should be emphasized that most of these studies were carried out at pH 9.0 (i.e., a pH that inactivates the vesicles), and that D-LDH activity is not inhibited by anti-D-LDH at this pH. Moreover, when ML 308-225 membrane vesicles are treated with antiserum against ATPase at pH 8.0 or below, there is no significant inhibition of ATPase activity. Futai (58) has stated, however, that he observes inhibition of ATPase activity by antiserum at pH 8.0 and below. The reason for this apparent discrepancy may be related to differences in the methods used to prepare membrane vesicles (36). Thus, homogenization of membrane vesicles at pH 8.0 or below sensitizes ATPase activity to antiserum, although D-lactate-dependent oxygen consumption and active transport in the homogenized vesicles are still completely resistant to the effects of anti-D-LDH γ-globulin. Finally, it has been demonstrated (36) that ATPase is not firmly bound to the membrane, and becomes easily dissociated during vesicle preparation. When the specific activities of ATPase, cytochrome b_1 and D-LDH are compared in sonicated cells, spheroplasts and membrane vesicles, that of ATPase remains constant, while the specific activities of cytochrome b_1 and D-LDH increase 5- and 3-fold, respectively. It is apparent therefore that 60-80% of the ATPase activity of the cell is lost during the preparation of membrane vesicles. The findings indicate that ATPase is loosely bound to the inner surface of the vesicle membrane, and that it may be dislocated from the inside to the outside surface or even lost entirely under relatively mild conditions. In any case, it seems apparent that ATPase activity *per se* or its inhibition by antibody is not a very useful means of determining membrane sidedness.

The flavin moiety of the holoenzyme appears to be critically involved in binding D-LDH to the membrane (55). Treatment with [1-^{14}C] hydroxybutynoate leads to inactivation of D-LDH by modification of the flavin adenine dinucleotide coenzyme bound to the enzyme (cf. Fig. 2). Enzyme labeled in this manner does not bind to dld^- membrane vesicles. The findings suggest that the flavin coenzyme itself may mediate binding or alternatively, that covalent inactivation of the flavin may result in a conformational change that does not favor binding. It is tempting to speculate on the relevance of this finding to the synthesis of membrane-bound dehydrogenases in the intact cell. Possibly, the apoprotein moiety of D-LDH is synthesized on cytoplasmic ribosomes, but is not inserted into the membrane until coenzyme is bound. If this is so, D-LDH mutants which are defective in the flavin binding site should exhibit soluble material which cross-reacts immunologically with native D-LDH.

MECHANISM OF ACTIVE TRANSPORT

There are certain fundamental observations that any proposed mechanism for energy-coupling in the vesicle system must take into account: (i) The apparent affinity of the carriers for substrate is much higher on the external surface of the membrane than on the internal surface, while the rate of "movement" from outside to inside and *vice versa* appears to be identical. (ii) There is no correlation between rates of oxidation of various electron donors and their relative effects on transport. Moreover, the cytochrome chain of *E. coli* vesicles is completely reduced by each substance that is oxidized by the vesicles, and essentially none of the vesicles in the preparations is inverted. (iii) Respiration-dependent transport is completely blocked by a variety of proton conductors, although these agents do not inhibit D-lactate oxidation. In a similar vein, a number of mutants have been isolated which exhibit pleiotropic transport defects, and vesicles prepared from some of these mutants exhibit increased permeability to protons (61,62). (iv) Although all inhibitors of D-lactate oxidation block uptake, only those that act after the site of energy-coupling in the electron transfer chain cause efflux when added to preloaded vesicles. (v) Reduction of the respiratory chain from either side of the vesicle membrane is able to drive active transport. (vi) There is a large excess of carriers relative to D-LDH. (vii) Appropriate ion gradients result in binding of dansyl- and azidophenylgalactosides in the absence of D-lactate oxidation. Similarly, these gradients also result in uptake of lactose and amino acids.

An initial model proposed by Kaback and Barnes (20) depicted the carriers as electron transfer intermediates in which a change from the oxidized to the reduced state results in translocation of the carrier-substrate complex to the inner surface of the membrane and a concomitant decrease in the affinity of the carrier for substrate. This model was intended merely to provide a working hypothesis that could account primarily for the observation that only certain electron transfer inhibitors cause efflux of accumulated solutes. However, in view of a large number of observations which have appeared since the presentation of this model, it seems obvious that the model is insufficient to explain many of the other basic properties of the transport system.

A very different type of hypothesis, one that emphasizes the positioning of respiratory chain components within the matrix of the membrane, was proposed by Mitchell (24,25,63-65). As visualized by this chemiosmotic model, oxidation of electron donors is accompanied by expulsion of protons into the external medium, leading to a pH gradient and/or electrical potential across the membrane. This electrochemical gradient is postulated to be the driving force for inward movement of transport substrates by way of passive diffusion in the case of lipophilic cations such as dibenzyldimethylammonium ion (66), via facilitated diffusion in the case of positively charged substrates such as lysine or potassium ions or via coupled movement of protons with a neutral substrate such as lactose or proline (i.e., "symport"). In instances where sodium efflux is observed (67), the chemiosmotic model invokes the concept of sodium-proton "antiport" which is postulated to catalyze electroneutral exchange of internal sodium with external protons, and *vice versa* (68). Moreover, the inhibitory effects of uncoupling agents on transport are attributed to the ability of these compounds to conduct protons across the membrane, thus short-circuiting the "proton-motive force" that drives transport (69).

Recent evidence has provided direct support for the Mitchell hypothesis with regard to the vesicle system. During D-lactate or reduced PMS oxidation, lipophilic cations such as dimethyldibenzylammonium (in the presence of tetraphenylboron)(70,72), triphenylmethylphosphonium (73), safranine-O (73), and rubidium (in the presence of valinomycin)(67) are accumulated. Moreover, there is a quantitative correlation between the steady-state level of accumulation of rubidium (in the presence of valinomycin), triphenylmethylphosphonium and dibenzyldimethylammonium (in the presence of tetraphenylboron)(73). These observations are consistent with the interpretation of Harold and co-workers (70,71) that D-lactate oxidation generates a membrane potential, interior negative.

Accumulation of these compounds is relatively specific for D-lactate or reduced phenazine methosulfate as electron donors. Since few, if any, of the vesicles in the preparations are inverted, it is unlikely that the relative inability of other electron donors to drive uptake of these compounds can be attributed to an artifact of this nature (see previous discussion). Thus, it appears likely that the machinery which is responsible for the generation of the membrane potential is located relatively specifically within the portion of the respiratory chain between D-LDH and cytochrome b_1.

Although all inhibitors of D-lactate oxidation block D-lactate-dependent solute accumulation, only those electron transfer inhibitors which block the respiratory chain after the energy-coupling site cause efflux of accumulated solutes. These observations have been interpreted as being inconsistent with a chemiosmotic mechanism for active transport because it is not immediately obvious why only certain electron transfer inhibitors should collapse the membrane potential and/or the pH gradient generated by D-lactate oxidation. Since precisely the same effects have been observed with a number of compounds which presumably equilibrate with the membrane potential and with D-lactate-induced proton gradients (73), it seems reasonable to suggest that inhibition of electron flow in a manner which leads to reduction of the energy-coupling site results in dissipation of the membrane potential, while inhibition of electron flow in a manner which leads to oxidation of the energy-coupling site does not result in collapse of the potential. In other words, one explanation that would account for the data is that the membrane potential may be in equilibrium with the redox state of the respiratory chain at that site between D-LDH and cytochrome b_1 which generates the membrane potential. Assuming that the carriers are in equilibrium with a membrane potential which, in turn, is in equilibrium with the redox potential of the energy-coupling site, a number of observations could be explained by a chemiosmotic mechanism.

Evidence is also available which provides a strong indication that the generation of a membrane potential, interior negative, is intimately related to the ability of the vesicles to catalyze active transport. As discussed above, binding of dansyl- and azidophenylgalactosides is induced by artificially imposed ion gradients that lead to a membrane potential, interior negative. Moreover, steady-state levels of lactose, proline, tyrosine, glutamic acid and glycine accumulation in the presence of D-lactate and ascorbate-PMS are directly related to the magnitude of the membrane potential as determined by triphenylmethylphosphonium accumulation under identical conditions (73). Finally, addition of lactose to membrane vesicles containing the β-galactoside transport system inhibits the uptake of both proline and triphenylmethylphosphonium (73). A reasonable explanation for the last observation is that lactose transport partially dissipates the membrane potential in a manner similar to that described for glucose in *Neurospora crassa* by Slayman and Slayman (74) who measured the potential directly. Despite the attractiveness of lactose/proton symport (75) as a mechanism for explaining these effects, however, all efforts to demonstrate this phenomenon directly in isolated membrane vesicles have been negative thus far.

Despite convincing evidence supporting a chemiosmotic mechanism for active transport in isolated membrane vesicles, a few inconsistencies remain which are not readily explained by the available data. According to the chemiosmotic theory, oxidation of any electron donor would presumably be associated with a unique value of the membrane potential, and this prediction is borne out by studies of triphenylmethylphosphonium and rubidium (in the presence of valinomycin) uptake (67,73). Accordingly, the order of effectiveness of the various electron donors in stimulating steady-state uptake would be expected to be the same for all respiration-linked transport systems, this order reflecting their relative effectiveness in generating a membrane potential. However, the order of effectiveness of various electron donors in stimulating uptake varies widely, depending on the transport system under study (72). Moreover, in chemiosmotic coupling, the logarithm of the steady-state concentration ratio attained in the presence of a given electron donor relative to the logarithm of the ratio established in the presence of D-lactate would be expected to be constant for all the transport systems. These ratios, however, exhibit enormous variation among the various systems.

Membrane vesicles and ethylenediaminetetraacetic acid-treated cells catalyze lactose accumulation at least 3 to 5 times better than triphenylmethylphosphonium (73). Based on a mechanism in which there is one positive charge (i.e., one proton) taken up per mole of lactose (75), a potential of approximately -120 mV is required to achieve a lactose concentration gradient of 100. However, the vesicles generate a potential of only about -75 mV (73). It seems apparent, therefore, that although the vesicles generate a membrane potential of the appropriate polarity, the magnitude of the potential is not sufficient to account for the phenomena observed. A similar situation has also been described in mitochondria (76) and in chloroplasts (77, 78). In addition, triphenylmethylphosphonium accumulation in response to an artificially generated membrane potential approximates that observed during D-lactate or reduced PMS oxidation, while lactose accumulation in response to the artificially generated potential is considerably less than that observed with D-lactate or ascorbate-PMS (73). These inconsistencies could be explained in a number of ways. First and most obvious, the accumulation of lipophilic cations or rubidium (in the presence of valinomycin) may not represent a truly quantitative estimate of the membrane potential. Second, there may be more than one positive charge taken up per mole of solute. Third, there may be another high-energy intermediate in addition to the membrane potential which is generated by D-lactate or reduced PMS oxidation, but not by an artificially imposed membrane potential. One such candidate is the chemical gradient of protons; however, no significant pH gradient is observed when the vesicles are incubated in the presence of D-lactate or ascorbate-PMS. This observation is not surprising in view of the fact that the transport assays are carried out in highly buffered reaction mixtures. Hopefully, some of these problems will be resolved by more quantitative estimates of the membrane potential and by the development of techniques which allow an assessment of solute/proton symport in the vesicles system. In any case, it seems apparent that definitive answers regarding the intimate mechanism of active transport will have to await the isolation of functionally active carrier molecules.

REFERENCES

1. Kaback, H.R. (1972) *Biochim. Biophys. Acta 265*, 367-416.
2. Kaback, H.R. and Hong, J.-s. (1973) In *CRC Critical Reviews in Microbiology* (Laskin, A.I., ed.) Vol. II, pp. 333-376, CRC Press, Ohio.
3. Kaback, H.R. (1973) In *Bacterial Membranes and Walls* (Leive, L., ed.) Vol. I, pp. 241-292, Marcel Dekker, New York.
4. Kaback, H.R. (1974) *Science 186*, 882-892.
5. Prezioso, G., Hong, J.-s., Kerwar, G.K. and Kaback, H.R. (1973) *Arch. Biochem. Biophys. 154*, 575-582.
6. Konings, W.N. and Freese, E. (1972) *J. Biol. Chem. 247*, 2408-2418.
7. Short, S.A., White, D.C. and Kaback, H.R. (1972) *J. Biol. Chem. 247*, 298-304.
8. Konings, W.N. and Kaback, H.R. (1973) *Proc. Nat. Acad. Sci. (U.S.A.) 70*, 3376-3381.
9. Boonstra, J., Huttunen, M.T., Konings, W.N. and Kaback, H.R. (in press) *J. Biol. Chem.*
10. Konings, W.N. and Boonstra, J. (in press) In *Current Topics in Membranes and Transport.*
11. Short, S.A. and Kaback, H.R. (1974) *J. Biol. Chem. 249*, 4275-4281.
12. Lombardi, F.J. and Kaback, H.R. (1972) *J. Biol. Chem. 247*, 7844-7857.
13. Short, S.A., White, D.C. and Kaback, H.R. (1972) *J. Biol. Chem. 247*, 7452-7458.
14. Kaback, H.R. (1971) In *Methods in Enzymology* (Jakoby, W.B., ed.) Vol. XXII, pp. 99-120, Academic Press, New York.
15. Kaback, H.R. (1974) In *Methods in Enzymology* (Fleischer, S. and Packer, L., eds.) Vol. XXXI, pp. 698-709, Academic Press, New York.
16. Kaback, H.R. (1970) *Annu. Rev. Biochem. 39*, 561-598.
17. Kaback, H.R. (1970) In *Current Topics in Membranes and Transport* (Bronner, F. and Kleinzeller, A., eds.) pp. 35-99, Academic Press, New York.

18. *The Lactose Operon* (1970)(Beckwith, J.R. and Zipser, D., eds.) Cold Spring Harbor Laboratories.

19. Kennedy, E.P. (1970) In *The Lactose Operon* (Beckwith, J.R. and Zipser, D., eds.) pp. 49-92, Cold Spring Harbor Laboratories.

20. Kaback, H.R. and Barnes, E.M., Jr. (1971) *J. Biol. Chem. 246*, 5523-5531.

21. Kaback, H.R. and Milner, L.S. (1970) *Proc. Nat. Acad. Sci. (U.S.A.) 66*, 1008-1015.

22. Barnes, E.M., Jr. and Kaback, H.R. (1970) *Proc. Nat. Acad. Sci. (U.S.A.) 66*, 1190-1198.

23. Barnes, E.M., Jr. and Kaback, H.R. (1971) *J. Biol. Chem. 246*, 5518-5522.

24. Mitchell, P. (1973) *J. Bioenergetics 4*, 63-91.

25. Harold, F.M. (1972) *Bacteriol. Rev. 36*, 172-230.

26. Hare, J.B., Olden, K. and Kennedy, E.P. (1974) *Proc. Nat. Acad. Sci. (U.S.A.) 71*, 4843-4846.

27. Altendorf, K.H. and Staehelen, L.A. (1974) *J. Bacteriol. 117*, 888-899.

28. Konings, W.N., Bisschop, A., Voenhuis, M. and Vermeulen, C.A. (1973) *J. Bacteriol. 116*, 1456-1465.

29. Stroobant, P. and Kaback, H.R. Manuscript in preparation.

30. Reeves, J.P., Lombardi, F.J. and Kaback, H.R. (1972) *J. Biol. Chem. 247*, 6204-6211.

31. Rosen, B.P. and McClees, J.S. (1974) *Proc. Nat. Acad. Sci. (U.S.A.) 71*, 5042-5046.

32. Hong, J.-s. and Kaback, H.R. (1972) *Proc. Nat. Acad. Sci. (U.S.A.) 69*, 3336-3340.

32a. Grau, F., Hong, J.-s. and Kaback, H.R. Unpublished information.

33. Wolfson, E.B. and Krulwich, T.A. (1974) *Proc. Nat. Acad. Sci. (U.S.A.) 71*, 1739-1742.

34. Thompson, J. and MacLoed, R.A. (1974) *J. Bacteriol. 117*, 1055-1064.

35. Short, S.A., Hawkins, T., Kohn, L.D. and Kaback, H.R. (in press) *J. Biol. Chem.*

36. Short, S.A., Kaback, H.R. and Kohn, L.D. (in press) *J. Biol. Chem.*

37. Short, S.A., Kaback, H.R., Kaczorowski, G., Fisher, J., Walsh, C.T. and Silverstein, S. (1974) *Proc. Nat. Acad. Sci. (U.S.A.) 71*, 5032-5036.

38. Walsh, C.T., Schonbrunn, A., Lockridge, O., Massey, V. and Abeles, R. (1972) *J. Biol. Chem. 247*, 6004-6006.

39. Walsh, C.T., Abeles, R.H. and Kaback, H.R. (1972) *J. Biol. Chem. 247*, 7858-7863.

40. Walsh, C.T. and Kaback, H.R. (1974) *Ann. N.Y. Acad. Sci. 235*, 519-541.

41. Walsh, C.T. and Kaback, H.R. (1973) *J. Biol. Chem. 248*, 5456-5462.

42. Shaw, L., Grau, F., Kaback, H.R., Hong, J.- s. and Walsh, C.T. (1975) *J. Bacteriol. 121*, 1047–1055.

43. Reeves, J.P., Schechter, E., Weil, R. and Kaback, H.R. (1973) *Proc. Nat. Acad. Sci. (U.S.A.) 70*, 2722–2726.

44. Schuldiner, S., Kerwar, G.K., Weil, R. and Kaback, H.R. (1975) *J. Biol. Chem. 250*, 1361–1370.

45. Schuldiner, S., Kung, H., Weil, R. and Kaback, H.R. (in press) *J. Biol. Chem.*

46. Schuldiner, S., Spencer, R.D., Weber, G. and Kaback, H.R. Manuscript in preparation.

47. Schuldiner, S., Weil, R. and Kaback, H.R. Manuscript in preparation.

48. Rudnick, G., Weil, R. and Kaback, H.R. (1975) *J. Biol. Chem. 250*, 1371–1375.

49. *Ibid.* In press.

50. Rudnick, G., Weil, R. and Kaback, H.R. (1975) *Fed. Proc. 34*, 491.

51. Rudnick, G., Weil, R. and Kaback, H.R. Manuscript in preparation.

52. Kohn, L. and Kaback, H.R. (1973) *J. Biol. Chem. 248*, 7012–7017.

53. Futai, M. (1973) *Biochemistry 12*, 2468–2474.

54. Reeves, J.P.. Hong, J.- s. and Kaback, H.R. (1973) *Proc. Nat. Acad. Sci. (U.S.A.) 70*, 1917–1921.

55. Short, S.A., Kaback, H.R. and Kohn, L.D. (1974) *Proc. Nat. Acad. Sci. (U.S.A.) 71*, 1461–1465.

56. Futai, M. (1974) *Biochemistry 13*, 2327–2333.

57. Konings, W.N. (1975) *Arch. Biochem. Biophys. 167*, 570–580.

58. Futai, M. (1974) *J. Membr. Biol. 15*, 15–28.

59. Van Thienen, G. and Postma, P.W. (1973) *Biochim. Biophys. Acta 323*, 429–440.

60. Weiner, J.H. (1974) *J. Membrane Biol. 15*, 1–14.

61. Rosen, B.P. (1973) *Biochem. Biophys. Res. Commun. 53*, 1289–1296.

62. Altendorf, K., Harold, F.M. and Simoni, R.D. (1974) *J. Biol. Chem. 249*, 4587–4593.

63. Mitchell, P. (1963) *Biochem. Soc. Symp. 22*, 142.

64. Mitchell, P. (1966) *Biol. Rev. (Cambridge) 47*, 445–502.

65. Mitchell, P. (1967) *Fed. Proc. 26*, 1370–1379.

66. Bakeeva, L.E., Grinius, L.L., Jasaitis, A.A., Kuliene, V.V., Levitsky, D.O., Liberman, E.A., Severina, I.I. and Skulachev, V.P. (1970) *Biochim. Biophys. Acta 216*, 13–21.

67. Lombardi, F.J., Reeves, J.P. and Kaback, H.R. (1973) *J. Biol. Chem. 248*, 3551–3565.

68. West, I.C. and Mitchell, P. (1974) *Biochem. J. 144*, 87–90.

69. Mitchell, P. and Moyle, J. (1967) *Biochem. J. 104*, 588–600.

70. Hirata, H., Altendorf, K. and Harold, F.M. (1973) *Proc. Nat. Acad. Sci. (U.S.A.) 70*, 1804–1808.
71. Altendorf, K., Hirata, H. and Harold, F.M. (1975) *J. Biol. Chem. 249*, 4587–4593.
72. Lombardi, F.J., Reeves, J.P., Short, S.A. and Kaback, H.R. (1974) *Ann. N.Y. Acad. Sci. 227*, 312–327.
73. Schuldiner, S. and Kaback, H.R. *J. Biol. Chem.*, submitted for publication.
74. Slayman, C.L. and Slayman, C.W. (1974) *Proc. Nat. Acad. Sci. (U.S.A.) 71*, 1935–1939.
75. West, I.C. and Mitchell, P. (1973) *Biochem. J. 132*, 587–592.
76. Rottenberg, H. (1970) *Eur. J. Biochem. 15*, 22–28.
77. Schuldiner, S., Rottenberg, H. and Avron, M. (1972) *Eur. J. Biochem. 25*, 64–70.
78. Rottenberg, H., Grunwald, T. and Avron, M. (1972) *Eur. J. Biochem. 25*, 54–63.

THE ACTIVE EXTRUSION OF WEAK AND STRONG ELECTROLYTES IN INTACT MITOCHONDRIA

Giovanni Felice Azzone, Stefano Massari, Tullio Pozzan
and Raffaele Colonna

C.N.R. Unit for the Study of Physiology of Mitochondria
and
Institute of General Pathology
University of Padova
Padova, Italy

SUMMARY

Two energy linked ion transport processes are analyzed in intact mitochondria: (a) the extrusion of Pi coupled to the extrusion of weak bases, and (b) the extrusion of strong acids coupled to the extrusion of strong bases. These processes are discussed in terms of three models: (a) the mechano-enzyme, (b) the electrogenic proton pump, and (c) the proton-driven cation-anion pump.

INTRODUCTION

Three ion species seem to be actively transported in mitochondria: protons, strong bases, and strong acids (1-3). In intact mitochondria protons and strong acids are extruded while strong bases are taken up. The scientific controversy concerns two aspects (1-3): (a) the mechanism of H^+ transport and (b) the force responsible for coupling ion fluxes. As to H^+ transport, the H^+ ions are supposed to move either through a special arrangement of the respiratory chain components, denoted as loops, or through specific membrane components denoted as H^+ carriers. As to the force responsible for coupling ion fluxes, this has been supposed to involve either long-range or short-range interactions, the former denoted as electrogenic and the latter as electroneutral proton pump. The role of long-range interactions, and thus of the membrane potential, has been accepted by the majority of workers mainly on the basis of the argument that the accumulation of K^+ in the presence of valinomycin can hardly be other than electrophoretic. The argument has been further strengthened by the observation that the energy linked movement of the strong acids and bases is completely unspecific in that any lipid permeable species can be either taken up or released (4).

In the present paper we will show that intact mitochondria may also catalyze active transport with an apparent inverted polarity (5,6), namely: (a) energy-linked extrusion instead of uptake of weak acids; (b) energy-linked uptake instead of extrusion of strong acids; and (c) energy-linked extrusion instead of uptake of strong bases. We will show that these processes shed new light on the mitochondrial high energy state and active transport.

THE ENERGY—LINKED EXTRUSION OF WEAK ACIDS

Fig. 1 shows that addition of nigericin to anaerobic liver mitochondria incubated in K-phosphate resulted in swelling due to penetration of K-phosphate. Swelling occurred also in ammonium phosphate without addition of ionophores. Addition of succinate at the end of the swelling phase caused shrinkage with restoration of the initial mitochondrial volume. The active shrinkage in the phosphate medium occurs under a narrow range of conditions. First, it has a maximal

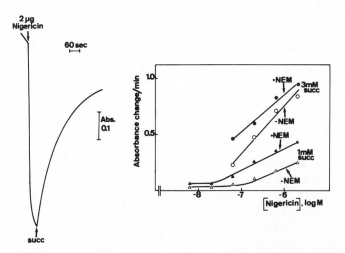

Fig. 1. Effect of various nigericin concentrations on active shrinkage in K-phosphate in the absence and presence of NEM. The medium contained 30 mM K-phosphate, 1 mM EDTA, pH 7.5, 1 μM rotenone and 1.8 mg of rat liver mitochondrial protein. Final volume, 2 ml. Swelling was initiated by 2 μg nigericin and shrinkage by 1 mM succinate. In the right part the concentrations of nigericin and succinate were varied and the ordinate refers to the rate of active shrinkage. Absorbance was recorded at 600 nm. NEM, N-ethylmaleimide.

rate at 20-30 mM phosphate and it decreases both at lower and higher electro-lyte concentrations. Second, it occurs at a negligible rate below pH 7, while it reaches its maximum at pH 7.5.

The process of active shrinkage in the phosphate medium has the following additional properties. First, as shown in Fig. 1, the rate of shrinkage in K-phosphate was markedly dependent on the concentrations of nigericin and of succinate. Below 10^{-7} M nigericin and 1 mM succinate the rate of shrinkage was very low. The rate increases proportionally to the nigericin concentration. Second, as also shown in Fig. 1, the process was markedly enhanced in rate as well as in extent, by the addition of the SH reagent, NEM (7-9). The question arises as to why SH reagents, which inhibit the Pi carrier, enhance the Pi extrusion. The answer is found by studying the effect of various SH inhibitors on the passive influx and efflux of Pi. Fig. 2 shows that NEM, as well as other SH inhibitors, are powerful inhibitors to the influx of Pi and the inhibition corresponds to the titration of the Pi carrier. On the other hand, the passive efflux of Pi from Pi-loaded swollen mitochondria is not inhibited by SH reagents. Thus the enhancing effect on active shrinkage corresponds to an asymmetric effect of the SH inhibitors which inhibit the influx but not the efflux of Pi.

THE MOLECULAR MECHANISM OF EXTRUSION OF WEAK ELECTRO–LYTES

The Pi extrusion could be due to the mechano-enzyme proposed by Lehninger (10,11). Energy would cause a conformational change with increase of hydro-

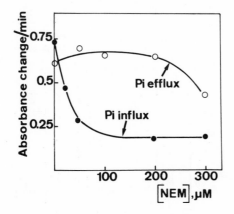

Fig. 2. Effect of N-ethylmaleimide on the passive influx and efflux of Pi. The experimental conditions were as in Fig. 1. The Pi influx was initiated by 2 μg nigericin and N-ethyl-maleimide (NEM) was added before nigericin. The Pi efflux was induced by adding 20 mM sucrose, and NEM was added before sucrose.

static pressure. The hydrostatic pressure causes a movement of water while the ions flow passively down concentration gradients. The mechanoenzyme model is in contrast with the observation of the enhancing effect of Pi extrusion by SH inhibitors. If the efflux of ions is a passive process, it is difficult to see why inhibition of Pi influx would enhance shrinkage. In effect the Pi efflux caused by osmotic shrinkage is insensitive to SH inhibitors. The effect of SH inhibitors suggests that during active shrinkage Pi is under the effect of two different forces at the two sides of the membrane.

In Fig. 3 is depicted schematically the process of Pi extrusion according to the electrogenic proton pump (12,13) and the proton-driven Pi pump (5). In both models the Pi extrusion may be considered as the primary process resulting in the formation of a ΔpH, alkaline inside. The ΔpH is used either to promote a reuptake of Pi or a release of K^+. The former process inhibits the extrusion while the latter enhances it. Thus inhibition of Pi influx by SH inhibitors favours shrinkage while acceleration of H^+/K^+ exchange by nigericin enhances shrinkage.

The fundamental question however concerns the mechanism of Pi extrusion. According to the electrogenic proton pump, the extrusion of the anion is an electrophoretic process driven by the membrane potential. The alternative model is that the Pi extrusion is driven by an energy linked binding of H^+ ions at the matrix side of the membrane. The H^+ drives the electroneutral extrusion of Pi. In both cases the overall process is electroneutral. However, the coupling between H^+ and Pi fluxes involves in the former case long-range, and in the latter case short-range, interactions.

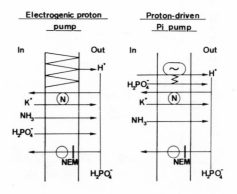

Fig. 3. Electrogenic proton pump and proton driven Pi pump. Schematic representation of the mechanism of active Pi extrusion according to the electrogenic proton pump and the proton-driven Pi pump.

THE ENERGY LINKED UPTAKE OF STRONG ACIDS

In this section we shall analyze the uptake of cations when there is: (a) a high concentration of permeant cations, and (b) strong acids.

It is well known that the membrane of intact mitochondria is impermeable to Cl^- (14,15). However, Brierley (16-18) has shown that respiration causes a large Cl^- uptake in valinomycin-treated mitochondria. According to Brierley (16) the increased permeability to Cl^- is due to H^+ extrusion with alkalinization of the matrix or of the membrane, while the penetration of KCl is a passive process. The interpretation of Brierley (16-18), within the frame of the electrogenic proton pump, leads, however, to a dilemma. If the membrane is de-energized, the permeability to Cl^- is low and the KCl penetration is limited by the low membrane Cl^- conductance. If the membrane is energized, the permeability is high but the membrane possesses also a membrane potential which tends to extrude Cl^-. Several laboratories have in fact reported that energization leads to extrusion of strong acids from the mitochondria.

Fig. 4 shows a correlation between H^+ ion extrusion and absorbance change during incubation of mitochondria in 50 mM KCl, pH 6.5, and 2 mM succinate. H^+ ion extrusion and swelling were initiated by 0.001 μg valinomycin. It is seen that H^+ ion extrusion and swelling went parallel during the whole process. Addition of FCCP caused a reuptake of H^+ ion and inhibited the continuation of swelling. The swelling described in Fig. 4 consists of about 50% of H^+/K^+ exchange and 50% of a penetration of KCl. At higher valinomycin concentra-

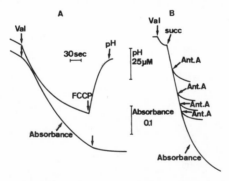

Fig. 4. Dependence of uptake of strong anions on H^+ ion extrusion and steady supply of energy. In the left part the medium contained 50 mM KCl, 1 mM Tris-Cl, pH 6.5, 0.5 mM EDTA, 2 mM succinate-Tris and 4.9 mg mitochondrial protein. Final volume, 2 ml. The reaction was initiated with 0.001 μg valinomycin and blocked with 1 μM FCCP. pH was measured with a H^+ ion electrode. In the right part the medium was the same. When indicated were added 0.003 μg valinomycin and 1 mM succinate. Antimycin was then added after 10, 15, 25 and 30 min, respectively. 1.8 mg protein.

tion the uptake of KCl is five times the H^+/K^+ exchange. In the right part of Fig. 4 it is seen that addition of valinomycin to mitochondria incubated at pH 6.5 caused no swelling because of absence of energy supply, while subsequent addition of succinate initiated swelling. Addition of antimycin inhibited swelling at any stage. Thus the process described in Fig. 4 appears as an uptake of KCl dependent on a steady extrusion of H^+ on one side and a supply of energy on the other.

A close correlation exists between rates of respiration and of KCl uptake. Increase of the concentration of valinomycin caused on one side an increase of the rate of KCl penetration and on the other an increase of the rate of respiration. Furthermore, Fig. 5 shows that inhibition of respiration due to increasing malonate concentrations was paralleled by a decrease of the rate of KCl uptake. In the former case it is the rate of electrolyte penetration to limit the respiratory rate. In the latter case it is the respiratory rate to limit the electrolyte penetration. However, rates of energy supply and of ion uptake appear to be strictly coupled.

Experiments 4 and 5 indicate that energy supply to mitochondria, incubated in the presence of permeant cations and the absence of weak acids, leads to an' apparent inversion of polarity in respect to the movement of strong acids, since they are accumulated instead of extruded.

THE ENERGY LINKED EXTRUSION OF STRONG BASES.

Fig. 6 shows that addition of valinomycin to aerobic mitochondria in 50 mM KCl at pH 7.5 resulted first in an extrusion of H^+ ion accompanied by a KCl

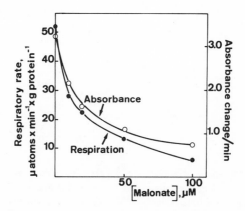

Fig. 5. Correlation between respiratory rate and KCl penetration. Experimental conditions as in Fig. 4. Rates of absorbance change and of respiration were measured in parallel experiments. 0.3 μg Valinomycin and variable amount of malonate. 2 and 4 mg protein.

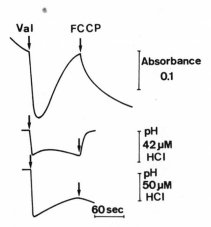

Fig. 6. Active swelling and shrinkage with strong electrolytes. The medium contained 30 mM KCl, 1 mM Tris-Cl, 0.5 mM EDTA, 2 mM succinate Tris, pH 7.5. The reaction was initiated by the addition of 0.1 μg valinomycin; 1 μM FCCP added where indicated. Upper trace, absorbance. Middle trace, pH measured with electrode. Bottom trace, pH measured with phenol red. Amount of protein, 3.3 mg. Final volume, 2 ml.

penetration. However, the rapid burst of H^+ ion ejection was followed by a levelling off and then by a slow reuptake. The reuptake of H^+ ion was followed by an extrusion of electrolytes.

Fig. 7 shows the passive swelling in 30 mM KCl due to valinomycin and the shrinkage due to succinate added 30, 60 and 120 min after valinomycin. The process of active shrinkage initiated only after the membrane had reached a certain degree of stretching whether through a passive or an active process.

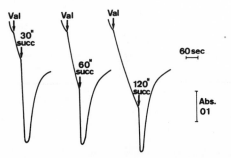

Fig. 7. Correlation between initiation of shrinkage and membrane stretching. The medium contained 30 mM KCl, 10 mM Tris-Cl, 1 mM EDTA, pH 7.5, 1 μM rotenone and 2 mg protein. The reaction was initiated by 0.1 μg valinomycin. 1 mM succinate was added after 30, 60 and 120 min, respectively.

The uptake of strong acids and extrusion of strong bases raises the question as to whether the swelling does not result in a fragmentation of the inner membrane and formation of sonic fragments. This is excluded for two reasons. First, the polarity of H^+ ion movement is always that of intact mitochondria. Second, addition of weak acids causes a shift from cation extrusion to uptake. Fig. 8 shows a passive swelling in 30 mM KI due to valinomycin and shrinkage due to succinate; however, if 5 mM acetate was present succinate caused a further swelling. Fig. 8 shows a similar experiment carried out with 30 mM tetrapropylammonium nitrate. Tetraphenylboron accelerated the penetration of the organic cation and succinate initiated a shrinkage which was inhibited by 3 mM Pi. The subsequent addition of an SH inhibitor restored the active shrinkage. It is therefore apparent that the process of active cation uptake and of active cation extrusion are not opposite but complementary processes, occurring in a membrane possessing the same polarity of active transport. The predominance of one process in respect to the other depends on: (a) the availability of H^+ ion donating weak acids, and (b) the degree of stretching of the membrane.

SITES FOR ACTIVE ANION TRANSPORT

In Fig. 2 it was shown that NEM, like other SH inhibitors, inhibited the passive influx but not the passive efflux of Pi. It was then concluded that the enhancing effect of the active Pi extrusion was due to rendering faster the rate of

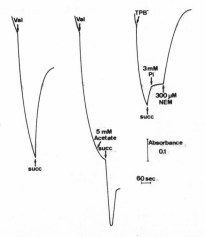

Fig. 8. Restoration of swelling by weak acids. The medium contained in the first and second trace 30 mM KI, 10 mM Tris-Cl, 1 mM EDTA, pH 7.5, and 1 μM rotenone. When indicated were added 0.3 μg valinomycin, 5 mM acetate and 1 mM succinate. In the third trace, KI was replaced by 30 mM tetrapropyl ammonium nitrate, and where indicated were added 5 μM tetraphenylboron, 1 mM succinate, 3 mM Pi and 300 μM NEM. 2 mg protein.

Pi efflux in respect to that of the Pi influx.

Fig. 9 shows that mersalyl caused a marked inhibition of the active shrinkage both in the system involving Pi plus nigericin and that involving strong anions plus valinomycin. In the case of the inhibition of Pi extrusion this was preceeded by a phase of enhancement. The experiment indicates that mersalyl has two effects: one at the level of the Pi influx, which shows up as an enhancement of the active extrusion, and the other at the level of the efflux of both Pi and the strong acids. It must be added that in the case of Pi, mersalyl inhibits both the active and the passive efflux of Pi, while in the case of the strong acids mersalyl inhibits only the active and not the passive efflux.

This complex pattern may be explained by assuming that the transport of anion involves at least two kinds of sites: (a) one whereby the anion overcomes a hydrophobic barrier and which is involved only in Pi transport; and (b) another whereby all anions interact with energy linked high affinity sites, which involve all anions actively extruded whether weak or strong acids.

THE MOLECULAR MECHANISM FOR THE ENERGY LINKED MOVEMENT OF STRONG ELECTROLYTES

While the mechanoenzyme model (10,11) is incompatible with the effect of the SH inhibitors in the case of the extrusion of weak electrolytes, it is compatible with all observations in the case of the extrusion of strong anions (15,19). Recently, Brierley (16,17) and Nichols (20) have proposed a model for the

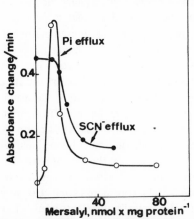

Fig. 9. Comparison between inhibitory effects of mersalyl on active extrusion of Pi and of strong acids. In the case of the Pi efflux the medium contained 30 mM K-phosphate, 2 μg nigericon, 1 mM EDTA. In the case of the SCN⁻ efflux the medium contained 30 mM KSCN, 10 mM Tris-Cl, 1 mM EDTA, pH 7.0, and 0.3 μg valinomycin. Mersalyl was added in various amounts after completion of swelling, and shrinkage was initiated by 1 mM ascorbate plus 100 μM tetramethylenephenylenediamine (TMPD). 2 mg protein.

active extrusion of strong electrolytes, which is an extension of the electrogenic proton pump. Their model predicts an electrophoretic diffusion of anion driven by a membrane potential and an extrusion of cation, via a neutral H^+/K^+ antiporter or via nigericin, driven by a ΔpH. This model is identical to that discussed in Fig. 3A for the active extrusion of Pi. Nichols considered a strong support for the model the fact that extrusion of electrolytes occurred only when the movement of cation was electroneutral and that of anions was electrical (20).

The application of the model of Brierley (16,17) and Nichols (20) to the extrusion of strong bases gives rise to two problems. First, the mechanism of K^+ release. The rate of KCl extrusion in the system analyzed in Fig. 6 may amount to $200 \ \mu mol \ x \ min^{-1} \ x \ g \ protein^{-1}$. However, the rate of the H^+/K^+ antiporter, which should transport K^+, is extremely low. In accord with Douglas and Cockrell (21), we have found a rate of about $10 \ \mu mol \ x \ min^{-1} \ x \ g \ protein^{-1}$. This is less than ten times the rate of K^+ extrusion. Furthermore, as shown in Fig. 1, the active extrusion of K^+ is negligible at low nigericin concentrations. Second, the relative permeabilities to anions and K^+. Since the membrane is permeant to both cations and anions, in order for the membrane potential to always be utilized for extruding anions and not for taking up cations, it is necessary for the rate of anion permeation to be much higher than that of cation permeation. This of course cannot be excluded, but is unlikely to occur in the presence of high valinomycin concentrations. We have found that high valinomycin concentrations do not inhibit the extrusion of K^+.

There are two experiments which argue against the model of Brierley (16,17) and of Nichols (20). First, the process described in Fig. 6 is inhibited by nigericin while the electrogenic proton pump model predicts an enhancement of the shrinkage rate due to an additional H^+/K^+ exchange. Second, the process is independent of the chemical structure of the cation. Fig. 10 shows that mitochondria incubated in 30 mM tetrapropylammonium nitrate underwent a large swelling phase. Addition of succinate at the end of the swelling caused shrinkage. It was possible to replace tetrapropylammonium with a variety of organic cations in the same manner as described by Skulachev's group (4,22) for the energy linked uptake of cations. Tetraphenylboron enhanced markedly the rate of uptake while it had practically no effect on the rate of extrusion. The experiment shows that there is no antiporter or $H^+/$cation exchange involved, and that the process of cation extrusion is to be considered as electrophoretic in nature.

In Fig. 11 are reported schematically the electrogenic proton pump and the proton-driven, cation-anion pump which is an alternative model for the energy linked uptake of strong anions and extrusion of strong cations. The latter model is based on two assumptions. First, the inner mitochondrial membrane possesses a proton-driven pump which may be coupled either to the uptake of cations or to the extrusion of anions. The latter coupling occurs in the stretched mem-

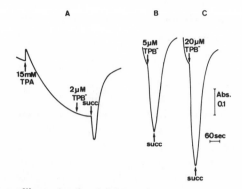

Fig. 10. Passive swelling and active shrinkage with organic cations. In A the medium contained 30 mM KNO₃, 10 mM Tris-Cl, 1 mM EDTA, pH 7.5, and 1 μM rotenone. At the points indicated were added 15 mM tetrapropylammonium (TPA), 2 μM tetraphenylboron (TPB⁻) and 1 mM succinate. In B the medium contained 30 mM TPA NO⁻₃, 10 mM Tris-Cl, 1 mM EDTA, pH 7.5, and 1 μM rotenone. Where indicated were added 2 μM TPB⁻ and 1 mM succinate. In C, conditions as in B except that TPB⁻ was 20 μM.

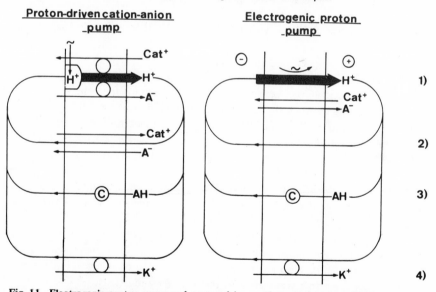

Fig. 11. Electrogenic proton pump and proton-driven cation-anion pump. (1) The electrogenic proton pump assumes an electrical coupling between primary proton movement and diffusion of permeable cations and anions. The proton-driven, cation-anion pump assumes a direct coupling between proton movement and cation and anion movement. (2) The electrogenic proton pump assumes that passive H⁺ permeation leads to decrease of $\Delta\bar{\mu}_{H^+}$. The proton-driven cation-anion pump assumes that passive H⁺ permeation leads under certain conditions to uptake of strong acids and extrusion of strong bases. (3) Indicates in both models the anion transport via carriers. (4) Indicates in both models the cation transport via natural antiporters or nigericin.

brane when there is a high rate of anion permeation. Second, the damaged membrane possesses a high rate of H^+ ion permeation (23).

Consider first the uptake of strong acids. The energy-linked proton-cation exchange leads to the formation of a ΔpH. The ΔpH drives a passive penetration of H^+ ion down the concentration gradient. This in turn drives the passive diffusion of strong anions down the electrical gradient. The molecular process consists of two electrical fluxes of H^+ and Cl^-, respectively. The system studied by us is not far from that analyzed by Nichols (20) for the passive uptake of KCl in brown adipose tissue mitochondria, which also relies on a high permeability to H^+ and Cl^-. There are, however, two important differences. First, here the damage of the membrane is due to a metabolism-linked membrane stretching and not to fatty acids. Second, in the Nichols model the driving force is the anion concentration gradient, while here it is the active proton-cation exchange.

Consider now the extrusion of strong bases. The energy linked proton-anion extrusion also leads to the formation of a ΔpH. The ΔpH drives the H^+ influx with formation of a diffusion potential. The potential, positive inside, drives the efflux of the cations down the electrical gradient.

SHORT RANGE AND LONG RANGE INTERACTIONS IN COUPLING ION FLUXES

The electrophoretic nature of the energy linked uptake of strong cations   or the release of strong anions has been considered to support the conclusion of an electrical coupling between the respiration driven H^+ extrusion and the fluxes of cations and anions. By reversing the argument the electrophoretic nature of the energy linked uptake of strong anions and of the release of strong cations supports the view of an electrical coupling between the passive H^+ ion influx and the fluxes of cations and anions.

The view that the primary reaction of the proton pump generates a large transmembrane potential, negative inside, has difficulties in explaining the uptake of strong acids and the extrusion of strong bases.

On the other hand, the assumption of the proton-driven, cation-anion pump, that charges move across the membrane at small and finite distances, without originating large transmembrane potentials, explains the coupling of ion fluxes at the level of the primary as well as of the secondary reactions.

The proton-driven cation-anion pump represents an extension of our early electroneutral proton pump (24). The binding of protons to a membrane protein, due to energization, is thought of as responsible for a dipolar state of the protein. Negatively and positively charged sites determine active transport of cations and anions through short range electrostatic interactions. The accessibility of cations and anions for the charged sites depends on various factors:

(a) presence of intrinsic carriers as for divalent cations, (b) intrinsic property of the ions as hydrophobic charge screening or charge delocalization, (c) addition of extrinsic ionophores as for valinomycin, and (d) membrane damage as for the strong acids. The interaction of the ions actively transported with the membrane sites is a dynamic process. This means a coupling of flows through pairing of charges, where the charges may be situated in different environments.

CONCLUSION

The present results indicate that mitochondrial active transport is much more complex than hitherto conceived. The energy linked extrusion of Pi, the energy linked uptake of strong acids, and the energy linked extrusion of strong bases represent phenomena which are difficult to explain within the frame of any of the existing models. It is likely that the basic metabolic force is exerted at the level of the proton transport, while the flux of protons is subsequently linked to the cation uptake or the anion extrusion. The fundamental question concerns the nature of the link, whether it involves short range or long range interactions. This is related to the other, equally fundamental, question as to whether direct or indirect types of interactions exist between respiratory chain and ATPase systems. The experimental tests to answer these questions are not easy. It is, however, important to recognize that the present state of knowledge renders unrealistic any dogmatic attitude in favour of either of the two opposing views.

REFERENCES

1. Azzone, G.F. and Massari, S. (1973) *Biochim. Biophys. Acta 301*, 195-226.
2. Azzone, G.F., Massari, S., Colonna, R., Dell'Antone, P. and Frigeri, L. (1973) *Ann. N.Y. Acad. Sci. 227*, 337-347.
3. Azzone, G.F., Massari, S., Colonna, R., Dell'Antone, P., Frigeri, L. and Beltrame, M. (1974) In *Dynamics of Energy-Transducing Membranes* (Ernster, L., Estabrook, R.W. and Slater, E.C., eds.) pp. 405-415, Elsevier, Amsterdam.
4. Lieberman, E.A. and Skulachev, V.P. (1970) *Biochim. Biophys. Acta 216*, 30-42.
5. Azzone, G.F., Massari, S. and Pozzan, T. In preparation.
6. Azzone, G.F., Massari, S. and Pozzan, T. In preparation.
7. Fonyo, A. (1968) *Biochem. Biophys. Res. Commun. 32*, 624-628.
8. Tyler, D. (1969) *Biochem. J. 111*, 665-668.
9. Guerin, B., Guerin, M. and Klingenberg, M. (1970) *FEBS Letters 10*, 265-268.
10. Lehninger, A.L. (1962) *Physiol. Rev. 42*, 467-517.

11.   Lehninger, A.L. (1961) *Biochim. Biophys. Acta 48*, 324-331.
12.   Brierley, G.P. (1973) *Ann. N.Y. Acad. Sci. 227*, 398-411.
13.   Brierley, G.P., Jurkowitz, M. and Scott, K.M. (1973) *Arch. Biochem. Biophys. 159*, 742-756.
14.   Azzi, A. and Azzone, G.F. (1967) *Biochim. Biophys. Acta 131*, 468-478.
15.   Azzone, G.F. and Piemonte, G. (1969) In *The Energy Level and Metabolic Control of Mitochondria* (Papa, S., Tager, J.M., Quagliariello, E. and Slater, E.C., eds.) pp. 115-124, Adriatica, Bari.
16.   Brierley, G.P. (1970) *Biochemistry 9*, 697-708.
17.   Brierley, G.P. and Stoner, C.D. (1970) *Biochemistry 9*, 708-713.
18.   Brierley, G.P., Jurkowitz, M. and Scott, K.M. (1971) *J. Biol. Chem. 246*, 2241-2249.
19.   Azzi, A. and Azzone, G.F. (1967) *Biochim. Biophys. Acta 135*, 444-453.
20.   Nichols, D. (1974) *Eur. J. Biochem. 49*, 585-593.
21.   Douglas, M.G. and Cockrell, R.S. (1974) *J. Biol. Chem. 249*, 5464-5471.
22.   Barkeeva, L.E., Grinius, L.L., Jasaitis, A.A., Kuliena, V.V., Levitsky, D.O., Liberman, E.A., Severina, I.I. and Skulachev, V.P. (1970) *Biochim. Biophys. Acta 216*, 13-21.
23.   Massari, S. and Azzone, G.F. (1972) *Biochim. Biophys. Acta 283*, 23-29.
24.   Massari, S. and Azzone, G.F. (1970) *Eur. J. Biochem. 12*, 300-309; *ibid.*, 310-319.

ION TRANSPORT THROUGH THE GRAMICIDIN A CHANNEL

E. Bamberg, H.-A. Kolb and P. Läuger

Department of Biology
University of Konstanz
D-775 Konstanz, Germany

The molecular mechanisms of ion permeation through the cell membrane are as yet not well understood, but it is likely that in many cases built-in proteins are involved which offer to the ion a hydrophilic pathway through the apolar core of the membrane. Such a structure may be called a channel (or a pore), i.e., a sequence of coordination sites at which the potential energy of the ion is drastically reduced in comparison with the extremely high value in a hydrocarbon environment.

A possible approach to the study of ion channels is to isolate the channel protein from the biological source and to incorporate the purified protein into an artificial lipid membrane. As far as planar bilayer membranes are concerned, this approach has met only with limited success so far. It is therefore fortunate that a number of relatively simple channel-forming molecules exists from which we may hope to get some insight into the mechanisms of ion transport through channels.

Such a molecule is gramicidin A, a linear pentadecapeptide (Fig. 1) which has been isolated from *Bacillus brevis* (1,2). This molecule has several remarkable structural features. With the exception of glycine in position 2, it consists of strongly hydrophobic amino acids. As both end groups are blocked (by a formyl residue at the amino terminal and by an ethanolamino residue at the carboxyl terminal), the molecule is electrically neutral. Furthermore, the optical configuration of the amino acids is alternately D and L (glycine in position 2 may stand for a D amino acid).

If gramicidin A is introduced into a biological membrane (3-6) or into an artificial (7-21) lipid bilayer, the membrane becomes cation permeable. In this respect gramicidin A is similar to the macrocyclic ion carriers such as valinomycin or monactin. It has become clear, however, that the action mechanism

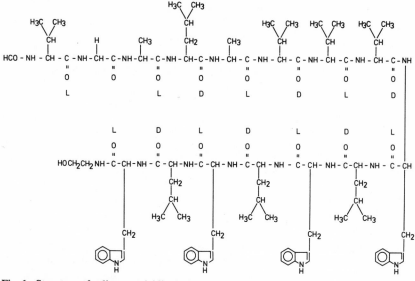

Fig. 1. Structure of valine-gramicidin A.

of gramicidin is quite different from that of the macrocyclic carriers, and that gramicidin probably forms channels in a lipid membrane. One piece of evidence came from the experiments of Krasne, Eisenman and Szabo (14). They were able to freeze lipid bilayers by reducing the temperature, and observed that below the freezing point of the membrane the valinomycin conductance dropped to a very low value, but the gramicidin conductance still persisted in the frozen state. This finding is in favor of the view that valinomycin acts as a mobile carrier but that gramicidin creates channels in the lipid membrane. Further support for a channel mechanism comes from the observed transport rates (see below) as well as from bulk phase experiments (34).

A detailed model of the gramicidin A channel has been proposed by Urry (22-24), and independently by Ramachandran and Chandrasekaran (25), on the basis of conformational energy considerations. The model of Urry consists of a helical dimer that is formed by head-to-head (formyl end to formyl end) association of two gramicidin monomers and which is stabilized by intra- and inter-molecular hydrogen bonds (Fig. 2). The central hole along the axis of the π^6 (L,D)- helix has a diameter of about 4 Å and is lined with the peptide C-O moieties, whereas the hydrophobic residues lie on the exterior surface of the helix. The total length of the dimer is about 20-30 Å, which is the lower limit of the hydrophobic thickness of a lipid bilayer. The length of the dimer would thus be sufficient to bridge the membrane if a local thinning of the lipid structure is assumed. The dimer hypothesis is supported by the finding that the

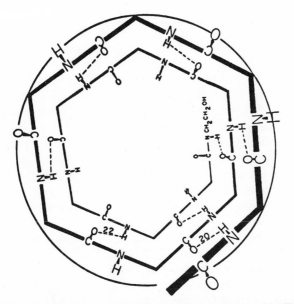

Fig. 2. Structure of the π^6 (L,D)- helix of gramicidin A, as proposed by Urry (15). The hydrophilic hole along the helix axis is lined with the oxygen atoms of the amide carbonyls which alternately point up and down. The hydrophobic residues which are located at the periphery of the helix are not shown.

dimer formed by chemical linkage of two gramicidin molecules at the formyl ends has a similar membrane activity as the monomer, but at much lower concentrations (15). Dimerization of gramicidin A in organic solvents has recently been studied by fluorescence and circular dichroism methods (27). Further support for the dimer model comes from kinetic experiments which are described below.

Recently, Veatch and Blout (26-28) proposed an alternative model of the gramicidin channel. They postulated a new class of double-helical dimers in which the two chains are coiled about a common axis and in which all hydrogen bonds are intermolecular. These double helices have a cylindrical hole along the axis with a diameter of greater than 3 Å. The rise per turn is about twice that of the π(L,D)- helices of comparable diameter. Undoubtedly, the model of Veatch and Blout is a possible candidate for the structure of the gramicidin channel. The observation that desformylation of gramicidin A leads to a virtually complete loss of activity (22) is easily explained on the basis of Urry's model, but is less obvious for a double-stranded helix. At the moment the available experimental data do not seem sufficient, however, to exclude one or the other model.

KINETICS OF CHANNEL FORMATION

Many details about the nature of the gramicidin-mediated cation permeability have been revealed by studying the electrical properties of gramicidin-doped black lipid films. An instructive experiment, which was first performed by Hladky and Haydon, consists of adding extremely small amounts of gramicidin A to the membrane and recording the time course of the electric current at a given voltage (12,13,19). Under these circumstances the current shows discrete fluctuations which have more or less the same amplitude (Fig. 3). If increasing amounts of gramicidin are added to the membrane, more and more fluctuations build up on top of each other and eventually fuse into an average macroscopic current. This experiment can be explained by the assumption that the single current fluctuation comes about by the formation and the disappearance of a single channel. The conductance of the channel, which is calculated from the current fluctuation, is of the order of $1 \times 10^{-11} \ \Omega^{-1}$ in 1 M NaCl at 25° C (19). This corresponds to a transport of about 10^7 sodium ions per second through the single channel at a voltage of 100 mV. This value is larger by a factor of 10^3 than the turnover number of a translational carrier such as valinomycin (19). The high transport rate through the single conductance unit of gramicidin A would therefore be difficult to reconcile with a carrier model but can easily be explained by a pore mechanism.

An obvious test for the hypothesis that the channel is a dimer would consist of measuring the dependence of membrane conductance λ on the aqueous gramicidin concentration c. Provided that a partition equilibrium of gramicidin between aqueous phase and membrane exists, the dimer model requires that λ should be proportional to c^2. Such conductance measurements turned out to be very inaccurate, however, apparently because a true partition equilibrium between membrane and water could never be reached (13). This difficulty may be circumvented by studying the kinetics of the gramicidin system (18-21).

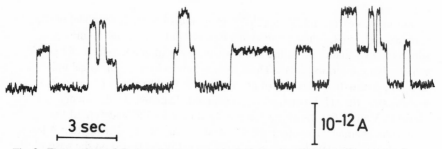

3 sec

10^{-12} A

Fig. 3. Fluctuations of the membrane current in the presence of very small amounts of gramicidin A (19). 1 M NaCl, 25° C, V_m = 90 mV. The membrane has been formed from dioleoyl-lecithin in n-decane. The base line corresponds to the membrane conductance without gramicidin.

If the dimer hypothesis is correct, then the channel-formation process may be described in the same way as a bimolecular chemical reaction between two gramicidin monomers G:

$$G + G \underset{k_D}{\overset{k_R}{\rightleftharpoons}} G_2 \tag{1}$$

where k_R and k_D are the rate constants of association and dissociation. Information about the kinetics of such a reaction may be obtained by relaxation experiments (18). The system is disturbed by a sudden displacement of an external parameter, and the time required for the system to reach a new stationary state is measured. In the case of a bilayer membrane, it is convenient to choose as the variable external parameter the electric field strength in the membrane (Fig. 4). Immediately after a sudden displacement of the voltage, a capacitive current is observed which decays with a time constant equal to the product of

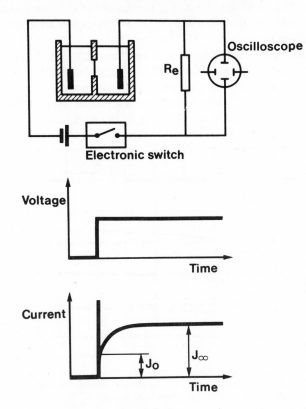

Fig. 4. Principle of the electric relaxation method.

the circuit resistance and the membrane capacitance. This capacitive spike limits the time resolution of the method, which under favorable conditions is of the order of 1 μs. After the decay of the capacitative transient, the membrane current approaches a stationary value J_∞ with a time constant τ which contains information about the conductance mechanism under study.

As the conductance of the bare membrane is negligibly small and as the monomers are likely to be non-conducting, the current is proportional to the instantaneous concentration of dimers G_2 in the membrane. For the above case of an association-dissociation reaction, the relaxation time τ_C of the membrane current is then related to the rate constants k_R and k_D, the single-channel conductance Λ, and the stationary membrane conductance λ_∞ in the following way (18):

$$\frac{1}{\tau_C} = k_D + 4 \left(\frac{k_R k_D \lambda_\infty}{L\Lambda}\right)^{\frac{1}{2}} \tag{2}$$

(L is Avogadro's constant). Thus, if τ_C is measured at different gramicidin concentrations, and $1/\tau_C$ is plotted as a function of the square root of λ_∞, a straight line should result.

The resultant of a typical relaxation experiment is shown in Fig. 5. After the voltage jump at time zero, the membrane current starts at a relatively low level and increases asymptotically towards a new stationary value. According to the channel model, this increase of current reflects a rise in the equilibrium concentration of channels under the action of the electric field.

The mechanism by which the electric field influences the number of con-

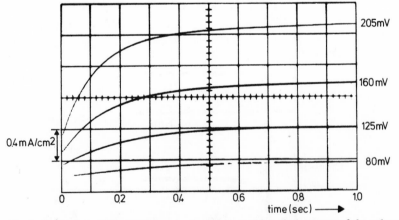

Fig. 5. Relaxation of the membrane current after a sudden displacement of the voltage (18). 1 M NaCl, 2×10^{-9} M gramicidin, 25° C, dioleoyl-lecithin membrane. The amplitude of the voltage jump is indicated at the right side of the figure. The capacitative transient of the current is not resolved.

ducting channels is not known with certainty. One possibility is that the decrease of membrane thickness which is brought about by the compressing force of the electric field tends to stabilize the dimeric channel (18). This explanation is suggested by the fact that the length of the dimer is less than the thickness of the membrane. There is evidence, however, that in addition to this purely geometrical effect the electric field may also act in a direct way on the gramicidin molecule in the membrane (30).

If the transients shown in Fig. 5 are analyzed, it is found that the time course of the current is described by a single relaxation time τ_c. A plot of $1/\tau_c$ as a function of $(\lambda_\infty)^{\frac{1}{2}}$ gives a straight line with positive slope in accordance with equation 2. The results of the relaxation experiments are thus consistent with the dimer model. This, of course, does not mean that more complicated models may be excluded at this stage of analysis. For instance, channel formation may be a two-step process, the first step being the formation of an inactive dimer which in the second step is transformed into the conducting channel by a monomolecular reaction. Or there may be a pre-equilibrium between unreactive and reactive monomers (24), only the latter being capable of forming dimers. While these possibilities require further investigation, the main result of the relaxation experiments is the conclusion that the overall reaction in channel formation has a molecularity that is higher than one.

Accepting for the moment the simple dimer model, we may analyze the experimental result in terms of equation 2 and determine k_D and k_R from the intercepts and the slope of the $1/\tau_c$ vs $(\lambda_\infty)^{\frac{1}{2}}$ plot, using the Λ value which has been obtained from the single-channel experiments. For a dioleoyl-lecithin membrane at 25° C this gives $k_D = 1.6$ s^{-1} and $k_R = 2.0$ x 10^{14} cm^2 x mole^{-1} s^{-1} (18,19). According to the dimer model, $1/k_D$ should be equal to the mean lifetime τ^* of the conducting channel which can be independently determined from a single-channel experiment. τ^* And $1/k_D$ have been measured at 10, 25 and 40° C. As seen from Table I, τ^* varies about 20-fold in this temperature range, but $1/k_D$ and τ^* agree within a factor of two at all temperatures.

The value of the association rate constant k_R may be illustrated in the following way. If the monomers are present in the membrane in such a concentration that the mean distance is 10^3 Å, then a given monomer will associate with another monomer with a frequency of 2 per second. As may be expected, this rate is considerably less than the limiting value of a diffusion-controlled reaction (18). It may be instructive to calculate from the equilibrium constant of association, $K_A = k_R/k_D = 1.2$ x 10^{14} cm^2 mol^{-1} (25° C), the ratio of monomer and dimers in the membrane under the condition of the conductance experiments. If the actual numbers of monomers and dimers in a membrane of area A are n_m and n_d, respectively, then the ratio n_m/n_d is given by:

TABLE I

Temperature Dependence of Gramicidin A Channel Parameters (19)

T	k_R	k_D	Λ	τ^*	$1/k_D$
°C	$cm^2 \times mole^{-1} \times s^{-1}$	s^{-1}	Ω^{-1}	s	s
10	2.3×10^{13}	0.25	0.65×10^{-11}	3.8	4.0
25	20×10^{13}	1.6	1.2×10^{-11}	1.1	0.63
40	68×10^{13}	4.5	2.3×10^{-11}	0.15	0.25

Dioleoyllecithin in n-decane, 1 M NaCl, V_m = 135 mV.

$$\frac{n_m}{n_d} = (\frac{LA}{n_d K_A})^{1/2} \tag{3}$$

For a membrane of area A = 1 mm^2, the ratio n_m/n_d in a single channel experiment (n_d = 1) is calculated to be about 7 x 10^3. The finding that a large number of monomers is present in the membrane under the conditions of a single-channel experiment is consistent with the observation (13) that at low conductance levels the probability for the simultaneous occurrence of exactly n channels follows the Poisson statistics. On the other hand, at high conductance levels, 10^8 or more channels may be present in a membrane of A = 1 mm^2, and under these conditions, n_m/n_d becomes of the order of unity.

ANALYSIS OF CHANNEL NOISE

In the past the kinetics of fast reactions (including transport processes in membranes) has been studied mainly by relaxation methods. An entirely different method is based on the principle that any reacting system exhibits random fluctuations around the stationary state. By analyzing the time- correlation of the fluctuations, information about the kinetic parameters of the system may be obtained. The fluctuation method has the advantage that the measurement is carried out while the system is in an equilibrium or stationary state. Recently, Feher and Weissman (31) studied conductance fluctuations of an electrolyte solution and deduced from the frequency spectrum of the fluctuations the relaxation time of the association-dissociation reaction of the electrolyte.

The method of fluctuation analysis may also be applied to the kinetic analysis of gramicidin channels in lipid bilayer membranes (20,21). A thin lipid membrane is, in the thermodynamic sense, a small system and is thus subject to relatively large fluctuations. In the case of a gramicidin-doped membrane, conductance fluctuations arise from the random formation and decay of channels. As described above, from experiments with only one or a few channels present in the membrane, the single channel conductance and the mean lifetime of the channel may be obtained. It is not possible, however, to evaluate also the formation rate constant of the channel from single-channel experiments. A more complete analysis, yielding (in terms of the dimer model) both k_D and k_R, becomes possible if the fluctuations of the macroscopic membrane conductance at different gramicidin concentrations are measured. Thus, the main difference between single-channel experiments and the study of macroscopic conductance fluctuations is that in the latter case additional information may be obtained from the variation of a further experimental parameter, the gramicidin concentration in the membrane.

If a constant voltage is applied to a bilayer membrane containing gramicidin A, the current J is found to fluctuate in time t around a mean value $\bar{J}$:

$$J(t) = \bar{J} + \delta J(t) \qquad (4)$$

A straightforward way to analyze the fluctuating component $\delta J(t)$ of the membrane current is to compute the so-called autocorrelation function $C(\tau)$:

$$C(\tau) = \overline{\delta J(0) \times \delta J(\tau)} \qquad (5)$$

$C(\tau)$ describes the average rate by which a current fluctuation decays that has occurred at time $\tau = 0$. The fluctuation-dissipation theorem states that the decay of a spontaneous fluctuation follows (on the average) the same time-law as the relaxation from a macroscopic perturbation. This means that for the dimerization reaction considered here, the autocorrelation function has the form:

$$C(\tau) = \overline{(\delta J)^2} \, e^{-\tau/\tau_c} \qquad (6)$$

where τ_c is the macroscopic relaxation-time given by equation 2.

The autocorrelation function may be directly computed from a record of J(t) with the aid of suitable instruments (20,21). Fig. 6 gives an example of the current fluctuations observed with a dioleoyl-lecithin membrane in the presence of gramicidin A. The autocorrelation function $C(\tau)$ of the current noise (Fig. 7) is found to be very nearly exponential with a single time constant τ_c. From here on the analysis proceeds in the same way as in the case of the relaxation experiments. The correlation time τ_c is plotted as a function of the square root of the mean membrane conductance λ; again, it is found that τ_c varies linearly with $\lambda^{1/2}$. The rate constants k_R and k_D that are obtained from the τ_c vs $\lambda^{1/2}$ plot agree within a factor of two with the values calculated from the relaxation experiments (Table II).

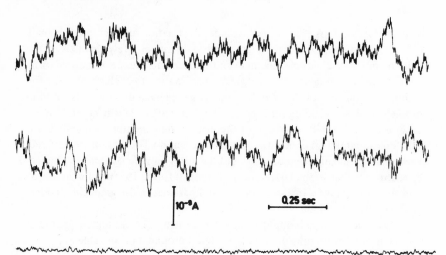

Fig. 6. Current noise from a dioleoyl-lecithin membrane in the presence of gramicidin A (upper traces). The mean current $\bar{J}$ was 3.9×10^{-7} amps, corresponding to a mean membrane conductance $\lambda = 3.1 \times 10^{-3}$ Ω^{-1} cm^{-2} (membrane area A = 6.8×10^{-3} cm^2, external voltage V = 18.5 mV). The aqueous phase contained 1 M NaCl. The lower trace represents a control experiment in which the noise was recorded in the same way as above but from a gramicidin-free membrane with an external resistor of 47 kΩ simulating the gramicidin-induced conductance (21).

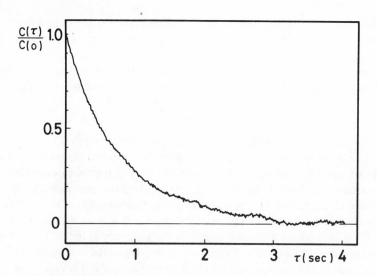

Fig. 7. Autocorrelation function $C(\tau)$ of the current noise, divided by the initial value $C(0)$ $\cong 7.8 \times 10^{-22}$ amps^2. Membrane area A = 7.0×10^{-3} cm^2, external voltage V = 18.5 mV. The aqueous phase contained 1 M NaCl (21).

TABLE II

Kinetic Parameters of the Gramicidin A Channel:
Comparison Between Noise Analysis, Relaxation Method
and Single-Channel Experiments (21)

Method	k_R	k_R	k_D	k_D	Λ
	18.5 mV	98 mV	18.5 mV	98 mV	
Noise analysis	0.62×10^{14}	1.4×10^{14}	1.7	1.9	0.9×10^{-11}
Relaxation method	1.3×10^{14}	1.5×10^{14}	1.6	1.6	---
Single-channel experiments	---	---	---	0.91	1.2×10^{-11}

Dioleoyl-lecithin in n-decane, 1 M NaCl, 25° C, k_R in cm^2 $mole^{-1}$ s^{-1}, k_D in s^{-1}, Λ in Ω^{-1}.

As a theoretical analysis shows, the mean square of the current fluctuations is related to the single-channel conductance Λ. Thus, from $C(0) = \overline{(\delta J)^2}$ information on Λ may be obtained. As shown in Table II, the values of Λ that are calculated from noise analysis and from single-channel experiments are in satisfactory agreement.

For gramicidin A in a dioleoyl-lecithin membrane the analysis of the current noise is rather straightforward and yields essentially the same information as the relaxation method. Such an agreement is indeed expected in the case of the simple dimerization reaction considered here. For more complicated mechanisms, however, a simple relationship between relaxation data and noise measurements no longer exists, in general. For instance, if the formation of a channel is a two-step reaction that is described by two relaxation times, it may occur that only one relaxation is excited by an external perturbation such as a voltage jump, whereas both processes appear in the fluctuation experiment. With membranes made from certain monoglycerides, it has been observed recently that the gramicidin channel noise shows more complicated spectral properties that are not contained in the model considered above. Such systems are currently under study (32).

ION SPECIFICITY

An efficient channel should combine high transport rates with the ability to discriminate between different ion species. For instance, the Na^+ channel of the nerve membrane has a $Na^+:K^+$ selectivity of 20:1 and, at the same time, shows transport rates of the order of 10^8 Na^+ ions per sec (33). It is an important goal of the theoretical analysis of channel mechanisms to explain how these seemingly conflicting properties of high selectivity and high transport rates are achieved. In addition, from experiments with model compounds such as gramicidin A we may hope to get direct insight into the relationship between chemical structure and transport properties of ion channels.

The gramicidin A channel is virtually impermeable to anions but barely discriminates between the different alkali ions, the selectivity ratios being similar to the mobility ratios in free aqueous solution (7, 13, 16). Correspondingly, the activation energy of the single-channel conductance in 1 M NaCl (13,19) is rather low (4.5 Kcal/mole in glycerylmonooleate and 7.3 Kcal/mole in dioleoyllecithin) and not much larger than the activation energy of free diffusion of Na$^+$ in water (3.6 Kcal/mole).

For a theoretical analysis of ion specificity, the channel may be considered as a sequence of binding sites that are separated by energy barriers (35,36) (see Fig. 8). The passage of an ion through the channel consists of a series of thermally activated jumps between successive potential minima. At a potential minimum (or binding site) the ion is in an energetically favorable position with respect to one or several ligands (oxygen atoms of the peptide carbonyl groups). The energy barrier between two binding sites may be of purely electrostatic origin or may contain contributions from the change in conformational energy of the channel associated with the passage of the ion from one site to the next.

The ion specificity of a channel depends both on thermodynamic and kinetic parameters (35). The essential thermodynamic quantity is the change of free energy (ΔF) associated with the transfer of an ion from the aqueous phase into a binding site inside the channel (for simplicity we assume here that all binding sites are energetically identical). ΔF determines the equilibrium constant K for

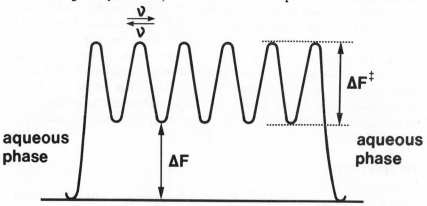

Fig. 8. Ion channel as a sequence of potential minima (binding sites) that are separated by energy barriers. For simplicity, the binding sites have been assumed to be energetically identical.

the occupancy of the channel, which relates the number N_0 of occupied channels in the membrane to the number N_e of empty channels:

$$K = \frac{N_0}{cN_e} \qquad (7)$$

c is the ion concentration in the aqueous phase, which is assumed to be sufficiently small so that a given channel is either empty or singly-occupied. K may be written as:

$$K = K_0 \, e^{-\Delta F/RT} \qquad (8)$$

(K_0 is a constant, T the absolute temperature, and R the gas constant). The energy ΔF may be evaluated in terms of ion-solvent and ion-ligand interactions on the basis of Eisenman's theory of ion specificity which has been successfully applied to ion carriers (37).

A second important quantity besides the equilibrium constant K is the rate constant v, which describes the frequency of jumps over an energy barrier in the channel. According to the theory of absolute reaction rates, v is related to the free energy of activation, $\Delta F^{\ddagger}$ (Fig. 8):

$$v = f \left(e^{-\Delta F^{\ddagger}/RT} \right) \qquad (9)$$

(see ref. 38). As the frequency factor f is of the order of $6 \times 10^{12} \, s^{-1}$, the overall transport rates may be quite high, provided that the free energy of activation, $\Delta F^{\ddagger}$, is not too large. A small value of $\Delta F^{\ddagger}$ means little interaction between ion and binding site, and if the interaction is small, the selectivity will be poor. Obviously, there must be a compromise between transport rates and selectivity.

The selectivity of a channel with respect to two ion species A and B may be defined as the ratio Λ_A/Λ_B of the single-channel conductances in the presence of A or B. Λ_A/Λ_B is obtained from the theory as a product of a thermodynamic term times a kinetic term that depends on the jump-rate constants (35). In the simple case of a channel consisting of a sequence of identical energy barriers, the selectivity ratio is given by the ratios of the equilibrium constants K_A and K_B of ion species A and B (equation 7) times the ratio of the jump-rate constants v_A and v_B:

$$\frac{\Lambda_A}{\Lambda_B} = \frac{K_A}{K_B} \times \frac{v_A}{v_B} \qquad (10)$$

The statement that the selectivity ratio depends both on thermodynamic as well as on kinetic factors also applies, in principle, to carriers, but in carrier-mediated ion transport there exists a wide range of conditions under which the specificity is determined by equilibrium constants alone (37). The reason is that in many cases the rate-limiting step is the translocation of the loaded carrier through the membrane and that the rate of this process depends little on the nature of the ion. The overall transport rate is then proportional to the equilibrium concentration of loaded carrier molecules in the membrane. On the other hand, the specificity of a channel is always influenced by kinetic parameters, as equation (10) shows.

REFERENCES

1. Hotchkiss, R.D. (1944) *Adv. Enzymol. 4*, 153-200.
2. Sarges, R. and Witkop, B. (1965) *J. Am. Chem. Soc. 87*, 2011-2020.
3. Pressman, B.C. (1965) *Proc. Nat. Acad. Sci. (U.S.A.) 53*, 1076-1083.
4. Henderson, P.J.F., McGivan, J.D. and Chappel, J.B. (1969) *Biochem. J. 111*, 521-535.
5. Podleski, T. and Changeux, J.-P. (1969) *Nature 221*, 541-545.
6. Wenner, C.E. and Hackney, J.H. (1969) *Biochemistry 8*, 930-938.
7. Mueller, P. and Rudin, D.O. (1967) *Biochem. Biophys. Res. Commun. 26*, 398-404.
8. Liberman, E.A. and Topaly, V.P. (1968) *Biochim. Biophys. Acta 163*, 125-136.
9. Tosteson, D.C., Andreoli, T.E., Tiefenberg, M. and Cook, P. (1968) *J. Gen. Physiol. 51*, 373S-384S.
10. Goodall, M.C. (1970) *Biochim. Biophys. Acta 219*, 471-478.
11. Goodall, M.C. (1971) *Arch. Biochem. Biophys. 147*, 129-135.
12. Hladky, S.B. and Haydon, D.A. (1970) *Nature 225*, 451-453.
13. Hladky, S.B. and Haydon, D.A. (1972) *Biochim. Biophys. Acta 274*, 294-312.
14. Krasne, S., Eisenman, G. and Szabo, G. (1971) *Science 174*, 412-415.
15. Urry, D.W., Goodall, M.C., Glickson, J.S. and Mayers, D.F. (1971) *Proc. Nat. Acad. Sci. (U.S.A.) 68*, 1907-1911.
16. Myers, V.B. and Haydon, D.A. (1972) *Biochim. Biophys. Acta 274*, 313-322.
17. Goodall, M.C. (1973) *Arch. Biochem. Biophys. 157*, 514-519.
18. Bamberg, E. and Läuger, P. (1973) *J. Membrane Biol. 11*, 177-194.
19. Bamberg, E. and Läuger, P. (1974) *Biochim. Biophys. Acta 367*, 127-133.
20. Zingsheim, H.P. and Neher, E. (1974) *Biophys. Chem. 2*, 197-207.
21. Kolb, H.-A., Läuger, P. and Bamberg, E. (1975) *J. Membrane Biol. 20*, 133-154.
22. Urry, D.W. (1971) *Proc. Nat. Acad. Sci. (U.S.A.) 68*, 672-676.

23. Urry, D.W. (1972) *Biochim. Biophys. Acta 265*, 115-168.
24. Urry, D.W. (1972) *Proc. Nat. Acad. Sci. (U.S.A.) 69*, 1610-1614.
25. Ramachandran, G.N. and Chandrasekaran, R. (1972) *Prog. Peptide Res. 2*, 195-215.
26. Veatch, W.R. and Blout, E.R. (1974) *Biochemistry 13*, 5249-5256.
27. Veatch, W.R. and Blout, E.R. (1974) *Biochemistry 13*, 5257-5264.
28. Fossel, E.T., Veatch, W.R., Ovchinnikov, Y.A. and Blout, E.R. (1974) *Biochemistry 13*, 5264-5275.
29. Läuger, P. (1972) *Science 178*, 24-30.
30. Bamberg, E. and Benz, R. To be published.
31. Feher, G. and Weissman, M. (1973) *Proc. Nat. Acad. Sci. (U.S.A.) 70*, 870-875.
32. Kolb, H.-A. and Bamberg, E. To be published.
33. Hille, B. (1970) *Prog. Biophys. Mol. Biol. 21*, 3-32.
34. Byrn, S.R. (1974) *Biochemistry 13*, 5186-5193.
35. Läuger, P. (1973) *Biochim. Biophys. Acta 311*, 423-441.
36. Läuger, P. and Frehland, E. (1974) *J. Theoret. Biol. 47*, 189-207.
37. Eisenman, G., Szabo, G., Ciani, S., McLaughlin, S. and Krasne, S. (1973) *Prog. Surface Membrane Sci. 6*, 139-241.
38. Läuger, P. and Neumcke, B. (1973) In *Membranes − A Series of Advances* (Eisenman, G., ed.) Vol. 2, pp. 1-59, Marcel Dekker, New York.

IV
Energy Transduction
in Membranes

ENERGY TRANSDUCTION IN MEMBRANES

E.C. Slater

Laboratory of Biochemistry
B.C.P. Jansen Institute
University of Amsterdam
Amsterdam, The Netherlands

The energy made available by electron transfer between redox couples of the appropriate redox potential present in the energy-transducing membranes of mitochondria, bacteria and chloroplasts can be utilized for energy-requiring reactions, namely the vectorial translocation of ions across the membrane against the *electrochemical potential* or non-vectorial or scalar chemical reactions proceeding outside the membrane against the *chemical potential*.

The electromotive force or potential acting on an ion moving across the membrane is the sum of the electrical potential difference across the membrane ($\Delta\psi$) and the Nernst term relating the ratio of the thermodynamic activities of the ion on the two sides of the membrane. For protons, Mitchell (1) has introduced the term protonmotive force which is given by Equation 1.

$$\Delta p = \Delta\psi - \frac{RT}{F} \Delta pH \tag{1}$$

where the Δ sign refers to the difference between the two sides of the membrane.

The chemical potential of a chemical reaction $A + B \rightleftharpoons C + D$ is represented by the value of ΔG, which is given by Equation 2.

$$\Delta G = \Delta G^O + RT \ln \frac{[C]\,[D]}{[A]\,[B]} \tag{2}$$

where $-\Delta G^O$ is the standard free energy change in the reaction. At equilibrium, $\Delta G = 0$, so that $\Delta G^O = -RT \ln K$. When ΔG is positive, energy has to be supplied to the reaction by coupling it with an energy-yielding reaction. For example, it has been found that respiring mitochondria at pH 7.7 and in the absence of added Mg^{2+} can still drive the reaction $ADP + P_i \rightleftharpoons ATP + H_2O$

when the concentration term $\dfrac{[ATP]}{[ADP]\ [P_i]}$ is equal to $3 \cdot 10^5\ M^{-1}$ (2). Since, under these conditions, $\Delta G^O = 8.7$ kcal/mol (3), the chemical potential (termed in this context the phosphate potential, ΔG_p) is given by $\Delta G_p = 8.7 + 1.36 \log 3 \cdot 10^5 = 16.1$ kcal/mol.

A second reaction that can be driven against the chemical potential by energy-transducing membranes is $NADH + NADP^+ \rightleftharpoons NAD^+ + NADPH$ whose ΔG^O is close to zero (+ 0.14 kcal/mol (4)). The maximum chemical potential reported for this reaction may be calculated from the data of Rydström *et al.* to be equal to 3.8 kcal/mol (5). A third type of reaction in respiring mitochondria or particles is the reduction of low-potential substrates by higher potential, for example the reaction

succinate + 2-oxoglutarate + $NH_3 \rightleftharpoons$ fumarate + glutamate.

Data required for the calculation of the chemical potential against which this reaction can be driven are not available.

The two processes that are coupled to electron-transfer in energy-transducing membranes — vectorial ion translocation against the electrochemical potential and scalar chemical reactions against the chemical potential — may be conceivably either (i) independent processes, (ii) linked in series, or (iii) linked in parallel. The universal association of the two processes in different types of membranes makes it unlikely that they are independent consequences of the electron transfer, and discussion now centres on two hypotheses. The Mitchell chemiosmotic hypothesis (1), which received support earlier from those working with the chloroplast than with the mitochondrion, is now accepted by many of the latter as well as probably a majority of those working with chloroplasts and bacteria. According to this hypothesis, the function of the electron-transfer chain is to translocate protons across the membrane, and the proton-motive force drives either the translocation of other ions against the concentration gradient or chemical reactions against the chemical potential. Proton translocation, which is the primary event in this hypothesis, is brought about by the alternate transfer of hydrogen atoms from one side of the membrane to the other, and of electrons back across the membrane, in a series of 'loops' (1). The protons are fed back through a proton 'well' (6) in the membrane to the protein (Coupling Factor 1 or F_1 (7)) that finally catalyses the synthesis of ATP from ADP and P_i.

An alternative view, favoured by Paul Boyer (8) and some others (e.g. ref. 9), is that the energy of electron transfer is primarily conserved in a conformationally strained high-energy form of the protein bearing the electron-accepting group concerned and is transmitted to the ATP-synthesizing protein. Specifically, it is proposed that the conformational change leads to the dissociation of ATP firmly bound to it, this ATP having been formed on the enzyme by reversal of reactions involved in the catalysis of added ATP (see Fig. 1). According to this view, proton translocation is incidental to protein conformation changes in the same way as protons are produced by the oxygenation of haemoglobin:

$$Hb + O_2 \rightleftharpoons HbO_2 + 2.8\ H^+ \text{ at pH 7.4.}$$
$$\quad T \qquad\qquad R$$

Deoxyhaemoglobin is in a strained (or T) conformation and becomes relaxed (in the R state) on oxygenation (10). At pH 7.0 and 20° C in the presence of 0.1 M phosphate, the free energy of this reaction is 8 kcal per mol less than expected from the reactivity of the individual polypeptide chains in haemoglobin (10). This free energy represents the energy of those interactions between and within the chains in the T state that are broken in the R state. This example shows that quite large amounts of energy may be stored in specific conformations of proteins. The protons are liberated during the transformation of the T to the R state as a consequence of the lowering of the pK of acidic groups. Indeed, binding of protons to specific nitrogen atoms in the amino acids (terminal amino group of the α- chains and histidine-122α and -146β side-chains (10)) are necessary for the stabilizing salt bridges in the T state.

By postulating that conformation-dependent acidic groupings present in the electron-transferring proteins take up protons from one side of the membrane and liberate them to the other, it is possible to construct a proton-translocating system as an alternative to the Mitchell loops and to link this with cation translocation (see Fig. 2). On the basis of this model one could either accept the

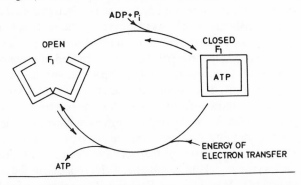

Fig. 1. Proposed hypothesis of ATPase-catalysed synthesis of ATP linked to electron transfer. It is proposed that the ATPase can exist in two conformations, an open (which binds aurovertin with a high fluorescence yield and binds the ATPase inhibitor weakly) and a closed. The open conformation can bind ADP and phosphate on the catalytic sites, with the result that dehydration takes place on the enzyme, with formation of ATP bound to the open conformation, which is converted spontaneously to the closed conformation which binds ATP more firmly. Electron transfer favours transformation into the open conformation with consequent dissociation of the ATP, and rebinding of the ADP and P_i, which are bound more firmly to this than to the closed conformation.

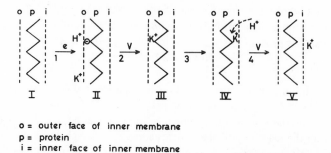

o = outer face of inner membrane
p = protein
i = inner face of inner membrane

Fig. 2. Possible mechanism whereby electron-transfer-induced conformation change in protein could lead to appearance of protons outside the mitochondrion and, in the presence of valinomycin (V), to exchange of K^+ with H^+. Reaction 1 represents a conformation change in a membrane protein (represented by a change of the zig-zag to its mirror image) induced by electron flow along the respiratory chain, resulting in dissociation of an acidic group near the outer phase of the membrane. A negative charge is left on the membrane, and a proton appears in the medium. In Reaction 2 valinomycin permits K^+ to enter the membrane where it combines with the negatively charged group on the protein. Combination of K^+ results in relaxation of the strained structure in II and III to the relaxed structure in IV (as in I) whereby K^+ now appears on the inner face. In Reaction 4 the K^+ is replaced by H^+, since the pK of conformation IV (= I) is higher than that of II + III.

Mitchell view that ATP synthesis is driven by the protonmotive force, or favour a direct transfer of conformational energy from the strained electron-carrying protein to the ATPase. An important difference between the Mitchell loops and protein conformational changes as the source of the protons translocated is that the first requires a specific stoicheiometry of protons translocated to electrons transferred, whereas the latter does not. Indeed, if protein conformation changes supplied the protons one might expect the stoicheiometry to be sensitive to change of pH. It is important, then, to resolve the experimental discrepancies between those reporting a strict stoicheiometry of 2 H^+ translocated per pair of electrons transferred through a coupling site or loop (e.g. ref. 12), and those who find a different stoicheiometry (e.g. ref. 13).

Jagendorf showed some years ago that ATP synthesis could be driven by a pH gradient across the thylakoid membrane of chloroplasts, in the absence of electron transfer (14). More recently, Racker and Stoeckenius (15) have reported ATP synthesis in illuminated artificial vesicles constructed from phospholipids, the ATPase complex isolated from mitochondria and a photoactive rhodopsin isolated from a halobacterium. Since illumination of the bacterial rhodopsin in both intact cells and vesicles has been shown to be accompanied by proton translocation, the ATP synthesis is presumably driven by the pH gradient. Although the ATPase complex used in these experiments was not free from electron-carrying components, experiments with intact cells where the purple membrane bearing the rhodopsin is quite separate from the electron-transfer chain (16) make it very likely that a pH gradient can drive ATP synthesis by the ATPase complex. It seems unlikely, however, that in this case the proton gradient is built up by proton translocation by means of a Mitchell 'loop'.

These experiments with the purple membrane of the halobacterium provide strong support for Mitchell's standpoint that the protonmotive force generated by electron transfer in mitochondria, chloroplasts and bacteria is also the driving force for ATP synthesis in these membranes. They can, however, also be accomodated in the conformational hypothesis since, just as protons favour the strained conformation (T state) of haemoglobin (see above), they may drive the equilibrium between two conformational states of the ATPase complex towards the one that promotes ATP synthesis (for example, by bringing about dissociation of firmly bound ATP). Fig. 3 illustrates such a model.

The expectation that changes in protein conformation will change the pK of dissociating groups makes it *a priori* difficult to devise experiments that will give a clear-cut answer to the question whether ATP synthesis is normally driven directly by protein conformational changes or a protonmotive force. Experience has shown that such a situation is liable to lead to controversies that can be sometimes more disruptive than productive. The younger experimentalist would be well advised to avoid these controversies, to stay uncommitted and to do experiments suggested to him on the basis of either hypothesis. Nature knows the answer, and this remains the same whoever wins the debate.

In any case, there are sufficient problems to be tackled independent of the coupling mechanism. I shall list some of them.

(1) Why do energy-transducing membranes contain so many electron accepting groups, in particular why are there so many of the same redox potential? Do electrons have to pass through all of them in optimally coupled membranes?

(2) What are the precise sources of the protons translocated across energy-transducing membranes?

(3) Are there extensive changes of protein conformation in the electron carriers or components of the ATPase complex during oxidative phosphorylation? (It is interesting to know this, whether or not these conformational changes are central to the coupling mechanism.)

(4) What are the properties of the coupling ATPase (F_1), both when dissociated from the membrane or attached to it? In particular, how do its substrates (ADP, P_i and ATP) bind to it, and what is the mechanism of the hydrolysis of ATP that it catalyses, in the absence of an external source of energy?

The last two problems are being actively tackled by many groups at the moment. It is clear from several lines of evidence that extensive changes in the ATPase complex do occur during electron transfer. This was shown first by

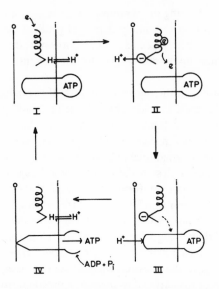

Fig. 3. Possible model of dissociation, induced by conformation change in ATPase, of firmly bound ATP. Electrons enter an electron-transferring protein represented by a helix (I). The uptake of an electron by this protein yields II and results in a conformation change whereby: (1) an acidic group originally in contact with the inner (i) phase now comes in contact with the outer (o) phase and (2) the pK of this group is lowered so that protons dissociate from the acid group leaving a negatively charged area (cf. Fig. 2). The electron is then lost to the following electron acceptor in the respiratory chain, so the electron-transferring protein becomes oxidized but in III it remains in the conformation characteristic of the reduced protein (II). By interactions (represented by the dotted arrow) between this polypeptide and polypeptides in the ATPase complex, the electron-transferring protein reverts in IV to the original conformation in I. These interactions result in a conformation change in the ATPase complex resulting in dissociation of ATP firmly bound to the ATPase (cf. Fig. 1). It is postulated that this conformation change results in an increase in the pK of a dissociating group in contact with the outer phase, whereby protons are taken up from the outer phase. ADP and P_i then bind to the open form of the ATPase, water is eliminated (cf. Fig. 1) and the ATPase reverts to its original conformation in I, with ATP firmly bound to it.

Jagendorf (17) who found an increased exchangeability of hydrogen atoms in the chloroplast ATPase when electron transfer takes place. Three separate lines of evidence have given further support to this view. First, the parameters of binding of aurovertin to the mitochondrial ATPase and the quantum yield of the bound aurovertin respond to both addition of the substrates of the ATPase (and to the cofactor Mg^{2+}) and to electron transfer (18-20). Secondly, electron transfer in the presence of ADP, P_i and Mg^{2+} leads to the dissociation from the ATPase of one of its subunits that has been shown to act as an inhibitor of its ATPase activity (20). Thirdly, electron transfer in chloroplasts promotes the incorporation of N-ethylmaleimide into the ATPase (21) and makes the adenine nucleotides that are firmly bound to all energy-transducing ATPases (22) exchangeable with added nucleotides (23).

The isolated ATPase and membrane contain weaker binding sites for ATP and ADP in addition to the firm binding sites occupied by adenine nucleotide in the isolated enzyme. These are presumably the sites primarily involved in the catalytic reaction. It has proved technically difficult to determine the binding constants of the relative weak binding sites accurately, but they are probably of the order of 10^6 M^{-1}, which is at least two orders of magnitude less than that of the strong binding sites. The increased exchangeability of the bound nucleotides during electron transfer in chloroplasts (23) raises the possibility, however, that under these conditions there is an interchange between strong and weak binding sites by a sort of flip-flop mechanism (cf. ref. 24).

However, we need to know much more about the chemistry and mechanism of action of the isolated ATPase, which has six different subunits, the intact enzyme (molecular weight 350,000) containing about ten subunits. It is a formidable but feasible problem to tackle by the techniques of protein chemistry and enzymology. A promising start has been made by Radda's group in Oxford, who find that one tyrosine (out of the 85 present) is specifically attacked by 7-chloro-4-nitrobenzo-2-oxa-1,3-diazole, with complete loss of ATPase activity (25).

REFERENCES

1. Mitchell, P. (1966) *Chemiosmotic Coupling in Oxidative Phosphorylation*, Glynn Research, Ltd., Bodmin.
2. Slater, E.C., Rosing, J. and Mol, A. (1973) *Biochim. Biophys. Acta 292*, 534-553.
3. Rosing, J. and Slater, E.C. (1972) *Biochim. Biophys. Acta 267*, 275-290.
4. Kaplan, N.O., Colowick, S.P. and Neufeld, E.F. (1953) *J. Biol. Chem. 205*, 1-15.
5. Rydström, J., Teixera da Cruz, A. and Ernster, L. (1970) *Eur. J. Biochem. 17*, 56-62.
6. Mitchell, P. and Moyle, J. (1974) *Biochem. Soc. Spec. Publ. 4*, 91-111.
7. Pullman, M.E., Penefsky, H.S., Datta, A. and Racker, E. (1960) *J. Biol. Chem. 235*, 3322-3329.
8. Boyer, P.D. (1974) In *Dynamics of Energy-Transducing Membranes* (Ernster, L., Estabrook, R.W. and Slater, E.C., eds.) pp. 289-301, Elsevier, Amsterdam.
9. Slater, E.C. (1974) In *Dynamics of Energy-Transducing Membranes* (Ernster, L., Estabrook, R.W. and Slater, E.C., eds.) pp. 1-20, Elsevier,

Amsterdam.

10. Perutz, M.F. (1970) *Nature (London)* 228, 726-739.
11. Ogawa, S. and Shulman, R.G. (1972) *J. Mol. Biol.* 70, 315-336.
12. Mitchell, P. and Moyle, J. (1968) *Biochem. J.* 105, 1147-1162.
13. Azzone, G.F. and Massari, S. (1973) *Biochim. Biophys. Acta* 301, 195-226.
14. Jagendorf, A.T. and Uribe, E. (1966) *Proc. Nat. Acad. Sci. (U.S.A.)* 55, 170-177.
15. Racker, E. and Stoeckenius, W. (1974) *J. Biol. Chem.* 249, 662-663.
16. Oesterhelt, D. and Stoeckenius, W. (1973) *Proc. Nat. Acad. Sci. (U.S.A.)* 70, 2853-2857.
17. Ryrie, I.J. and Jagendorf, A.T. (1972) *J. Biol. Chem.* 247, 4453-4459.
18. Bertina, R.M., Schrier, P.I. and Slater, E.C. (1973) *Biochim. Biophys. Acta* 305, 503-518.
19. Chang, T.M. and Penefsky, H.S. (1973) *J. Biol. Chem.* 248, 2746-2754.
20. Van de Stadt, R.J. and Van Dam, K. (1974) *Biochim. Biophys. Acta* 347, 240-252.
21. McCarty, R.E. and Fagan, J. (1973) *Biochemistry* 12, 1503-1507.
22. Slater, E.C., Rosing, J., Harris, D.A., Van de Stadt, R.J. and Kemp, A., Jr. (1974) In *Membrane Proteins in Transport and Phosphorylation* (Azzone, G.F., Klingenberg, M.E., Quagliariello, E. and Siliprandi, N., eds.) pp. 137-147, North-Holland, Amsterdam.
23. Harris, D.A. and Slater, E.C. (in press) *Biochim. Biophys. Acta.*
24. Repke, K.R.H. and Schön, R. (1974) *Acta Biol. Med. Germ.* 33, K27-K38.
25. Ferguson, S.J., LLoyd, W.J., Lyons, M.H. and Radda, G.K. (in press) *Eur. J. Biochem.*

RECENT ADVANCES ON THE CHEMISTRY OF THE MITOCHONDRIAL OXIDATIVE PHOSPHORYLATION SYSTEM*

Y. Hatefi, Y.M. Galante,** D.L. Stiggall and L. Djavadi-Ohaniance***

Department of Biochemistry
Scripps Clinic and Research Foundation
La Jolla, California 92037

INTRODUCTION

The inner membrane of bovine heart mitochondria is a highly specialized system for oxidative phosphorylation and energy-linked ion translocation. These energy transduction processes are extremely efficient and well-regulated, and constitute one of the most challenging problems of membrane biology today. Energy capture, transduction and utilization are achieved in a number of reactions which are energetically interlocked and intercommunicating. These reactions are: (a) electron transfer at three stages of the respiratory chain corresponding to coupling sites 1, 2 and 3; (b) ATP hydrolysis and synthesis; (c) ion translocation; and (d) hydride ion transfer from NADH to NADP. Thus, the mitochondrial inner membrane appears to be a system *par excellence* for study of the structural basis of membrane function.

The enormous complexity of this machinery made it necessary to develop a systematic fractionation procedure as a result of which discrete segments with well-defined composition and enzymatic activities could be isolated and studied. It was also important to accomplish this fractionation in such a way that the isolated segments would retain, if at all possible, the ability to interact and reconstitute larger functional units comparable to their parent particles. To a considerable extent, this goal, as well as extensive studies on the composi-

* The studies reported here were supported by USPHS grants AM08126 and CA13609, and NSF grant GB43470.
** Postdoctoral Fellow of the San Diego County Heart Association.
*** Permanent address: Faculty of Science, University of Tehran, Tehran, Iran.

tion and enzymatic properties of the isolated segments, has been achieved, and the following is a brief account of our progress to date.

The mitochondrial electron transport-oxidative phosphorylation system appears to be composed of five enzyme complexes, which can be isolated from the same batch of starting material by a relatively simple procedure (Fig. 1) (1,2). The composition of these enzyme complexes in terms of the known components of the respiratory chain is given in Table I. Complexes I, II, III and IV are segments of the electron transport system (3). They were first isolated in 1961 and shown to interact stoichiometrically to reconstitute a complete respiratory chain identical to the unfragmented parent particles in terms of electron transfer activity from NADH and succinate to molecular oxygen and response to site specific respiratory chain inhibitors (3-8). More recently, evidence has been obtained that preparations of complex I and complex IV (cytochrome oxidase) are also capable of energy conservation and transfer (9-11). Whether complex III has such capability has not been reported.

The well-defined components of the respiratory chain as found in the four electron transfer complexes are shown in the kinetic scheme of Fig. 2. Previous studies on the enzymatic and reconstitution properties of these enzyme complexes have been reviewed (2,3,12). Recent work of interest concerns the finding of multiple b-type cytochromes and iron-sulfur centers; the isolation of a new, membrane-bound iron-sulfur flavoprotein; the discovery of a direct pathway for NADPH oxidation by the respiratory chain; the isolation of complex V, which is capable of uncoupler-sensitive energy conservation and ATP-

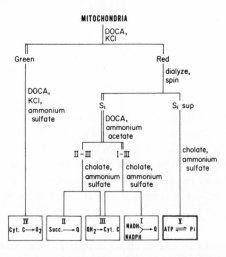

Fig. 1. Scheme showing the fractionation of beef-heart mitochondria into enzyme complexes I, II, III, IV and V with the use of cholate, deoxycholate (DOCA), ammonium acetate and ammonium sulfate. (From Hatefi et al. (1)).

TABLE I

Components of Complexes I, II, III, IV and V

Complex	Component	Concentration
		per mg protein
I. NADH-Q reductase	FMN	1.4-1.5 nmol
	nonheme iron	23-26 ng atom
	labile sulfide	23-26 nmol
	ubiquinone	4.2-4.5 nmol
	lipids	0.22 mg
II. Succinate-Q reductase	FAD	4.6-5.0 nmol
	nonheme iron	36-38 ng atom
	labile sulfide	32-38 nmol
	cytochrome b	4.5-4.8 nmol
	lipids	0.2 mg
III. QH_2-cytochrome c reductase	cytochrome b	8.0-8.5 nmol
	cytochrome c_1	4.0-4.2 nmol
	nonheme iron	10-12 ng atom
	labile sulfide	6-8 nmol
	ubiquinone	2-4 nmol
	lipids	0.4 mg
IV. Cytochrome c oxidase	cytochromes aa_3	8.4-8.7 nmol
	copper	9.4 ng atom
	lipids	0.35 mg
V. ATP-Pi exchange complex	cytochrome b	0.1-0.2 nmol
	cytochromes $c + c_1$	0.05-0.1 nmol
	cytochromes aa_3	nil
	flavin	~1 nmol
	iron	3-4 ng atom
	labile sulfide	3-4 nmol
	lipids	0.18 mg

Pi exchange; and a new approach to the study of the mechanism of uncoupling.

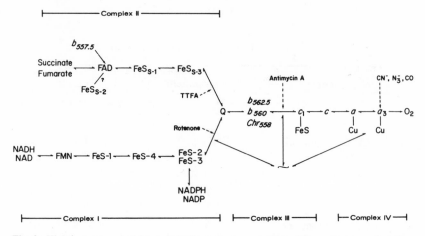

Fig. 2. Kinetic sequence of the well-defined electron carriers of the respiratory chain in complexes I, II, III and IV. The approximate sites for energy coupling are shown by arrows leading to $\sim$, and those for inhibition by rotenone (or piericidin A and barbiturates), TTFA (2-thenoyltrifluoroacetone), cyanide, azide, and carbon monoxide are shown by dashed arrows. FeS, iron-sulfur center.

THE b-TYPE CYTOCHROMES

There are three, and possibly four, b-type cytochromes in the electron transport system. One of these, designated cytochrome $b_{557.5}$ (λ_{max} in the α region is at 557.5 nm at 77° K), is found in complex II (13). This cytochrome appears to have a negative reduction potential and is not detectably reduced by succinate in complex II or submitochondrial particles. However, its reduced form is rapidly oxidized by either ubiquinone or fumarate. The role of $b_{557.5}$ in the electron transport system is not known. It might be an entry point for an unknown electron tributary of the respiratory chain, or it might undergo energy-linked reduction by reverse electron transfer and serve as a low-potential electron donor for a reductive process in the matrix.

The other b-type cytochromes are cytochromes $b_{562.5}$ and b_{560} (also designated b_T and b_K, respectively, by Chance and coworkers (see refs. 14-16)), and a cytochrome b-like chromophore provisionally designated chromophore-558 (13). They are located in complex III, and appear to be involved in electron transfer and modulation of electron flow from reduced ubiquinone to cytochromes c_1 and c. Cytochrome $b_{562.5}$ is reduced by substrates only when the system is energized or treated with antimycin A (13-16). Even under these conditions, the reduction of $b_{562.5}$ by substrates requires that cytochrome c_1 be in the oxidized state, and the rate of electron flow from the substrates be fast. Cytochrome b_{560} is reduced by substrates in the absence or presence of antimycin A or energized conditions. Unlike $b_{562.5}$, its reduction is not prevented when the rate of electron flow from substrates is slow, or when cytochrome c_1 is reduced. However, in the latter case, its reduction by substrates becomes very sluggish. Chromophore-558 exhibits a relatively small α peak at 558 nm at 77° K. It was discovered by fourth derivative analysis of the absorption spectra of b_{560} and $b_{562.5}$ under various conditions (13). Similar to b_{560}, chromophore-558 is reduced by substrates in the presence or absence of antimycin A. Otherwise, it behaves like $b_{562.5}$ in that it is not detectably reduced when the rate of electron flow from substrates is slow or when cytochrome c_1 is in the reduced state.

The b cytochromes have a unique position in the respiratory chain. They are located at or near coupling site 2, and at the juncture of incoming electrons from NADH, NADPH, succinate, fatty acyl coenzyme A, α-glycerol phosphate, choline, etc., and of outgoing electrons by way of cytochromes c_1, c and aa_3. Befitting this scheme of multiple electron tributaries and a single electron withdrawal pathway (at least insofar as cytochrome c_1 is concerned), the half-time for electron transfer from the c-type cytochromes through cytochrome oxidase to oxygen is close to two orders of magnitude shorter than the half-time of ubiquinone reduction by the fastest substrate oxidation route (NADH). Thus, the ability of the b cytochromes to respond to the redox states of the preceding (ubiquinone) and following (cytochrome c_1) electron carriers of the respiratory chain appears to be a device for modulation of electron flow and efficient energy capture at this important juncture in the chain. The details of the kinetic sequence of b_{560}, $b_{562.5}$ and chromophore-558, and the molecular mechanism by which they recognize and respond to the energized state of the membrane and the redox states of ubiquinone and cytochrome c_1 have yet to be worked out. However the kinetics of b_{560} and $b_{562.5}$ in the absence and presence of antimycin A under conditions of slow and fast electron flux from substrates suggest that the respiratory chain is branched at the level of these cytochromes.

IRON–SULFUR CENTERS

An iron-sulfur center may be defined as a protein prosthetic group composed of 2 or 4 iron atoms in chelation with 2 or 4 atoms of inorganic sulfur and 2 or 4 cysteine sulfur atoms of the protein. The inorganic sulfur atoms are released in acid in the −2 oxidation state, and the iron atoms as a mixture of ferrous and ferric. A protein molecule may contain more than one such center. Similar to flavoproteins and cytochromes, iron-sulfur proteins have a wide range of reduction potential. Each center takes up one electron, and the reduced center exhibits epr resonance, generally in the region of the magnetic field corresponding to $g = 2$.

The mitochondrial respiratory chain contains at least 9 well-defined iron-sulfur centers, most of which are associated with the early events of succinate and NADH oxidation (for recent reviews, see refs. 12 and 17). Succinate dehydrogenase was recently purified by selective extraction of complex II with chaotropic ions (18,19). The enzyme has a molecular weight of 97,000 ± 4%, and contains one mole of covalently-bound FAD, 7-8 g atoms of iron and 7-8 moles of inorganic sulfide per mole. It is composed of two unlike subunits with the molecular weights and composition shown in Table II. Ohnishi and collaborators (20,21) have recently identified in succinate dehydrogenase three iron-sulfur centers, designated centers S-1, S-2 and S-3, with E_m values at pH 7.4 of $0 ± 10$ mV, $-400 ± 15$ mV and $+60 ± 15$ mV, respectively. Centers S-1 and S-2 are believed to contain one Fe_2S_2 cluster each, while center S-3 is thought to contain an Fe_4S_4 core. Centers S-1 and S-3 are implicated in electron transfer from succinate to the respiratory chain, but the role of the low-potential center S-2 which is not reduced by succinate is not known.

The NADH-ubiquinone reductase segment of the respiratory chain (complex I) contains four well-defined iron-sulfur centers, designated centers 1, 2, 3 and 4 (12,22,23). These iron-sulfur centers occur in complex I at concentrations comparable to that of the NADH dehydrogenase flavin, and appear to be kinetically competent for electron transfer from NADH to ubiquinone (23). The epr signals of centers 3 and 4 overlap, but center 3 appears to have a more positive reduction potential than center 4. It can be reduced together with center 2, but without reduction of centers 1 and 4, when complex I is treated with small amounts of NADH. According to Ohnishi and coworkers (17,24),

TABLE II

Molecular Properties of Succinate Dehydrogenase

Molecule	Molecular weight	Flavin	Iron	Labile sulfide
		nmoles (ng atoms)/mg protein		
Succinate dehydrogenase	97,000 ± 4%	10.3[a]	70-80	70-80
Large subunit	70,000 ± 7%	12-13	45-55	45-55
Small subunit	27,000 ± 5%	<0.5[b]	95-110	90-100

[a] Average of 12 preparations. (From Davis and Hatefi (18)).
[b] From the small subunit obtained by chaotrope-induced resolution of succinate dehydrogenase (18). The small subunit obtained by resolution and column chromatography in presence of SDS is completely free of flavin.

the E_m values of these centers at pH 7.2 are: center 1, component a, $-380 \pm$ 20 mV, component b, -240 ± 20 mV; center 2, -20 ± 20 mV; center 3, -240 ± 20 mV; center 4, -410 ± 20 mV. However, the work of Beinert and his colleagues (22,23) suggests that the reduction potential of center 3 is close to that of center 2. As determined by the rapid-freeze epr technique, the kinetic sequence of the reduction of these centers by NADH is 1, 3+4, 2 (22,23).

In addition to the above, other iron-sulfur centers have been reported to exist in the respiratory chain (17,24-28). Two of these have been isolated as iron-sulfur containing proteins and will be described below. The existence of the others as *bona fide* components of the respiratory chain has yet to be ascertained, because at present satisfactory data regarding their concentration and kinetic competence are not available. One of the two centers mentioned above occurs in complex III, and is associated with a protein of 26,000 molecular weight. This protein has been isolated from complex III in its succinylated form (27). It contains 2 g atoms of Fe and 2 moles of inorganic sulfide per mole, and exhibits an E_m = +220 mV at pH 7.0 and 28°. The oxidoreduction properties of this iron-sulfur protein suggest that it is located in the respiratory chain at the level of cytochrome c_1. Whether it is an obligatory electron carrier between the b cytochromes and cytochrome c is not known.

The second center is associated with a new iron-sulfur flavoprotein, which has been recently discovered independently by Ruzicka and Beinert (private communication) and by ourselves. Its low field epr resonance resembles that of "center-5" of Ohnishi *et al.* (28). The flavin is acid extractable, and its chromatographic mobility and ability to activate the D-amino acid oxidase apoenzyme indicated that it is FAD. Ruzicka and Beinert have shown that the iron-sulfur center of a partially purified preparation is reduced by reduced ETF (electron transfer flavoprotein) with a half-time of about 5 msec, which suggests that the new iron-sulfur flavoprotein might serve as a link for electron transfer from fatty acyl CoA dehydrogenases and ETF to the respiratory chain. However, this possibility is not in accord with the facts that the above iron-sulfur flavoprotein is present in submitochondrial particles, yet ETF oxidation by these particles is still sluggish.

We first detected the new iron-sulfur flavoprotein in complex V (cf. Table I), since the flavin of complex V was found to be largely acid-extractable FAD, thus excluding NADH and succinate dehydrogenases as possible sources of this flavin. Fractionation of complex V yielded a preparation containing per mg protein approximately 5 nmoles of flavin, 20 ng atoms of Fe and 20 nmoles of inorganic sulfide. Thus the flavin:iron:labile sulfide ratio appeared to be 1:4:4, which agrees with the data of Ruzicka and Beinert for preparations containing close to 4 nmoles flavin per mg protein. The absorption spectrum of our preparation is shown in Fig. 3. This figure also shows the contributions of flavin

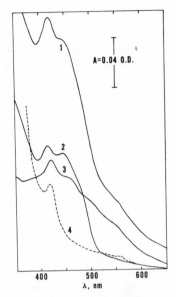

Fig. 3. Absorption spectrum of the new iron-sulfur flavoprotein of beef heart mitochondria containing acid-extractable FAD. Protein concentration, 2 mg per ml. Trace 1, oxidized iron-sulfur flavoprotein; trace 2, absorbance contribution of flavin plus minor cytochrome impurity to trace 1; trace 3, absorbance contribution of iron-sulfur chromophore to trace 1; trace 4, absorbance of cytochrome impurity in reduced form.

(trace 2) and the iron-sulfur chromophore (trace 3) to the total spectrum (trace 1) of the protein. The preparation, as shown in trace 4 of Fig. 3, contains a trace of cytochrome at its present stage of purity (probably about 30-40% pure). The Soret peak of the oxidized form of this cytochrome is seen in trace 2 at about 415 nm together with the flavin spectrum (peak at 450 nm).

THE NADPH OXIDASE PATHWAY OF THE RESPIRATORY CHAIN

Until recently, it was generally held that the mitochondrial electron transport system is unable to oxidize NADPH directly. The oxidation of this nucleotide was believed to occur exclusively by transhydrogenation to NAD, followed by subsequent oxidation of the NADH so formed by the normal NADH oxidase pathway. Two years ago, we discovered, however, that preparations of submitochondrial particles from beef heart could oxidize NADPH by molecular oxygen in the absence of effective amounts of NAD and without the apparent participation of NADPH → NAD transhydrogenase and NADH dehydrogenase (12,29-31). The rate of NADPH oxidation at pH 6.0, which is close to

the optimum, was $\geqslant 250$ nmoles$\cdot$min$^{-1}\cdot$mg^{-1} protein at 30°. The NADPH
dehydrogenase and NADPH → NAD transhydrogenase fractionated into complex
I. Preparations of complex I do not contain detectable amounts of bound NAD,
but were able to catalyze the oxidation of NADPH by ferricyanide or ubiquinone
(30). Epr and light absorption spectroscopy indicated that at pH 7-8 NADH
reduced the bulk of the flavin ($\geqslant$90%) and the iron-sulfur centers 1, 2, 3 and 4
of complex I, whereas NADPH reduced only centers 2 and 3 (12, 30, 31).

Subsequent studies showed that the transhydrogenase activity of submitochondrial particles and complex I can be selectively and completely inhibited
by treating the preparations with trypsin or with butanedione in the presence
of borate at pH 9.0 (12,32). Butanedione binds reversibly to protein arginyl
residues to form a 4,5-dimethyl-4,5-dihydroxy-2-imidazoline derivative, and
this *cis*-diol further reacts with borate to form a more stable complex. After
complete destruction of transhydrogenase activity from NADPH → NAD (Figs.
4 and 5), submitochondrial particles were still capable of full NADH oxidase
and better than 90% NADPH oxidase activity (i.e., $\geqslant$220 nmoles$\cdot$min$^{-1}\cdot$mg^{-1}
protein at pH 6.0 and 30°). Incubation of the particles with trypsin at 30°
(Fig. 4, right panel) further showed that the NADH oxidase, the NADPH oxidase and the NADPH → NAD transhydrogenase activities exhibited different
patterns of sensitivity to trypsin. These and other results described elsewhere
(32) have shown unambiguously that: (a) submitochondrial particles are capable of NADPH oxidation at appreciable rates in complete absence of transhydrogenase activity from NADPH → NAD; and (b) NADH dehydrogenation,
NADPH dehydrogenation and NADPH → NAD transhydrogenation are independent reactions in respiratory particles. Because of certain functional similarities, we had considered earlier that NADPH → NAD transhydrogenation and
NADPH dehydrogenation might be catalyzed by the same enzyme (30). The
above results do not exclude this possibility because trypsin and butanedione

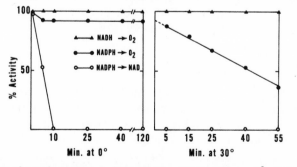

Fig. 4. Effect of trypsin treatment of submitochondrial particles at 0° and 30° C on the
rates of NADH and NADPH oxidation, and NADPH → NAD transhydrogenation. (From
Djavadi-Ohaniance and Hatefi (32)).

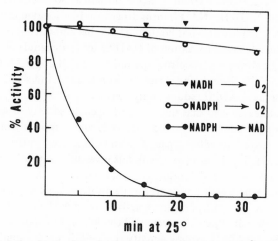

Fig. 5. Effect of preincubation of submitochondrial particles at 25° C with 10 mM butanedione in the presence of 50 mM borate, pH 9.0, on the rates of NADH and NADPH oxidation, and NADPH → NAD transhydrogenation. (From Djavadi-Ohaniance and Hatefi (32)).

might have modified only the NAD binding site of the enzyme without impairing NADPH binding and dehydrogenation.

The oxidation of NADH or NADPH by phosphorylating submitochondrial particles is associated with ATP synthesis at comparable yields (P/O of 2.4-2.9). Assuming that coupling sites 1, 2 and 3 are shared by the pathways for NADH and NADPH oxidation, the ATP yields and the epr data described above suggest that coupling site 1 is associated with the region of the electron transport system which includes iron-sulfur centers 2 and 3. Fig. 6 summarizes the above considerations with regard to the inter-relationships of NADH dehydrogenase, NADPH dehydrogenase, NADPH → NAD transhydrogenase, and coupling site 1 in the complex I segment of the respiratory chain.

COMPLEX V

This enzyme complex was recently isolated from mitochondria or submitochondrial particles (1,33) by the general procedure which also yields the electron transfer complexes I-IV (cf. Fig. 1). Preparations of complex V catalyze ATP-Pi exchange and ATP hydrolysis at the rates shown in Table III. The ATP-Pi exchange activity is specific for ATP as a nucleoside triphosphate, and is inhibited by various uncouplers, rutamycin, dicyclohexylcarbodiimide (DCCD), triethyl tin, arsenate, azide, adenylyl-imidodiphosphate (AMP-PNP), and valino-

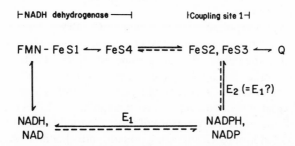

⊢NADH dehydrogenase ⊣   ⊦Coupling site 1⊣

FMN - Fe S1 ⟷ FeS4 ⇌===== FeS2, FeS3 ⟷ Q

NADH, $\underset{NAD}{\xrightarrow{\hspace{2cm} E_1 \hspace{2cm}}}$ NADPH, NADP

$E_2 (= E_1?)$

Fig. 6. Mitochondrial nicotinamide adenine dinucleotide cycle, showing the relationships of NADH dehydrogenase, NADPH dehydrogenase (E_2), NADPH $\rightleftharpoons$ NAD transhydrogenase (E_1), and coupling site 1 in the complex I segment of the respiratory chain. Dashed arrows indicate energy-requiring reactions. (From Hatefi and Stiggall (12)).

mycin + K^+ (1,33). The ATPase activity and the polypeptide composition (as examined by polyacrylamide gel electrophoresis in the presence of dodecyl sulfate and mercaptoethanol) of complex V are comparable to those of the oligomycin-sensitive ATPase (OS-ATPase) of Tzagoloff *et al.* (34). However, the latter preparation, isolated from mitochondria or phosphorylating submitochondrial particles, is essentially devoid of ATP-Pi exchange activity. It is not

TABLE III

Activities of Complex V

Activity	Rate	
	$nmoles \cdot min^{-1} \cdot mg^{-1}$ protein	
ATP-Pi exchange		
+ PL	270-310*	350-470**
- PL	100-130*	150-195**
ITP-Pi exchange	~20	
GTP-Pi exchange	~12	
ATPase	$8\text{-}10 \times 10^3$	

* As measured.
** Corrected for ATP hydrolysis during exchange.
PL, sonicated soybean phospholipids.

known whether this important difference is structural or due to lack of crucial factors in OS-ATPase.

Ryrie (35) has reported recently that yeast OS-ATPase has a molecular weight of 500,000 and acquires ATP-Pi exchange activity (25 nmoles·min^{-1}· mg^{-1} protein) when dialyzed in the presence of phospholipids and cholate, a procedure worked out by Racker and coworkers for vesicle formation (36, 37). The ATP-Pi exchange activity of Ryrie's preparation is only about 15% of that of yeast mitochondria (38). Assuming that the F_1 (ATPase) content of yeast mitochondria is in the range of those reported for heart (0.27 nmole/mg protein) and liver (0.12 nmole/mg protein) mitochondria (39, see also ref. 40 for cytochrome content of yeast mitochondria), and assuming that each molecule of yeast OS-ATPase contains one F_1, then the molecular weight of 500,000 would mean 7- to 17-fold purification. This in turn means that per mole of F_1 the ATP-Pi exchange activity of Ryrie's preparation is only 1-2% of that of yeast mitochondria. While as an OS-ATPase the preparation of Ryrie appears to be very interesting, the above calculations of its ATP-Pi exchange activity as compared to yeast mitochondria do not allow the implied conclusion that vesicle formation is necessary for ATP-Pi exchange. This is simply because more than 95% of the vesicles formed by application of the lipid-cholate-dialysis procedure could not have acquired ATP-Pi exchange activity.

By contrast to the yeast preparation, the ATP-Pi exchange activity of complex V per mg protein is 2-3 times the best activities obtained for phosphorylating submitochondrial particles. This agrees with the increase in complex V of the concentration of the mitochondrial uncoupler-binding site (see Table IV, below). Another important consideration is that, as shown in Table III, complex V is capable of considerable ATP-Pi exchange activity in the absence of any added phospholipids. Preparations of complex V contain less than 15% phospholipids, and are completely clear in solution at the concentrations used for ATP-Pi exchange assay. Whether under these conditions and in complete absence of added lipids they still form vesicles remains to be ascertained.

MECHANISM OF UNCOUPLING

We have recently conducted a detailed study of the mechanism of uncoupling in beef heart mitochondria, using the uncouplers 2-azido-4-nitrophenol (NPA) and 2,4,6-trinitrophenol (picrate, TNP)(41-44). The value of NPA, which was designed and synthesized in radioactive form by Dr. W.G. Hanstein, is two-fold. NPA is a potent, *water soluble* uncoupler (2-3 times more effective than its structural analog, 2,4-dinitrophenol, DNP) and, being an aromatic azide, it is capable of photodissociation into a nitrene which can insert into a neighboring molecule (e.g., a mitochondrial protein or lipid, cf. Fig. 7). The

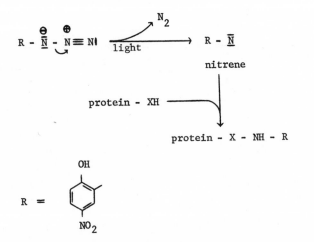

Fig. 7. Photoactivation of 2-azido-4-nitrophenol and insertion of the resultant nitrene into a protein. X can be C, N, S, O. (From Hanstein and Hatefi (41)).

value of trinitrophenol is that it, too, is a structural analog of the classical uncoupler, dinitrophenol, and similar to other substituted phenols would be expected to act as an uncoupler. However, picrate is a much stronger acid (pK_a = 0.8) than DNP (pK_a = 4.0). Consequently, TNP is largely ionized at the neutral pH of oxidative phosphorylation experiments, and the mitochondrial membrane would be expected to be relatively impermeable to picrate ions. Hence, in picrate we have the possibilities of a *membrane-impermeable* uncoupler. Our studies, which have been extensively discussed elsewhere (44) have led to the following important results and conclusions:

(a) Equilibrium binding of tritiated NPA to mitochondria in the dark has shown that mitochondria contain a specific uncoupler-binding site at a concentration comparable to that of F_1 (ATPase) or the individual electron carriers (e.g., cytochrome c_1). This site appears to bind various uncouplers (substituted phenols, phenylhydrazones, salicylanilides, even azide) in competition with NPA. It is located in the inner membrane, is associated mainly with protein, and fractionates into complex V (Table IV) as opposed to complexes I, III and IV, which in mitochondria contain coupling sites 1, 2 and 3, respectively. The concentration of the uncoupler-binding site in complex V, as determined at 4° C and pH 8.0, is 2-3 times that of whole mitochondria. Covalent labeling of mitochondria with tritiated NPA under photoactivating conditions has shown that a major NPA-binding component is a polypeptide of M_r = 30,000 ± 10%.

(b) When added to mitochondria or mitochondria denuded of outer membranes, trinitrophenol does not interact with the uncoupler-binding sites and

TABLE IV

NPA-Binding Capacity of Complexes I, III, IV and V

Complex	nmoles NPA bound/mg protein
I	⩽0.05
III	<0.01
IV	<0.01
V	0.81

Conditions: The NPA binding experiments were carried out at pH 8.0 and 4° according to ref. 41. Under these conditions (i.e., at pH 8.0) the NPA binding capacity of whole mitochondria was 0.35 nmole/mg protein. (From Hatefi et al. (33).

does not uncouple. By contrast, when it is added to sonicated submitochondrial particles, which have an inside-out orientation of the inner membrane with respect to the medium, then trinitrophenol interacts with the uncoupler-binding sites and uncouples various energy-linked functions of the particles (Table V). These and other results have established that picrate is, indeed, a membrane-impermeable uncoupler (42,44).

(c) In agreement with its membrane-impermeability, picrate is also a poor protonophore, and has little effect on the proton permeability of submitochondrial vesicles, even at 2-3 times the concentration needed for complete uncoupl-

TABLE V

Uncoupling of Energy-Linked Functions
of Submitochondrial Particles by Trinitrophenol

Function	$\phi_{1/2}$ (μM)
Oxidative phosphorylation	93
Succinate ⇆ NAD	41
NADH → NADP	59
Release of oligomycin-induced respiratory control (EDTA particles)	58

From Hatefi (44). $\phi_{1/2}$, concentration needed for 50% uncoupling.

ing. By comparison, dinitrophenol increases the proton permeability of these vesicles by 10- to 12-fold at 50% uncoupling concentration. Consequently, there is no correlation between the uncoupling and the protonophoric potencies of picrate and dinitrophenol. However, as shown in Table VI, there is an excellent agreement between the uncoupling potencies ($\phi_{1/2}$, the uncoupler concentration needed for 50% uncoupling) and the dissociation constants (K_D) of several water-soluble uncouplers, including picrate and the weak uncoupler, azide. Indeed, in all these cases $K_D/\phi_{1/2}$ was essentially constant. These results completely agree with the presence of specific uncoupler-binding sites in mitochondria, which when occupied by uncouplers would result in uncoupling. They do not agree, however, with a mechanism of uncoupling according to which uncouplers act as *mobile* protonophores in the mitochondrial membrane and uncouple by collapsing the transmembrane proton gradient.

TABLE VI

Comparison of Binding Affinities and Uncoupling Potencies

Uncoupler	K_D	$\phi_{1/2}$	$K_D/\phi_{1/2}$
	μM	μM	
NPA	15-29	5-10	3
DNP	48-73	15-25	3
TNP	121	40-60	2-3
PCP	22-44	4-8	5
Azide	$11\text{-}15 \times 10^3$	$3\text{-}4 \times 10^3$	3.7

K_D, dissociation constant; $\phi_{1/2}$ uncoupler concentration required for 50% uncoupling. K_D values for DNP, TNP, PCP (pentachlorophenol), and azide were determined from the competitive inhibition by each uncoupler of NPA binding to mitochondria and submitochondrial particles. The K_D values were determined at $3°$ and $\phi_{1/2}$ at $30°$. Assuming a ΔH for binding of DNP, TNP, PCP and azide comparable to that of NPA ($\Delta H = -8 \pm 1$ Kcal/mole) and using the van't Hoff equation, the K_D values given were calculated for $30°$ to correspond to the temperature at which the $\phi_{1/2}$ values were determined. (From Hatefi (2,44)).

CONCLUDING REMARKS

In view of the above results we feel that, in oxidative phosphorylation, energy generation, conservation, transfer and utilization for ATP synthesis are accomplished by a series of precise molecular interactions in and among the components of complexes I, II, III, IV and V (Fig. 8). Only by considering a molecular mechanism can we intuitively account for the existence of such an elaborate machinery packaged in five discrete enzyme complexes with precise stoichiometry of components. If ATP synthesis with the speed, efficiency, and regulation that is accomplished by mitochondria could have been managed by coupling a vectorial proton pump to a reversible ATPase, then presumably a pair of protein molecules such as the bacteriorhodopsin, discussed earlier in this Symposium by Stoekenius, and the mitochondrial F_1 (ATPase) would have sufficed. On the other hand, there is no doubt now that, in both mitochondria (45) and chloroplasts (see the contribution of Avron, this volume), an acid-base transition can result in ATP synthesis. As discussed by Avron and Lehninger in this volume, the $H^+/\sim$ ratio seems to be about 3 to 4 rather than 2, as originally advanced in the chemiosmotic hypothesis (46). This will require re-examination of the electron transport system and modification of Mitchell's loops to achieve 3 or 4 H^+ transfers per coupling site, especially at coupling site 1 where NADPH oxidation does not appear to involve flavin — the crucial respiratory chain component in Mitchell's first hydrogen translocation loop (46). However, a more important issue is how to reconcile ATP synthesis

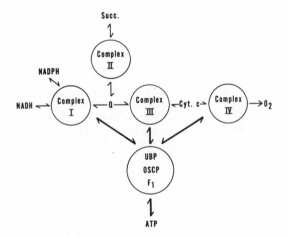

Fig. 8. Schematic representation of the functional relationship of the five enzyme complexes of the mitochondrial electron transport-oxidative phosphorylation system. UBP, uncoupler-binding protein. (From Hatefi et al. (33)).

coupled to acid-base transition of the particles with our findings described above (cf. particularly the preceding two sections).

We feel that these seemingly conflicting results can be explained by a mechanism such as that shown in the minimal scheme of Fig. 9. This scheme has been extensively discussed elsewhere (44). Its essence is two-fold: (1) It proposes that as a result of electron transfer from substrate (AH_2) to oxygen certain components of the membrane are energized (whether by conformation change or formation of a "high-energy" chemical intermediate remains to be seen), and concomitantly a transmembrane proton gradient (ΔpH) is produced by the redox reactions of respiratory chain components anisotropically located in the mitochondrial inner membrane. The energized state of the membrane can lead to ATP synthesis or be deenergized (reaction 9) by the interaction of uncouplers with an intermediate carrying the uncoupler-binding site. The ΔpH could be utilized for ion translocation; it could also have a regulatory role (e.g., for the pH sensitive transhydrogenase and NADPH dehydrogenase reactions), and serve as an energy storage form, as suggested by Avron elsewhere in this volume. (2) Fig. 9 further suggests that the energy yielding and requiring reactions 1, 2, 5 and 6 are in equilibrium with the transmembrane proton gradient and with forward and reverse electron flow (except for the final steps involving oxygen reduction), but that membrane energization by ΔpH (as shown by the dashed arrows 4 and 7) *is relatively slow.*

This scheme allows for the following crucial observations:

(a) Production of ΔpH by substrate oxidation or ATP hydrolysis.

(b) Uncoupling by membrane-impermeable uncouplers (reaction 9) such as TNP without concomitant collapse of the ΔpH.

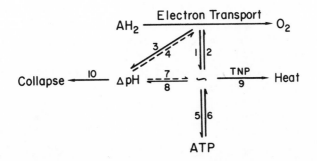

Fig. 9. Proposed scheme for oxidative phosphorylation, generation of transmembrane proton gradient (ΔpH), the inter-relationship of ΔpH and energized ($\sim$) membranes, and uncoupling by membrane-impermeable uncouplers, membrane perturbants and unspecific permeant ions. The dashed arrows suggest that reactions 4 and 7 are slower than the other reactions. (From Hatefi (44)).

(c) Uncoupling by permeant ions and membrane perturbants (e.g., deter-gents), which do not interact with the uncoupler-binding site, by collapsing the transmembrane proton gradient (reaction 10). Uncouplers with high protono-phoric potency might uncouple through both reactions 9 and 10, while the data of Table VI suggest that water-soluble uncouplers dissipate energy mainly through reaction 9 by interacting with the uncoupler-binding site.

(d) ATP synthesis by ΔpH. The assumption that this reaction is slower than ATP synthesis by the main pathway (reactions 1 and 5) is not in conflict with the recent work of Thayer and Hinkle (45), which elegantly demonstrates that ATP synthesis *after* transfer of the particles from acid to alkaline medium is fast. In these studies the particles were incubated at pH 5.0 and low K^+ con-centration for *60 sec*, then they were transferred to a medium of high pH (7.7) and high K^+ concentration and ATP synthesis was measured at roughly 20 msec intervals up to 120 msec. Therefore, the slow step could have occurred during the 60 sec acid treatment, as suggested by reactions 4 and 7 of Fig. 9.

REFERENCES

1. Hatefi, Y., Hanstein, W.G., Galante, Y. and Stiggall, D.L. (1975) *Fed. Proc. 34*, 1699-1706.
2. Hatefi, Y. (in press) In *Membrane-Bound Enzymes* (Martonosi, A., ed.)
3. Hatefi, Y. (1966) In *Comprehensive Biochemistry* (Florkin, M. and Stotz, E.H., eds.) Vol. 14, pp. 199-231, Elsevier, Amsterdam.
4. Hatefi, Y., Haavik, A.G. and Griffiths, D.E. (1961) *Biochem. Biophys. Res. Commun. 4*, 441-446 and 447-452.
5. Fowler, L.R. and Hatefi, Y. (1961) *Biochem. Biophys. Res. Commun. 5*, 203-208.
6. Hatefi, Y., Haavik, A.G. and Griffiths, D.E. (1962) *J. Biol. Chem. 237*, 1676-1680 and 1681-1685.
7. Fowler, L.R., Richardson, S.H. and Hatefi, Y. (1962) *Biochim. Biophys. Acta 64*, 170-173.
8. Hatefi, Y., Haavik, A.G., Fowler, L.R. and Griffiths, D.E. (1962) *J. Biol. Chem. 237*, 2661-2669.
9. Ragan, C.I. and Racker, E. (1973) *J. Biol. Chem. 248*, 2563-2569.
10. Ragan, C.I. and Widger, W.R. (1975) *Biochem. Biophys. Res. Commun. 62*, 744-749.
11. Racker, E. and Kandrach, A. (1971) *J. Biol. Chem. 246*, 7069-7071.
12. Hatefi, Y. and Stiggall, D.L. (in press) In *The Enzymes* (Boyer, P.D., ed.).
13. Davis, K.A., Hatefi, Y., Poff, K.L. and Butler, W.L. (1972) *Biochim. Biophys. Acta 325*, 341-356.
14. Chance, B., Wilson, D.F., Dutton, P.L. and Erecińska, M. (1970) *Proc. Nat. Acad. Sci. (U.S.A.) 66*, 1175-1182.

15. Sato, N., Wilson, D.F. and Chance, B. (1971) *Biochim. Biophys. Acta* *253*, 88-97.

16. Wilson, D.F. and Erecińska, M. (1975) *Arch. Biochem. Biophys. 167,* 116-128.

17. Ohnishi, T. (1973) *Biochim. Biophys. Acta 301*, 105-128.

18. Davis, K.A. and Hatefi, Y. (1971) *Biochemistry 10*, 2509-2516.

19. Hatefi, Y. and Hanstein, W.G. (1974) *Methods Enzymol. 31*, 770-790.

20. Ohnishi, T., Winter, D.B., Lim, J. and King, T.E. (1974) *Biochem. Biophys. Res. Commun. 61*, 1017-1025.

21. Ohnishi, T., Leigh, J.S., Winter, D.B., Lim, J. and King, T.E. (1974) *Biochem. Biophys. Res. Commun. 61*, 1026-1035.

22. Orme-Johnson, N.R., Orme-Johnson, W.H., Hansen, R.E., Beinert, H. and Hatefi, Y. (1971) *Biochem. Biophys. Res. Commun. 44*, 446-452.

23. Orme-Johnson, N.R., Hansen, R.E. and Beinert, H. (1974) *J. Biol. Chem. 249*, 1922-1927.

24. Ohnishi, T. (1975) *Biochim. Biophys. Acta 387*, 475-490.

25. Ohnishi, T., Leigh, J.S., Ragan, C.I. and Racker, E. (1974) *Biochem. Biophys. Res. Commun. 56*, 775-782.

26. DerVartanian, D., Baugh, R.F. and King, T.E. (1973) *Biochem. Biophys. Res. Commun. 50*, 629-634.

27. Rieske, J.S. (1965) In *Non-Heme Iron Proteins: Role in Energy Conservation* (San Pietro, A., ed.) pp. 461-467, Antioch Press, Yellow Springs.

28. Ohnishi, T., Wilson, D.F., Asakura, T. and Chance, B. (1972) *Biochem. Biophys. Res. Commun. 46*, 1631-1638.

29. Hatefi, Y. (1973) *Biochem. Biophys. Res. Commun. 50*, 978-984.

30. Hatefi, Y. and Hanstein, W.G. (1973) *Biochemistry 12*, 3515-3522.

31. Hatefi, Y. (1974) In *Dynamics of Energy-Transducing Membranes* (Ernster, L., Estabrook, R.W. and Slater, E.C., eds.) pp. 125-141, Elsevier, Amsterdam.

32. Djavadi-Ohaniance, L. and Hatefi, Y., submitted.

33. Hatefi, Y., Stiggall, D.L., Galante, Y. and Hanstein, W.G. (1974) *Biochem. Biophys. Res. Commun. 61*, 313-321.

34. Tzagoloff, A., Byington, K.H. and MacLennan, D.H. (1968) *J. Biol. Chem. 243*, 2405-2412.

35. Ryrie, I.J. (1975) *Arch. Biochem. Biophys. 168*, 704-711 and 712-719.

36. Kagawa, Y. and Racker, E. (1971) *J. Biol. Chem. 246*, 5477-5487.

37. Kagawa, Y., Kandrach, A. and Racker, E. (1973) *J. Biol. Chem. 248*, 676-684.

38. Groot, G.S.P., Kováč, L. and Schatz, G. (1971) *Proc. Nat. Acad. Sci. (U.S.A.) 68*, 308-311.

39.   Bertina, R.M., Schrier, P.I. and Slater E.C. (1973) *Biochim. Biophys. Acta 305*, 503-518.
40.   Burger, G., Lang, B., Bradlow, W. and Kaudewitz, F. (1975) *Biochim. Biophys. Acta 396*, 187-201.
41.   Hanstein, W.G. and Hatefi, Y. (1974) *J. Biol. Chem. 249*, 1356-1362.
42.   Hanstein, W.G. and Hatefi, Y. (1974) *Proc. Nat. Acad. Sci. (U.S.A.) 71*, 288-292.
43.   Hatefi, Y. and Hanstein, W.G. (1974) In *Membrane Proteins in Transport and Phosphorylation* (Azzone, G.F., Klingenberg, M.E., Quagliariello, E. and Siliprandi, N., eds.) pp. 187-200, North-Holland, Amsterdam.
44.   Hatefi, Y. (in press) *J. Supramolecular Struct.*
45.   Thayer, W.S. and Hinkle, P.C. (1975) *J. Biol. Chem. 250*, 5330-5335 and 5336-5342.
46.   Mitchell, P. (1966) *Chemiosmotic Coupling in Oxidative and Photosynthetic Phosphorylation*, Glynn Research Ltd., Bodmin.

PROBES OF ENERGY INPUT IN ATP FORMATION
BY OXIDATIVE PHOSPHORYLATION

Jan Rosing, Celik Kayalar and Paul D. Boyer

Molecular Biology Institute
and
The Department of Chemistry
University of California
Los Angeles, California 90024

INTRODUCTION

The energy transitions of oxidative phosphorylation may be conveniently divided into redox energy input, energy transmission, and phosphorylation output. This contribution focuses on the latter — how energy transmitted from oxidations may be used to make ATP from ADP and P_i. The hypothesis currently under evaluation in our laboratory is that ATP synthesis is accomplished by energy linked conformational changes at the catalytic site (1-4). The demonstration that the exchange of water oxygens with P_i ($P_i \rightleftharpoons$ HOH exchange) is much less sensitive to uncouplers of oxidative phosphorylation than are the $P_i \rightleftharpoons$ ATP and ATP $\rightleftharpoons$ HOH exchanges led to the suggestion that one important use of energy is to promote the release of ATP from the catalytic site (1). But it was recognized that additional energy input may occur at an earlier step or steps. The results presented in this paper indicate that an important use of energy occurs in the binding of P_i and/or ADP. They further indicate a remarkable insensitivity of the covalent bond forming and breaking step (ADP + $P_i \rightleftharpoons$ ATP + HOH) to lack of energization. Additional experimentation is necessary before any definitive conclusions can be made, but the results appear to be of sufficient potential importance to warrant presentation at this symposium for your consideration.

These important new findings result from the probes of the rates of different steps in the reaction made possible by isotope exchange techniques. One principal stimulus for the additional probes came from the observations on the effect of hexokinase additions on the $P_i \rightleftharpoons$ HOH exchange in the presence of submitochondrial particles, ADP, P_i and Mg^{++} (4). A hexokinase-glucose trap to remove ATP formed by adenylate kinase markedly reduced the observed oxygen exchange, indicative of ATP stimulation of the exchange. It was this result that suggested energy requirement in addition to that for ATP release. As one means of gaining information about where energy input may occur, we undertook evaluation of the two components of the incorporation of water oxygen into P_i, namely "medium" and "intermediate" exchange. "Intermediate" exchange is that occurring with the P_i formed by hydrolysis of ATP and prior to release of P_i from the catalytic site. "Medium" exchange is that occurring with P_i from the reaction medium. The gain of ^{18}O from $H^{18}OH$ into P_i (that present in the medium plus that formed from ATP), corrected for the one water oxygen that must be introduced into P_i by ATP hydrolysis, gives the total $P_i \rightleftharpoons$ HOH exchange, that is the sum of medium and intermediate exchanges. Medium exchange may be measured by the loss of ^{18}O from $^{18}O\text{-}P_i$. Subtraction of the medium exchange from the total exchange gives the intermediate exchange.

For probe of the site of the additional energy input, a careful assessment of the effect of uncouplers on both the medium and intermediate oxygens exchanges, and on the ATP $\rightleftharpoons$ HOH, $P_i \rightleftharpoons$ ATP exchanges as well as the ATPase seemed desirable. Results of these studies show that it is intermediate exchange that is resistant to uncouplers, and as mentioned above point to energy input for substrate binding.

Our findings also have other implications. Two possible modes of transmission of energy from the redox output to the phosphorylation input appear to us to warrant major consideration at this time. One is direct conformational transmission and the other is transmission by a potential gradient. Energy from conformational transmission could readily be used to drive the conformational changes at the catalytic site necessary for ATP synthesis. But how potential and/or proton gradients might be coupled to ATP synthesis is not as readily apparent. Mitchell has suggested a "chemiosmotic" mechanism for use of the energy of potential and/or H^+ gradients to make ATP (5). His suggestions have been regarded as untenable by Boyer (6). Also, our experimental findings are incompatible with the primary use of energy for making a covalent bond between ADP and P_i, as in his chemiosmotic mode of ATP formation. This does not, however, argue against possible transmission of energy by a membrane potential. An additional purpose of the present paper is to present briefly suggestions for how a membrane potential derived from oxidations could be used to drive the conformational transitions necessary for ATP synthesis.

It must be emphasized that the interpretations given in this paper are predicated on the assumption that the oxygen exchanges associated with oxidative phosphorylation result from the dynamic reversal of ATP formation at the catalytic site. If this is true, measurement of the oxygen exchanges gives us a powerful probe for delineation of catalytic events. An alternative mechanism for the exchange has been suggested by Young *et al.* (7), and although regarded as unlikely by us, has not been disproved. But millisecond mixing and quenching experiments show that dynamic reversal of ATP cleavage is at least one source of oxygen exchange and may be the only source (3). Extension of such experiments may allow more definitive conclusions.

MATERIAL AND METHODS

Materials

Water of approximately 9.49 or 26.11 atom % excess ^{18}O was obtained from Yeda Research and Development Company, Rehovoth, Israel. $[^{18}O]KH_2PO_4$ with about 33 atom % excess ^{18}O was prepared as described by Cohn and Drysdale (8). $[\beta-^{32}P]ADP$ was prepared as follows: 3 mg of rat liver mitochondria were incubated for 30 min at 20° C in a 2 ml reaction mixture containing 75 mM sucrose, 20 mM KCl, 50 mM Tris-Cl, 2.5 mM EDTA, 0.5 mM $[^{32}P]P_i$ (~10^7 cpm/nmole), 5 mM succinate, 0.3 mM AMP, 0.05 mM ATP and 5 mM $MgCl_2$ at pH 7.5. Because of the combined adenylate kinase and phosphorylation activities of the mitochondria the AMP was phosphorylated to a mixture of $[\beta-^{32}P]ADP$ and $[\beta,\gamma-^{32}P]ATP$. Mitochondria were removed by centrifugation and $[\beta,\gamma-^{32}P]ATP$ was converted to $[\beta-^{32}P]ADP$ by addition of hexokinase and glucose. After stopping the reaction with 1 ml of 1 M perchloric acid the nucleotides were adsorbed to 40 mg of acid treated charcoal and the charcoal filtered off by suction. The charcoal was washed with 2.5 ml of a solution containing 0.3 M perchloric acid, 25 mM PP_i and 0.1 M H_3PO_4 and then carefully rinsed with 10 ml H_2O. The adenine nucleotides were eluted from the charcoal with 6 ml ethanol/1 N ammonia 40:60 v/v and applied to a 0.5 x 3 cm column of anion exchange resin (Bio-Rad AG1-X4, 400 mesh) to separate the $[\beta-^{32}P]$-ADP from traces of $[^{32}P]P_i$ and $[\beta,\gamma-^{32}P]ATP$. The P_i was eluted with 5 ml of 0.2 M Tris pH 7.5, the ADP with 3 ml of 60 mM HCl and the ATP with 3 ml of 1 M HCl.

2,4-Dinitrophenol was purified by recrystallization from chloroform. 5-Chloro-3-*t*-butyl-2'-chloro-4'-nitrosalicylanilide (S-13) was the gift of Monsanto Chemical Co. Hexokinase, pyruvate kinase and lactate dehydrogenase were purchased from Boehringer Mannheim Co. and dialysed before use.

Rat liver mitochondria were prepared as described by Johnson and Lardy (9). Heavy beef heart mitochondria were prepared by a modification of the method of Smith (10) and used for preparation of submitochondrial particles (ETPH– $Mg^{2+}-Mn^{2+}$) essentially as described by Beyer (11).

Measurement of ATPase activity and exchange reactions

Reactions were stopped by addition of 1 M perchloric acid in amounts given in the legend to the figures and protein was removed by centrifugation. Measurement of P_i and of P_i release from ATP (ATPase activity) was carried out by the method of Sumner (12). In the experiments described in Fig. 1 and Table I, ATP hydrolysis was determined by measurement of $[^{32}P]P_i$ formation. ATP was determined by a coupled assay with hexokinase and glucose-6-phosphate dehydrogenase and ADP with pyruvate kinase and lactate dehydrogenase as described by Bergmeyer (13).

For estimation of the $P_i \rightleftharpoons HOH$ exchange P_i was isolated from the perchlorate extracts and its ^{18}O content measured as described by Boyer and Bryan (14). To estimate the ATP $\rightleftharpoons HOH$ exchange the adenine nucleotides present in the perchlorate extract were adsorbed to acid treated charcoal (14 mg charcoal/ μmole adenine nucleotide). The charcoal was filtered off by suction and washed with 5 ml of a solution containing 0.3 N perchloric acid, 25 mM PP_i and 0.1 M H_3PO_4 and then rinsed with 20 ml H_2O. The charcoal was suspended in 2 ml 2 M HCl and heated in a boiling water bath for 30 min to release the β- and γ-phosphoryl groups of ADP and ATP. Phosphate was isolated from the filtrate and analyzed for its ^{18}O content.

For determination of the amount of ATP $\rightleftharpoons P_i$ exchange, concentrated HCl and 50 mM aqueous ammonium molybdate were added to final concentrations of 1 M and 12 mM, respectively, and the phosphomolybdate complex was removed by repeated extraction with cold isobutyl alcohol-benzene (1:1 v/v). 0.5 ml of the lower layer was counted in aqueous solution using Cerenkov light (15).

Calculation of the exchange rates

When relatively extensive exchange occurred, correction for the approach to isotopic equilibrium was made (14). In the case of the ATP $\rightleftharpoons P_i$ and the $P_i(^{18}O) \rightleftharpoons HOH$ exchange, the specific radioactivity of the P_i and the atom % excess ^{18}O of the P_i were decreased in the course of the experiment by P_i released from ATP. The calculations were based on the mean P_i concentrations present during the experiment. Since the calculation of the rate of the $P_i(^{18}O) \rightleftharpoons H_2O$ exchange is based on measurements of the loss of ^{18}O from medium P_i, a correction was also made for the loss of ^{18}O from P_i to ATP via ATP $\rightleftharpoons P_i$ exchange. We assumed that about 50% of the oxygens of P_i introduced into ATP via ATP $\rightleftharpoons P_i$ originate from medium P_i in accord with earlier findings (8,16). Where exchange rates are given the incorporation rate was taken to be linear over the periods measured.

RESULTS

Effect of ATP on the $P_i \rightleftharpoons HOH$ exchange

The rapid $P_i \rightleftharpoons HOH$ exchange catalyzed by submitochondrial particles in the presence of P_i and ADP but in the absence of oxidations or an added energy donor gives evidence for a dynamic reversal of formation of tightly bound ATP at the catalytic sites of oxidative phosphorylation. However, as shown by Cross and Boyer (4) this exchange is inhibited considerably by addition of hexokinase and glucose indicating that the exchange is probably stimulated by ATP formed from the added ADP by adenylate kinase present in the submitochondrial particle preparations.

EFFECT OF HEXOKINASE ON THE $P_i \rightarrow HO^{18}H$ EXCHANGE AND ATP HYDROLYSIS

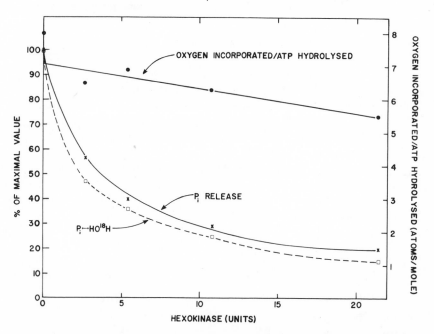

HEXOKINASE (UNITS)

Fig. 1. Effect of hexokinase on the $P_i \rightleftharpoons H^{18}OH$ exchange and ATP hydrolysis. Reaction mixtures contained, in a 1.0 ml volume at pH 7.4, and 30° C, 5 mM [β-^{32}P] ADP (208 cpm/nmole), 10 mM $MgCl_2$, 40 mM Tris-Cl, 10 mM P_i, 40 mM glucose, 1.9 mg beef heart submitochondrial particles, $H^{18}OH$ (0.76 atom % excess) and hexokinase as indicated. Reaction was started by addition of particles and stopped after 5 min by addition of 0.45 ml of 1 M perchloric acid. The samples were analyzed for ^{18}O incorporation into P_i and liberation of [^{32}P]P_i as described in the experimental procedure.
100% for $P_i \rightleftharpoons H^{18}OH$ exchange is 720 nmoles· min^{-1}·mg^{-1}
100% for ATP hydrolysis is 90 nmoles·min^{-1}·mg^{-1}.

To quantitate the amount of ATP formed by adenylate kinase and then hydrolyzed to yield P_i, the exchange experiments were carried out in the presence of $[\beta\text{-}^{32}P]$ ADP. Any ATP formed by adenylate kinase is thus labeled with $[^{32}P]P_i$ in both the β and γ positions and subsequent hydrolysis of this ATP can be measured as release of $[^{32}P]P_i$.

As shown in Fig. 1, the $P_i \rightleftharpoons HOH$ exchange stimulated by added ADP is indeed accompanied by a considerable ATP hydrolysis and both the exchange and ATP hydrolysis are inhibited to the same extent by addition of hexokinase and glucose. Because the number of oxygen atoms incorporated into P_i per mole of ATP hydrolyzed remains nearly constant, it appears that ATP hydrolysis might be necessary for the $P_i \rightleftharpoons HOH$ exchange.

That ATP formed by adenylate kinase is necessary for this exchange is shown from the data given in Table I. The $P_i \rightleftharpoons HOH$ exchange carried out in the absence of an added energy donor is inhibited about 90% by amounts of AMP that inhibit ATP formation from ADP by adenylate kinase nearly completely.

TABLE I

Effect of Hexokinase plus Glucose, AMP and Dinitrophenol
on the $P_i \rightleftharpoons H^{18}OH$ Exchange and ATP Hydrolysis

Addition	P_i Release		$P_i \rightleftharpoons H^{18}OH$ Exchange	
	nmoles· $\text{min}^{-1} \cdot \text{mg}^{-1}$	% inhibition	natoms· $\text{min}^{-1} \cdot \text{mg}^{-1}$	% inhibition
---	54	---	735	---
21.5 units hexokinase	10	81	72.5	90
10 mM AMP	4.9	91	58.5	92
10 mM AMP, 21.5 units hexokinase	2.1	96	14.5	98
100 μM DNP	58	---	161	78
100 μM DNP, 10 mM AMP, 21.5 units hexokinase	0.6	99	11	98.5

Reaction mixtures contained in a 1.0 ml final volume at pH 7.4 and 30°, 1 mM $[\beta\text{-}^{32}P]$ ADP (1250 cpm/nmole), 10 mM $MgCl_2$, 40 mM Tris-HCl, 10 ml P_i, 40 mM glucose, 2 mg beef heart submitochondrial particles, $H^{18}OH$ (0.76 atom % excess), and hexokinase and AMP as indicated. The reaction was started by addition of particles and stopped after 5 min by addition of 0.45 ml 1 M perchloric acid. The samples were analyzed for ^{18}O incorporation into P_i and $[^{32}P]P_i$ release as described in experimental procedures.

The almost complete abolishment of the $P_i \rightleftharpoons$ HOH exchange measured in the presence of hexokinase, glucose, AMP and the uncoupler, 2,4-dinitrophenol indicates that indeed some energy input is necessary to drive this medium $P_i \rightleftharpoons$ HOH exchange. These results can be explained by assuming some energy input in one of the steps leading to the formation of the ATP at the catalytic site.

Effect of uncouplers on exchange reactions

To further probe the reactions leading to the formation of tightly bound ATP and the sites of energy input in this process we studied in more detail the effect of uncoupler on the various exchanges. Fig. 2 shows the effect of increasing concentrations of the uncoupler, 2,4-dinitrophenol, on the exchange reac-

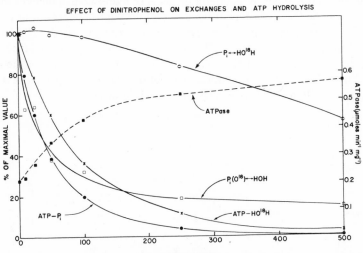

EFFECT OF DINITROPHENOL ON EXCHANGES AND ATP HYDROLYSIS

Fig. 2. Effect of 2,4-dinitrophenol on exchange reactions and ATP hydrolysis. 3 Mg of beef heart submitochondrial particles were incubated at 30° C in a reaction mixture containing 15 mM ATP, 3 mM ADP, 7.5 mM P_i, 15 mM $MgCl_2$, 20 mM Tris-Cl and 150 mM sucrose at pH 7.5. After 1 min 50 sec 2,4-dinitrophenol was added to a final concentration as indicated and 10 sec later a mixture of [^{32}P]P_i and $H^{18}OH$ was added to give a final concentration of 10 mM P_i and a final 1.96 atom % excess $H^{18}OH$. 5 Min later reactions were stopped by addition of 1 ml of 1 M perchloric acid and analyzed as described in experimental procedures. For the measurement of the $P_i(^{18}O) \rightleftharpoons$ HOH exchange an analogous incubation was carried out but at 2 min $P_i(^{18}O)$ was added to give a final concentration P_i of 10 mM containing 8.3 atom % excess ^{18}O.

100% for $P_i \rightleftharpoons H^{18}OH$ exchange is 1158 nmoles min^{-1}·mg^{-1}
100% for $P_i(^{18}O) \rightleftharpoons$ HOH exchange is 1300 nmoles min^{-1}·mg^{-1}
100% for ATP $\rightleftharpoons$ HOH exchange is 347 nmoles min^{-1}·mg^{-1}
100% for ATP $\rightleftharpoons P_i$ exchange is 212 nmoles min^{-1}·mg^{-1}.

tions. The results show that the total $P_i \rightleftharpoons HOH$ exchange is much less sensitive to the uncoupler than the $ATP \rightleftharpoons P_i$, $ATP \rightleftharpoons HOH$, and the medium $P_i \rightleftharpoons HOH$ exchanges. The difference in sensitivity of the total and medium $P_i \rightleftharpoons HOH$ exchange shows that it is in fact the intermediate part of the $P_i \rightleftharpoons HOH$ exchange that is insensitive to uncoupler.

Because the effects of 2,4-dinitrophenol on the exchange reactions may be partly due to nonspecific effects of this uncoupler the exchange reactions were also titrated with the powerful inhibitor, 5-Cl-3-t-butyl-2′-Cl-4′-nitrosalicylanilide or "S-13" (Fig. 3). The effects of this uncoupler were close to being the same as for 2,4-dinitrophenol and show again the remarkable insensitivity of the intermediate $P_i \rightleftharpoons HOH$ exchange to uncoupler.

The effect of the uncoupler "S-13" on the intermediate $P_i \rightleftharpoons HOH$ exchange is shown in Fig. 4. Also shown, as calculated from the data of Fig. 3, is the amount of "extra" oxygens incorporated per ATP hydrolyzed, a number that turned out to be fairly constant at the different uncoupler concentrations. This reflects the insensitivity of the reversible formation of bound ATP from bound ADP and P_i to uncoupler. The increase in rate of the intermediate exchange by addition of uncoupler is simply explained by the increase of the rate of ATP hydrolysis with incorporation of about one "extra" oxygen atom from H_2O into the γ-P_i released from ATP during each turnover.

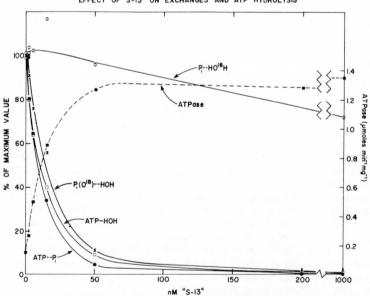

Fig. 3. Effect of "S-13" on exchange reactions and ATP hydrolysis. The experiment was carried out as described in the legend to Fig. 2, but with varying amounts of "S-13."

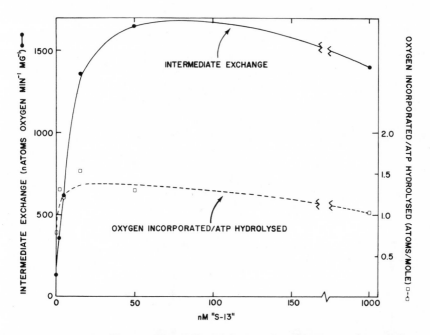

Fig. 4. Effect of "S-13" on the intermediate exchange and on the amount of oxygen incorporated per ATP hydrolyzed. The rate of the intermediate exchange is calculated from the difference in the rate of the $P_i \rightleftharpoons H^{18}OH$ and $P_i(^{18}O) \rightleftharpoons HOH$ exchanges given in Fig. 3. From the rate of ATP hydrolysis (Fig. 3) the number of O atoms incorporated into P_i per mole of ATP hydrolyzed was calculated. The rates for exchange are in addition to the one oxygen from H_2O required for each P_i formed by ATP hydrolysis.

The demonstration that the intermediate exchange is insensitive to uncouplers has been independently observed by Robert Mitchell (17). Agreement of the two laboratories gives additional confidence in the results.

DISCUSSION

The importance of uncoupler-insensitive intermediate $P_i \rightleftharpoons HOH$ exchange

Probably the most important part of the present data is the demonstration that the intermediate $P_i \rightleftharpoons HOH$ exchange is quite insensitive to uncouplers. The significance of this may not be readily apparent, and further comment is desirable. In the absence of uncouplers and during the net cleavage of ATP as typical for submitochondrial preparations, the fastest reaction being catalyzed is the medium $P_i \rightleftharpoons HOH$ exchange. This means that even though net ATP cleavage is proceeding, the binding of P_i readily occurs so that most of the bound

ATP at the catalytic site arises from P_i and a rapid medium exchange results. Such exchange would involve steps 1 and 3 of Fig. 5B. In addition, a rapid ATP $\rightleftharpoons$ HOH exchange is also occurring. This means that ATP is being rapidly bound and rapidly released from the catalytic site accompanied by the rapid and reversible cleavage of the bound ATP. As uncoupler concentration is increased, the ATP $\rightleftharpoons$ HOH, P_i $\rightleftharpoons$ ATP and the medium P_i $\rightleftharpoons$ HOH are decreased (Figs. 2,3). The pronounced decrease in the medium exchange was a somewhat

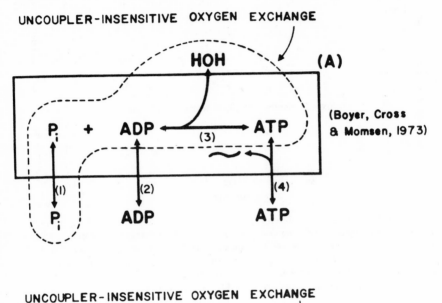

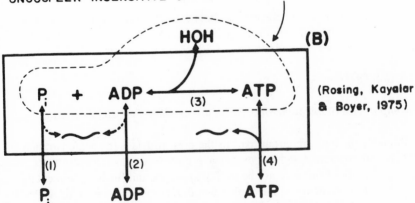

Fig. 5. Schemes for energy input in oxidative phosphorylation: (A) Input only for ATP release; (B) input for P_i and/or ADP binding and ATP release.

unexpected but quite significant finding. The remaining very prominent oxygen exchange is thus nearly all intermediate exchange. Further, as noted in Fig. 4, even though there is a pronounced increase in ATPase with increased uncoupler concentration, the extra oxygen incorporated by exchange into each P_i formed from ATP remains nearly constant. Thus the covalent bond cleavage of bound ATP to give bound ADP and bound P_i remains readily reversible even with high uncoupler concentration. Clearly this indicates that the energy requirement for formation of bound ATP from bound ADP and P_i at the catalytic site is low, but not necessarily zero.

The continuation of the phosphate oxygen exchange in the presence of uncouplers was one of the principal bases for the hypothesis advanced by Boyer *et al.* (1) that a prominent function of energy derived from electron transfer is to bring about release of ATP formed at the catalytic site by reversal of hydrolysis. These relationships are indicated in Fig. 5A. Our new experimental findings confining the uncoupler-insensitive oxygen exchange to the enzyme-bound reactants, point strongly to another effect of energy input, namely favoring the binding of substrate(s), ADP and/or P_i. The increased information confines the uncoupler-insensitive oxygen exchange as indicated in Fig. 5B.

Substrate binding and steps of energy input

Interrelations of P_i and ADP binding need clarification. The dramatic decrease in the medium exchange with continuation of the intermediate exchange as uncoupler concentration increases means that P_i is being released but not rebound. In the absence of uncouplers, the release and the rebinding of P_i were both rapid, as shown by the high rate of the medium exchange. The absence of energy caused by the presence of uncouplers has obviously interfered with P_i binding. This could occur solely because of energy requirement for a tight binding of P_i. However, bound ADP must also be present for oxygen exchange to occur. It is possible that ADP departs from the enzyme prior to, with, or shortly after the P_i. If ADP were not rebound because of energy lack, inhibition of the medium $P_i \rightleftharpoons$ HOH exchange would occur. Experiments on ADP $\rightleftharpoons$ ATP exchange may shed light on this possibility.

With regard to ATP binding and release as uncoupler concentration is increased, ATP obviously can still bind as shown by the ATPase activity. Further, the continuation of the intermediate $P_i \rightleftharpoons$ HOH exchange means that the ATP bound at the catalytic site must have an appreciable steady-state ^{18}O content. All the ^{18}O from $H^{18}OH$ that appears in the medium P_i above that arising from net ATP hydrolysis must be present at one time in the γ-phosphoryl group of ATP. Such results can be explained if ATP is being bound but not released. They are in agreement with the previous interpretation (1-4) of energy use for ATP release.

Another possibility also needs consideration. This is that energy input serves only to favor tight binding of ADP and/or P_i. Such a possibility, as depicted in Fig. 6, obviously explains the sensitivity of the medium $P_i \rightleftharpoons$ HOH exchange and the insensitivity of the intermediate oxygen exchange to uncouplers. It also offers an alternate way of achieving ATP release for net synthesis, namely by conversion of the catalytic site to a non-ATP binding form as soon as ATP dissociates. The scheme of Fig. 6, however, does not readily explain the marked sensitivity of the ATP $\rightleftharpoons$ HOH exchange to uncouplers. There is the possibility that such sensitivity could result if the steady-state of E·ATP containing ^{18}O were drastically reduced in the presence of uncoupler. This does not seem likely in view of the continued intermediate exchange. Additional experiments are underway for evaluation of the various possibilities.

Some analogies with myosin ATPase

It is of interest that the mitochondrial phosphorylation system in the presence of uncouplers, and with the membrane intact has exchange properties like myosin ATPase. With myosin ATPase there is a quite weak medium exchange and a prominent intermediate exchange (18,19,20). Also, there is direct evidence for the reversible cleavage of bound ATP to bound ADP and bound P_i at the catalytic site (21,22). Other proteins in the mitochondrial membrane are still exerting a prominent influence on the ATPase, however. With the purified ATPase, both intermediate and medium exchange are very low or absent. We interpret this to mean that conformational restraints may still be operative with the membrane ATPase even in the presence of uncouplers. These restraints are lost when the ATPase is purified.

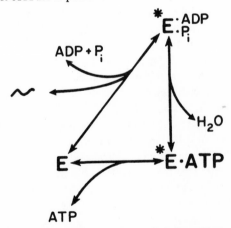

Fig. 6. An alternate scheme for oxidative phosphorylation with energy input for P_i and ADP binding.

The interactions of myosin and actin that occur in the muscle may even have further analogy for oxidative phosphorylation. Let us consider how acto-myosin could be made to synthesize ATP. Assume that in a mixture of acto-myosin, ADP and P_i, a mechanism existed for energy input to cause the conformational change that led to myosin and actin dissociation, and that the dissociated actin could be transitorily kept from the myosin. Removal of actin would promote ADP and P_i binding to the myosin to form bound ATP. Allowing recombination with the dissociated actin would release ATP. A similar series of events could occur in mitochondria, but with better design for ATP synthesis. In mitochondria, energy input could cause transitorial rupture of protein-protein interactions in the ATPase complex. These conformational changes could give the forms that form ADP and P_i binding with formation of bound ATP at the catalytic site. A reformation of the original protein interactions could promote release of the ATP. Thus, the synthesis of ATP would be accomplished, with the energy serving to favor both binding of reactant ADP and P_i and release of product ATP.

Other suggestions for ATP formation

The indication from our data that the formation of bound ATP from bound ADP and bound P_i at the catalytic site has little or no energy requirement is not in accord with the "chemiosmotic" mechanism of phosphorylation. In Mitchell's most recent and previous chemiosmotic suggestions (5) the primary input of energy serves to drive formation of the β-γ covalent bond of ATP. Also, our findings are not in harmony with any hypothesis where a covalent precursor, e.g., an I $\sim$ X or an acyl-S-, serves as a precursor to ATP. With uncoupler present, such a precursor should be dissipated and the continued intermediate oxygen exchange should not be observed.

As mentioned earlier, the requisite conformational changes for changing substrate affinities might be energized by oxidations through an interlocking protein network or through a membrane potential gradient. Also, it appears clear that ATP cleavage can be coupled to proton transport, reflecting either a step in the overall process of oxidative phosphorylation or a reversible side reaction. Mention has been made that membrane potential might be coupled to ATP synthesis by conformational changes (4), but it may not be readily apparent how a potential could be used for ATP synthesis correlated with a proton pump. Thus some amplification is presented here briefly.

One very logical thing for a membrane potential to do is to cause a charged group to move or change position, and this might be coupled to vital conformational changes. A quite hypothetical scheme depicting such movement driven by membrane potential is depicted in Fig. 7. This scheme is considered in some-

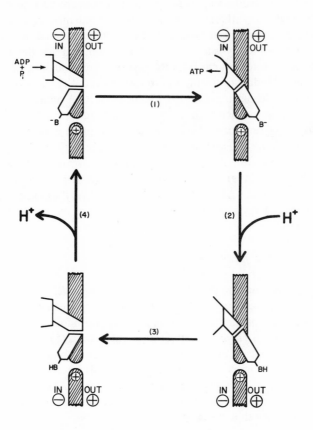

Fig. 7. An illustrative scheme for coupling of membrane potential to ATP synthesis.

what more detail elsewhere (23) but will be briefly discussed here. Binding of ADP and P_i at the phosphorylation site of the inner mitochondrial membrane is accompanied by a movement of a protein-attached group, B^- to the exterior. Driven by such movement, the bound ADP and P_i form bound ATP at the site, and the ATP is released to the interior. The B^- group is protonated, and in the uncharged form moves to the interior, where the proton dissociates and the cycle can be repeated. The negative potential inside has thus driven ATP formation and charge balance to compensate for the B^- group movement has been attained by a proton movement.

For simplicity, the figure depicts only one proton transported per ATP made.

Two or more properly aligned groups could move to account for observed stoichiometries (24,26). The figure indicates large movements — very small movements to expose B⁻protons in the aqueous phase on different sides would suffice. To avoid an energy barrier for dehydration, the group is regarded as moving in a channel where water can penetrate. A positive charge near this aqueous channel prevents proton passage, but the channel might serve for passage of permanent anions.

The arrows depict the direction of the transitions when a membrane potential is used to form ATP. By operation in the reverse direction, the cleavage of ATP would drive a proton transfer and establish a membrane potential, even in highly buffered media. An important point is that the scheme would serve for ATP synthesis coupled to membrane potential without any proton gradient. Changes in affinity of substrates at the phosphorylation site by energy-linked conformational change thus makes an attractive way of transducing energy of membrane potential to energy of ATP.

ACKNOWLEDGEMENTS

J. Rosing is recipient of a NATO Science Fellowship awarded by the Netherlands Organization for the Advancement of Pure Research (Z.W.O.).

C. Kayalar is recipient of a Science Fellowship awarded by The Scientific and Technical Research Council of Turkey.

The researches were supported in part by grant GM 11094 of the Institute of General Medical Sciences, U.S. Public Health Service, and Contract AT(04-3)-34 of the U.S. Atomic Energy Commission.

REFERENCES

1. Boyer, P.D., Cross, R.L. and Momsen, W. (1973) *Proc. Nat. Acad. Sci. (U.S.A.) 70*, 2837-2839.
2. Boyer, P.D. (1974) In *Dynamics of Energy-Transducing Membranes* (Ernster, L., Estabrook, R.W. and Slater, E.C., eds.) pp. 289-301, Elsevier, Amsterdam.
3. Boyer, P.D., Stokes, B.O., Wolcott, R.G. and Degani, C. (1975) *Fed. Proc. 34*, 1711-1717.
4. Cross, R.L. and Boyer, P.D. (1975) *Biochemistry 14*, 392-398.
5. Mitchell, P. (1974) *FEBS Letters 43*, 189-194.
6. Boyer, P.D. (1975) *FEBS Letters 50*, 91-94.
7. Young, J.H., Korman, E.F. and McKlick, J. (1974) *Bioorg. Chem. 3*, 1-15.
8. Cohn, M. and Drysdale, G.R. (1955) *J. Biol. Chem. 216*, 831-846.
9. Johnson, D. and Lardy, H. (1967) *Meth. Enzymol. 10*, 94-96.
10. Smith, A.L. (1967) *Meth. Enzymol. 10*, 81-86.

11. Beyer, R.E. (1967) *Meth. Enzymol. 10*, 186-194.
12. Sumner, J.B. (1944) *Science 100*, 413-414.
13. Bergmeyer, H.U. (1970) *Methoden der Enzymatischen Analyse*, Band I and II, Verlag Chemie, Weinheim/Bergstr.
14. Boyer, P.D. and Bryan, M. (1967) *Meth. Enzymol. 10*, 60-71.
15. Clausen, T. (1968) *Anal. Biochem. 22*, 70-73.
16. Boyer, P.D., Luchsinger, W.W. and Falcone, A.B. (1956) *J. Biol. Chem. 223*, 405-421.
17. Mitchell, R., Lamos, C.M. and Russo, J.A. (in press) *J. Supramolec. Struc.*
18. Levy, H.M. and Koshland, D.E., Jr. (1958) *J. Am. Chem. Soc. 80*, 3164-3165.
19. Dempsey, M.E., Boyer, P.D. and Benson, E.S. (1963) *J. Biol. Chem. 238*, 2708-2715.
20. Sartorelli, L., Fromm, H.J., Benson, R.W. and Boyer, P.D. (1966) *Biochemistry 5*, 2877-2884.
21. Bagshaw, C.R. and Trentham, D.R. (1973) *Biochem. J. 133*, 323-328.
22. Wolcott, R.G. and Boyer, P.D. (1974) *Biochem. Biophys. Res. Commun. 57*, 709-716.
23. Boyer, P.D. Submitted for publication.
24. Mitchell, P. and Moyle, J. (1968) *Eur. J. Biochem. 4*, 530-539.
25. Azzone, G.F. and Massari, S. (1971) *Eur. J. Biochem. 19*, 97-107.
26. Junge, W., Rumberg, B. and Schröder, H. (1970) *Eur. J. Biochem. 14*, 575-581.

BIOCHEMICAL GENETICS OF OXIDATIVE PHOSPHORYLATION

David E. Griffiths

Department of Molecular Sciences
University of Warwick
Coventry CV4 7AL, England

There has been a major time lag in the application of the classical techniques of biochemical genetics to the study of the mechanism of oxidative phosphorylation as compared with studies of other metabolic pathways. The experimental problems are formidable but many systems are now under investigation especially as the relationship of membrane lipoproteins to membrane structure and organisation and full knowledge of the components of the oxidative phosphorylation system is being elucidated.

Significant progress in the biochemical genetics of mitochondrial oxidative phosphorylation has been made in the past seven years and the subject has been extensively reviewed by Kovac (1). Extensive progress has also been made in studies of the biochemical genetics of oxidative phosphorylation in prokaryotes due to the work of Gibson (2) and Gutnick (3). Such studies are giving information on the structure, function and organisation of the mitochondrial inner membrane and the bacterial plasma membrane and the ATP-synthetase complex. Information on the sites of synthesis and regulation of synthesis and assembly of mitochondrial inner membrane components is now available from biochemical genetic studies and is of particular value in elucidating the nucleo-cytoplasmic interactions involved in mitochondrial biogenesis.

The known components of the electron transport chain and the ATP-synthetase complex have been elucidated by isolation of component complexes and correlation with function in the integrated membrane complex. The major components of the oxidative phosphorylation system can be isolated as five functional complexes.

COMPLEX I	NADH-ubiquinone reductase (4)
COMPLEX II	Succinate-ubiquinone reductase (5,6)
COMPLEX III	Ubiquinol-cytochrome c reductase (7,8)

COMPLEX IV   Cytochrome c oxidase (9,10)
COMPLEX V   Oligomycin-sensitive (O.S.) ATPase complex
                     or ATP synthetase (11,12)

The biogenesis of some of the components of these complexes has been discussed in recent reviews by Tzagoloff (13) and Schatz (4).

The following major conclusions emerge:

1. Mitochondrial protein complexes involved in energy conservation contain subunits synthesised on cytoribosomes (nuclear coded) and subunits synthesised on mitoribosomes (assumed to be mtDNA coded).

2. Three subunits of complex IV (cytochrome oxidase) and at least one subunit of complex III (ubiquinol-cytochrome c reductase) are mitoribosome synthesised (mtDNA coded). The correlation of these results with the properties of the classical mitochondrial mutation, the cytoplasmic 'petite' mutants which lack cytochromes aa_3, b and c_1, provides supporting evidence that mitoribosome synthesised components of complex III and complex IV are products of the mitochondrial genome.

3. To date, no mitochondrial gene or genetic locus has been identified with mitoribosome synthesised components of complexes III and IV.

4. To date, there is no evidence to indicate the presence of mitochondrially-coded components in complex I and complex II.

5. Four subunits of the ATP synthetase complex are mitoribosome synthesised and hence are assumed to be coded for by mtDNA. These include the membrane proteins which are the binding sites for oligomycin and DCCD.

6. There is tentative evidence that the ADP translocase complex contains a mitoribosome synthesised component.

7. Nuclear and mitochondrial mutations result in modification of components of the oxidative phosphorylation system.

8. Study of mitochondrial mutants may lead to the isolation and characterisation of previously unknown components of the mitochondrial energy conservation system.

9. Study of nuclear and mitochondrial mutants which lead to a modification of components of the inner membrane may give information as to the mechanism of oxidative phosphorylation and/or the structural organisation of the component complexes.

Biochemical genetic studies of yeast mitochondria which are underway in our laboratory will be discussed in relation to:

(a) the isolation and utilisation of mitochondrial drug resistant mutants;

(b) mitochondrial genes which specify components of the ATP-synthetase complex;

(c) genetic evidence for a relationship between ATP synthetase and ADP translocase;

(d) evidence that other cytoplasmic determinants not located on mtDNA specify mitochondrial inner membrane components.

MITOCHONDRIAL DRUG RESISTANT MUTANTS FOR ANALYSIS OF COMPONENTS OF OXIDATIVE PHOSPHORYLATION

The biochemical genetic studies of oxidative phosphorylation have been underway in our laboratory for the past five years and extensive use has been made of mitochondrial drug-resistant mutants.

The investigation is based on the following premises (15):

1. Drugs which affect mitochondrial energy conservation reactions (inhibitors, uncouplers, ionophores) have specific inhibitor sites associated with specific protein subunits of the oxidative phosphorylation complex located in the mitochondrial inner membrane.

2. Drug resistant mutants should exhibit modified sensitivity to the drug at the whole cell, mitochondrial, sub-mitochondrial and purified enzyme levels.

3. Demonstration that the mutation is cytoplasmically inherited and located on mtDNA is good *a priori* evidence that a mitochondrial gene product which is a component of the oxidative phosphorylation complex has been modified.

The drugs which show promise as specific inhibitors of mitochondrial oxidative phosphorylation are listed in Table I, and uncoupling agents, ATPase inhibitors, adenine nucleotide translocase inhibitors, and ionophores satisfy the general criteria outlined above (15,16).

Detailed biochemical genetic studies of oligomycin resistant, venturicidin-resistant and triethyltin-resistant mutants are now available and correlate well with modification of inhibitor binding sites on the mitochondrial oligomycin-sensitive (O.S.) ATPase complex.

Oligomycin-resistant mutants can be divided into two general classes (Class I and Class II) on the basis of cross resistance to other mitochondrial drugs (17). Class I mutants show cross resistance to aurovertin, Dio-9, venturicidin, triethyltin, uncouplers and other mitochondrial drugs. In contrast, Class II mutants are specifically resistant to oligomycin and structurally related antibiotics and show no cross resistance to venturicidin, triethyltin, DCCD, or uncoupling agents (Table II). All the Class II mutants exhibited typical cytoplasmic inheritance and the resistance determinants are located on mtDNA. Genetic analysis indicates that two loci (OLI and OLII) located on independent cistrons on mtDNA are involved (18,19). Biochemical studies (17,20) indicate a difference in oligomycin sensitivity, which is demonstrable at the mitochondrial, submitochondrial and purified ATPase levels, and are consistent with the modification of two of the mitochondrially synthesised components of the

TABLE I

Mitochondrial Drugs Useful in Genetic Studies

Drug	Biochemical locus of action	Resistant mutants isolated and locus
Antimycin A	Electron transport	Yes; nuclear
CCCP	Uncoupling agent	Yes; cytoplasmic
TTFB	Uncoupling agent	Yes; cytoplasmic
1799	Uncoupling agent	Yes; cytoplasmic
Oligomycin	O.S. ATPase	Yes; nuclear and mitochondrial
Venturicidin	O.S. ATPase	Yes; nuclear, mitochondrial and cytoplasmic
Trialkyltin	O.S. ATPase	Yes; nuclear and cytoplasmic
Aurovertin	F_1 ATPase	Yes; nuclear
Dio-9	F_1 ATPase	Yes; nuclear
Bongkrekic acid	ADP translocase	Yes; nuclear and cytoplasmic
Rhodamine 6G	ADP translocase?	Yes; cytoplasmic
Valinomycin	K^+ transport	Yes; mitochondrial

CCCP, carbonylcyanide m-chlorophenylhydrazone; TTFB, 4,5,6,7-tetrachloro-2-trifluoromethylbenzimidazole; 1799, α,α -bis(hexafluoroacetonyl)acetone.

TABLE II

Cross Resistance of Oligomycin Resistant Mutants

Class	Resistance level x parental strain				
	OLIGO	VEN	TET	1799	(MD)
I	>50	>50	10-20	>20	3-5
II	>50	1	1	1	1

MD = mitochondrial drugs, Chloramphenicol, Antimycin A, CCCP, etc. The resistance levels quoted are the values required for growth inhibition as compared to the sensitive parental strain (see refs. 17 and 21).

O.S. ATPase complex. Current studies are concerned with the isolation of mitochondrially synthesised subunits of the O.S. ATPase, and comparison of mutant and parental strains by peptide mapping and amino acid sequencing in order to establish a correlation between a mitochondrial mutation and a mitochondrial gene product.

Venturicidin resistant mutants also divide into two general classes (Table III). Two types of specific venturicidin-resistant mutants can be isolated (21). Class II (V,O) mutants show cross resistance to oligomycin and the resistance allele is located on mtDNA at a locus which is closely linked to OLI, termed OLIII (22). Class II (V,T) mutants are cytoplasmic mutants which show cross resistance to triethyltin, and are similar to cytoplasmic triethyltin mutants (Table IV). Genetic analysis indicates that these resistance alleles are essentially unlinked to other mitochondrial loci and are located on a different molecule of DNA (22,23). The genetic properties of these mutants will be discussed later, but there is sufficient evidence to indicate that the binding or inhibitory sites of oligomycin and triethyltin are not identical and that the triethyltin binding site is located on a different mitochondrial gene product to those which are involved in oligomycin binding.

Interaction and co-operative effects between different binding sites on the mitochondrial ATPase have been demonstrated in studies of the effect of the insertion of the TET^R phenotype into mitochondrial OLY^R mutants and provide an experimental basis for complementation studies at the ATP-synthetase level. These studies provide strong evidence for separate but interacting reaction sites for oligomycin, triethyltin and venturicidin and indicate that we are dealing with modification of drug receptor sites on the membrane bound sub-

TABLE III

Cross Resistance of Venturicidin Resistant Mutants

Class	Resistance level x parental strain				
	OLIGO	VEN	TET	1799	(MD)
I	>50	>50	10-20	>20	3-5
II (V,O)	100	100	1	1	1
II (V,T)	1	50	20	10	1

TABLE IV

Cross Resistance of Triethyltin Resistant Mutants

	Resistance level x parental strain				
Class	OLIGO	VEN	TET	1799	(MD)
I	>50	>50	>20	>20	5-10
IIa	1	>50	>20	1	1
IIb	1	>50	>20	10	1

units of the O.S. ATPase complex. The relationship of these mutations to the mitochondrially synthesised components of the O.S. ATPase complex described by Tzagoloff (13) is under investigation and should lead to the correlation of a mitochondrial mutation with modification of a mitochondrial gene product.

MITOCHONDRIAL GENES WHICH SPECIFY COMPONENTS OF THE O.S. ATPase COMPLEX

The biogenesis of the O.S. ATPase complex has been intensively investigated by Tzagoloff (13) and has been shown to consist of ten subunits, four of which are tightly associated with the mitochondrial inner membrane and are synthesised on mitoribosomes. These four mitoribosome synthesised subunits contain the sensitivity (inhibitor binding) sites for oligomycin, venturicidin, DCCD and triethyltin as we have demonstrated that the purified O.S. ATPase is sensitive to these inhibitors and contains the binding sites for these inhibitors, whereas the F_1 ATPase is not inhibited and does not bind these inhibitors (20,24). Genetic analysis indicates that three loci, OLI, OLII and OLIII, which are probably located on two separate genes on mtDNA, specify oligomycin and venturicidin resistance (21,22,23). The location of these resistance alleles on mt-DNA is strongly supported by evidence for linkage to other mitochondrial loci and ready loss of the resistance allele on petite induction with ethidium bromide (17,18,19,22). In contrast, the cytoplasmic TET^RVEN^R and VEN^RTET^R mutations do not appear to be linked to other loci on mtDNA, and the resistance alleles appear to be located on another DNA species. Recent experiments (25) have shown that the VEN^RTET^R resistance alleles are retained in ρ^o petites in which mtDNA has been deleted by ethidium bromide treatment. However, this cytoplasmic DNA species is closely associated with mtDNA as loss of the VEN^R or TET^R resistance alleles is always accompanied by loss of the ρ^+ state. This DNA species could be directly responsible for the maintenance of the ρ^+ state or indirectly via a necessity for the stability of mtDNA. This DNA species could be a second mtDNA molecule and may form part of the mitochondrial genome, and a possible candidate is the 2μ circular DNA species found to be associated with mitochondrial fractions by several workers and termed omicrom-DNA (o-DNA) by Clark-Walker (26).

The intra- or extra-mitochondrial location of o-DNA remains to be established and both possibilities must be considered together with the possible reversible integration into mtDNA in a fashion analogous to the bacterial plasmid situation. These possibilities have important consequences in studies of mitochondrial biogenesis which are summarised in Fig. 1. The assignment of protein components of petite mitochondria as products of the nuclear genome needs to be reinvestigated in view of the demonstration that cytoplasmic determinants which specify components of the ATPase complex are present in ρ^0 petites which have lost all mtDNA. The relationship of the ADP translocase system to components of the O.S. ATPase complex is a particular area of interest in this context.

Biochemical genetic studies have thus implicated two mitochondrial genes which code for two of the four mitoribosome synthesised subunits of O.S. ATPase. A third subunit represented by the triethyltin binding site is determined by an unknown cytoplasmic determinant which could be another species of mtDNA, possibly o-DNA.

GENETIC EVIDENCE FOR A RELATIONSHIP BETWEEN ATP–SYNTHETASE AND ADP TRANSLOCASE

We have previously presented evidence that $TET^R VEN^R$ mutants show in-

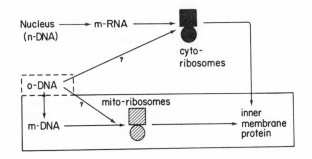

Fig. 1. Nucleo-cytoplasmic interactions involved in mitochondrial biogenesis.

creased resistance to bongkrekic acid (BA) and that these mutants exhibit a decreased sensitivity to BA as compared with the atractyloside sensitivity of mitochondrial ADP translocase in parental and mutant strains (27). On the basis of these observations we suggested that the ADP translocase complex contains a cytoplasmically determined subunit involved in the BA binding site and that this subunit could be a common subunit of the O.S. ATPase complex involved in the triethyltin binding site. The presence of a cytoplasmically determined subunit in ADP translocase has also been suggested by Haslam et al. (28), but has been questioned by Kolarov and Klingenberg (29) on the basis that a cytoplasmic petite (ρ^-) mutant contains a translocase system with normal sensitivity to atractyloside and bongkrekic acid. In view of the demonstration that components of the O.S. ATPase complex may be specified by cytoplasmic determinants which are not located on mtDNA, we suggest that the question remain open. We have investigated the translocase activity and atractyloside binding capacity of mitochondria from parental strains and from ρ^O petites which have retained ($\rho^O V^R T^R$) or lost ($\rho^O V^O T^O$) the $V^R T^R$ resistance alleles. In both petite strains the translocase activity is not detectable but binding studies with tritiated atractyloside indicate that an ADP sensitive atractyloside site is present but the affinity for atractyloside is 5-10 fold less than in the parental strain, and both petite strains contain less than 15% of the number of binding sites present in the parental strain. The results indicate that a full integrated ADP translocase complex requires the synthesis of a mitochondrially determined subunit or of mitochondrially determined components which serve a structural role in the ADP translocase complex. The isolation of a complex containing O.S. ATPase and the ADP translocase complex is now being investigated.

CYTOPLASMIC DETERMINANTS OTHER THAN mtDNA CODE FOR COMPONENTS OF THE INNER MEMBRANE

The demonstration of cytoplasmic determinants which code for components of the O.S. ATPase complex has already been discussed in relation to triethyltin resistant mutants, and a related series of cytoplasmic mutants specifying resistance to uncoupling agents are now of interest in view of recent developments from Hatefi's laboratory. Uncoupler resistant mutants isolated by selection against TTFB showed no cross resistance to other drugs and inhibitors, but showed cross resistance to other uncoupling agents, even though of a different molecular structure (15,16). These mutants are cytoplasmic mutants and the resistance allele is not lost on petite induction with ethidium bromide. There is no evidence that this mutation is located on mtDNA or has any genetic interaction with mtDNA. It was inferred that uncoupler resistance involved modification of a cytoplasmically determined binding site which was able to bind uncoupler molecules of different chemical structure. Supporting evidence has now been obtained by Hanstein and Hatefi in their studies of a mitochondrial uncoupler binding site (30). Ligand binding studies have shown that the mitochondria from uncoupler resistant strains have a modified binding site as evidenced by changes in the binding constant and the number of binding sites (Skipton, Griffiths and Hanstein, unpublished results). Hatefi's group has now shown that the uncoupler binding site fractionates with and concentrates in the O.S. ATPase complex (complex V). The relationship of the components involved in the uncoupler binding site to the mitochondrially determined binding sites specifying oligomycin and venturicidin resistance and the cytoplasmic determinants specifying triethyltin resistance is under active investigation. Of particular interest is the possibility that only two of the mitochondrially synthesised subunits of O.S. ATPase are determined by genes on mtDNA, and that the other two subunits (including the uncoupler binding site) are determined by genes on another species of cytoplasmic DNA.

ACKNOWLEDGMENTS

This work was supported by grants from the Science Research Council, the Medical Research Council and the Tin Research Institute.

REFERENCES

1. Kovac, L. (1974) *Biochim. Biophys. Acta. 346*, 101-135.
2. Gibson, F. and Cox, G.B. (1973) In *Essays in Biochemistry* (Campbell, P.N. and Dickens, F., eds.) Vol. 9, pp. 1-29, Academic Press, London.
3. Kanner, B.I. and Gutnick, D.L. (1972) *FEBS Letters 22*, 197-199.
4. Hatefi, Y., Haavik, A.G. and Griffiths, D.E. (1962) *J. Biol. Chem. 237*, 1676-1680.
5. Ziegler, D.M. and Doeg, K.A. (1959) *Biochem. Biophys. Res. Commun. 1*, 344-349.
6. Davis, K.A. and Hatefi, Y. (1972) *Arch. Biochem. Biophys. 149*, 505-512.
7. Hatefi, Y., Haavik, A.G. and Griffiths, D.E. (1962) *J. Biol. Chem. 237*, 1681-1685.
8. Rieske, J.S., Baum, H., Stoner, C.D. and Lipton, S.H. (1967) *J. Biol. Chem. 242*, 4854-4866.
9. Griffiths, D.E. and Wharton, D.C. (1961) *J. Biol. Chem. 236*, 1850-1862.
10. Fowler, L.R., Richardson, S.H. and Hatefi, Y. (1962) *Biochim. Biophys. Acta 64*, 170-173.
11. Tzagoloff, A. and Meagher, P. (1971) *J. Biol. Chem. 246*, 7328-7336.

12. Hatefi, Y., Stiggall, D.L., Galante, Y. and Hanstein, W.G. (1974) *Biochem. Biophys. Res. Commun. 61*, 313-321.

13. Tzagoloff, A., Rubin, M.S. and Sierra, M.F. (1973) *Biochim. Biophys. Acta 301*, 71-104.

14. Schatz, G. and Mason, T. (1974) *Ann. Rev. Biochem. 43*, 56-87.

15. Griffiths, D.E., Avner, P.R., Lancashire, W.E. and Turner, J.K. (1972) In *Biochemistry and Biophysics of Mitochondrial Membranes* (Azzone, G.F., Carafoli, E., Lehninger, A.L., Quagliariello, E. and Siliprandi, N., eds.) pp. 505-521, Academic Press, New York and London.

16. Griffiths, D.E. (1972) In *Mitochondria: Biogenesis and Bioenergetics* (Van den Borgh, S.G., Borst, P. and Slater, E.C., eds.) pp. 95-104, North Holland, Amsterdam.

17. Avner, P.R. and Griffiths, D.E. (1973) *Eur. J. Biochem. 32*, 301-311.

18. Avner, P.R. and Griffiths, D.E. (1973) *Eur. J. Biochem. 32*, 312-321.

19. Avner, P.R., Coen, D., Dujon, B. and Slonimski, P.P. (1973) *Mol. Gen. Genet. 125*, 9-52.

20. Griffiths, D.E. and Houghton, R.L. (1974) *J. Biol. Chem. 46*, 157-167.

21. Griffiths, D.E., Houghton, R.L., Lancashire, W.E. and Meadows, P.A. (1975) *Eur. J. Biochem. 51*, 393-402.

22. Lancashire, W.E. and Griffiths, D.E. (1975) *Eur. J. Biochem. 51*, 403-413.

23. Lancashire, W.E. and Griffiths, D.E. (1975) *Eur. J. Biochem. 51*, 377-392.

24. Griffiths, D.E. and Griffiths, E.J. Unpublished results.

25. Griffiths, D.E., Lancashire, W.E. and Zanders, E.D. (1975) *FEBS Letters 53*, 126-130.

26. Clark-Walker, G.D. (1973) *Eur. J. Biochem. 32*, 263-267.

27. Cain, K., Lancashire, W.E. and Griffiths, D.E. (1974) *Biochem. Soc. Trans. 2*, 215-218.

28. Haslam, J.M., Perkins, M. and Linnane, A.W. (1973) *Biochem. J. 134*, 935-947.

29. Kolarov, J. and Klingenberg, M. (1974) *FEBS Letters 45*, 320-323.

30. Hanstein, W.G. and Hatefi, Y. (1974) *J. Biol. Chem. 249*, 1356-1362.

ON THE MECHANISM OF ELECTRON FLOW AND CHARGE SEPARATION IN THE UBIQUINONE–CYTOCHROME C SPAN OF THE RESPIRATORY CHAIN

S. Papa, M. Lorusso and F. Guerrieri

Institute of Biological Chemistry, Faculty of Medicine
and
Institute of Biological Chemistry, Faculty of Science
University of Bari
Bari, Italy

The components of the respiratory chain of mitochondria are assembled in polymeric complexes. Much attention is today devoted to the stereochemical events taking place in the polypeptide subunits of these complexes and to orientation of their redox centers in the membrane.

Our group has undertaken a systematic study of the kinetics of redox processes and protonic reactions in cytochrome oxidase (complex IV)(1-3) and ubiquinol-cytochrome c oxido-reductase (complex III)(3,4). This approach gives a most direct insight into the topography of the redox complexes in the membrane and the molecular mechanism of generation of energy-linked, transmembrane electrochemical proton gradient.

Electrons are transferred from complex III to complex IV by cytochrome c at the outer side of the insulating layer of the membrane (see ref. 5 for review). Kinetic analysis with flow potentiometric and spectrophotometric techniques shows, on the other hand, that cytochrome oxidase reacts with oxygen at the inner side of the membrane (1,3, cf. refs. 6,7). In fact, rapid aerobic oxidation of cytochrome oxidase and c cytochromes is accompanied, in intact mitochondria, by synchronous proton release in the medium instead of the expected stoichiometric proton consumption (2). In the presence of proton conducting uncouplers, however, the oxidation of electron carriers is accompanied by delayed, stoichiometric proton consumption. In sonic particles which exhibit an inverted orientation of the mitochondrial membrane, the oxidation of oxygen-terminal electron carriers is accompanied by synchronous and stoichiometric consumption of protons (2). Valinomycin plus K^+ enhances the rate of aerobic oxidation of cytochrome c and of the accompanying proton consumption (3).

In conclusion, complex IV is oriented transversely across the mitochondrial membrane, in order to mediate fast vectorial flow of electrons from the outer to the inner side of the membrane. This functional arrangement of the cytochrome oxidase region will lead to generation of a net transmembrane electrochemical proton gradient if other segments of the chain are topographically arranged in order to lead to proton electron separation at the outer side of the membrane and give rise to alternation of outward hydrogen currents and inward electron currents.

Kinetics analysis shows that the aerobic oxidation of endogenous ubiquinol is accompanied by synchronous proton release from intact mitochondria and synchronous proton uptake by sonic particles (4). Thus electron flow along complex III is associated to proton translocation across the membrane. This proton pump appears to be electrogenic (4,8,9).

In the experiment in Fig. 1, the respiratory carriers of sonic particles were

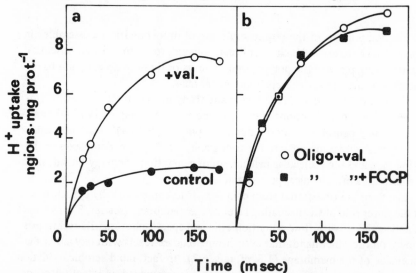

Fig. 1. Effect of valinomycin plus K^+ and FCCP on the kinetics of proton uptake associated to aerobic oxidation of endogenous ubiquinol in EDTA sonic particles. The kinetics of proton uptake were analyzed with a Roughton type continuous-flow pH meter. Mixing ratio 1:60 (v:v). The main syringe contained 250 mM sucrose, 3.5 mM potassium succinate, 3.5 mM potassium malonate, 30 mM KCl and EDTA particles, 2 mg protein/ml. Final pH 6.8. Temperature 20°. The smaller syringe contained an oxygen-saturated mixture consisting of 250 mM sucrose and 30 mM KCl. Where added: valinomycin 0.5 μg/mg protein; oligomycin 2 μg/mg protein and FCCP 0.5 μM.

reduced in anaerobiosis by succinate. Succinate oxidation was then inhibited by malonate so that subsequent oxygen-induced electron flow was limited to net oxidation of endogenous ubiquinol and reduced electron carriers. Fast proton uptake associated to aerobic oxidation of respiratory carriers stopped almost completely in 20 msec when about 2 ng ion H^+ per mg protein had been taken up. This inhibition was removed by valinomycin. Fast proton uptake associated to oxidation of endogenous respiratory carriers was unaffected by FCCP. Thus the proton pumping activity represents a primary energy-conserving act of the respiratory chain.

Flow analysis of the $H^+/2\ e^-$ stoichiometry for the aerobic oxidation of ubiquinol in sonic submitochondrial particles has revealed that, at least at pH 6.8, 4 g ion H^+ are taken up *per* mole ubiquinol oxidized (3,4). $H^+/2\ e^-$ ratio of 4 has also been found by Lawford and Garland (10) by pulsing intact mitochondria with small amounts of exogenous quinols.

The finding that 4 H^+ are translocated per 2 e^- flowing from ubiquinol to oxygen implies that one (or two) hydrogen carrier(s) exist in complex III. Additional evidence for the existence of a hydrogen carrier in complex III comes from studies on proton translocation in phospholipid liposomes inlaid with purified complex III (8,9). In this system reduction of added cytochrome *c* by duroquinol is accompanied by electrogenic proton translocation with an H^+/e^- ratio of approximately 2 (9).

The finding that electron transfer along complex III is associated to electrogenic proton translocation from the inner to the outer side of the membrane indicates that the components of the complex are indeed arranged in the membrane in such a way as to give rise to an inward vectorial flow of electrons and outward hydrogen flow (see Fig. 2).

Complex III delivers electrons to cytochrome *c* at the outer side of the membrane; we are however left with the following problems: (i) where is the site at which ubiquinol is oxidized by electron carriers of complex III and hydrogen atoms are split into protons and electrons; (ii) which electron carriers of

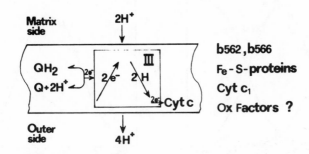

Fig. 2.

complex III mediate the inward electron flow; (iii) which is the hydrogen carrier involved in the outward hydrogen flow?

The electron carriers of complex III are two b cytochromes, b_{566} and b_{562}, cytochrome c_1, and two Fe-S-proteins (11). There are also factors not well identified (12-14).

No classical hydrogen carrier is known to exist in the complex. We have, however, proposed a model (4, 15,16) according to which electron carriers might act as effective hydrogen carriers and might be responsible for transmembrane proton transport through linkage of oxido-reduction of the metal to protonic equilibria of acidic groups in the apoprotein. This mechanism, which emphasizes the role of stereochemical events in the apoproteins of electron carriers, requires the midpoint redox potential of the carrier to be pH dependent. The midpoint potential of c_1 is pH independent in the physiological range. Prince and Dutton (17) have recently reported that the midpoint redox potential of the Rieske Fe-S-protein is pH independent in the range of pH 6.3 to pH 8.3. A unique position is that of b cytochromes. Their midpoint redox potential is pH dependent above pH 6.9 (see ref. 11 for review). The pH-dependence curve of the potential gives a value of 6.8 for the pK of the oxidized form and a tentative value of 9 for the pK of the reduced form (Fig. 3). Utilizing these pK values it is possible to construct the curves for the ionization state of the oxidized and reduced b cytochromes as a function of pH, and from these curves it is possible to calculate to what extent at a given pH the b cytochromes function as electron or hydrogen carriers. At pH 8 the b cytochromes function practically as hydrogen carriers, however the percentage of b cytochromes acting as hydrogen carriers sharply decreases as the pH is decreased or increased until below 5 and above 10 the b cytochromes function as pure electron carriers (see Fig. 3).

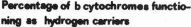

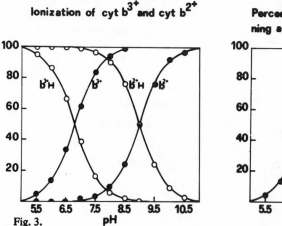

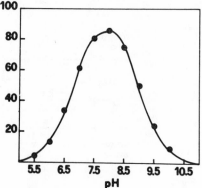

Fig. 3.

We have utilized these properties of the *b* cytochromes to design experiments to examine the topographical arrangement in the membrane of *b* cytochromes and their role in the redox proton pump of complex III (18).

Work from various laboratories had led to the proposal that cytochrome b_{566} is predominantly located at the outer and b_{562} at the inner side of the membrane (19,20). If this were the case the aerobic oxidation of b_{566}, but not that of b_{562}, should be depressed by the membrane potential and this depression should decrease as the b_{566} changes, upon increasing the pH, from an electron to a hydrogen carrier.

Fig. 4 illustrates the kinetics of the aerobic oxidation of succinate reduced *b* cytochromes in intact mitochondria. At pH 6.8 the rate of oxidation of cytochrome b_{566}, whose absorbance predominates over that of b_{562} at the wavelength pair 566-575 nm, is much lower than that of b_{562}. Addition of

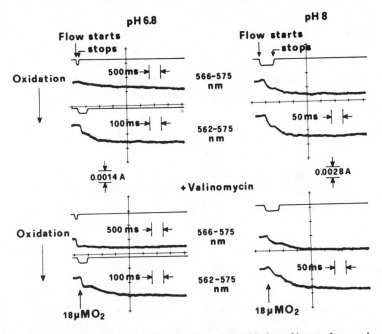

Fig. 4. Effect of valinomycin on the kinetics of aerobic oxidation of b cytochromes in rat liver mitochondria. The kinetics of oxidation of b cytochromes was analyzed with a Johnson Foundation dual wavelength, stopped-flow spectrophotometer. Mixing ratio 1:80 (v:v). The main syringe contained 200 mM sucrose, 10 mM potassium succinate, 10 mM potassium malonate, 30 mM KCl, 2 μg/mg protein oligomycin, 1 μg rotenone/mg protein and rat liver mitochondria, 2.2 mg protein /ml. Temperature, 20°. The smaller syringe contained an oxygen saturated mixture consisting of 200 mM sucrose and 30 mM KCl. Where indicated 0.1 μg/mg protein valinomycin was added. For other details see ref. 4.

valinomycin plus K^+, which collapses the membrane potential, causes a marked acceleration of the oxidation of cytochrome b_{566} but has only a minor effect on the rate of oxidation of b_{562}. At pH 8 the aerobic oxidation of b_{566} is much faster than at pH 6.8 and practically synchronous to the oxidation of b_{562}. At this pH, valinomycin plus K^+ has no detectable effect on the rate of oxidation of the two b cytochromes.

Table I summarizes data on the pH dependence of the kinetics of aerobic oxidation of b cytochromes in EDTA sonic particles (18). Oligomycin, which inhibits passive proton diffusion in these particles (21) and thus allows the establishment of aerobic transmembrane electrochemical proton gradient, causes, at pH 7, marked inhibition of the oxidation of cytochrome b_{566}, as revealed by the enhancement of the t½. This inhibition is completely released by valinomycin. The inhibitory effect of oligomycin and the releasing effect of valinomycin are much less marked at pH 7.7 and disappear completely at pH 8.5. Also, the oxidation of cytochrome b_{562} is inhibited by oligomycin and stimulated by the subsequent addition of valinomycin. However, in this case the effects are much less pronounced and practically pH independent. These results are consistent with the proposed location of cytochrome b_{566} at the outer and b_{562} at the inner side of the membrane. This arrangement of b cytochromes offers the possibility of various models. Two models are presented in Fig. 5.

TABLE I

pH Dependence of the Kinetics of Cytochrome b Oxidation
in the Anaerobic-Aerobic Transition of EDTA Particles

Additions	Cytochrome b oxidation, t½ (msec)					
	566-575 nm			560-540 nm		
	pH					
	7.0	7.75	8.5	7.0	7.75	8.5
None	200	140	170	147	162	200
Oligomycin	1080	580	236	316	225	270
Oligomycin + valinomycin	210	178	165	175	105	141
Oligomycin + valinomycin + nigericin	210	187	165	165	150	250

The reaction mixture contained: 250 mM sucrose, 10 mM Tris-succinate, 5 mM Tris-malonate, 20 mM Tris-HCl, 20 mM KCl, 1.5 mg particles protein/ml. (From Table IV of Papa et al. (18)).

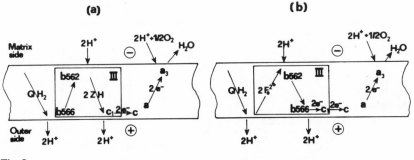

Fig. 5.

According to model (a)(cf. refs. 19 and 20) cytochrome b_{566} would be closer to the substrate than cytochrome b_{562}. According to model (b) cytochrome b_{562} would be closer to the substrate than b_{566}. According to model (a) the b cytochrome molecules would be spread across the membrane in order to give rise to vectorial flow of electrons or hydrogens — this depends upon the actual pH of the system — from the outer to the inner side of the membrane. The membrane potential impedes electron flow in the normal direction from cytochrome b_{566} at the outer to cytochrome b_{562} at the inner side. Specific collapse of the potential should, in this case, reveal a crossover point between the two types of b cytochromes. This crossover exerted by electrical forces should disappear when, experimentally increasing the pH, the b cytochromes change from electron to hydrogen carriers.

A second prediction of model (a) is that there exists a hydrogen carrier between the b cytochrome at the inner and c cytochromes at the outer side of the membrane. This carrier will mediate a vectorial outward flow of hydrogen which will be under the control of transmembrane ΔpH. Thus specific collapse of the ΔpH should reveal a crossover point between b and c cytochromes.

According to model (b), on the other hand, the b cytochromes will give rise to a vectorial flow of electrons (or hydrogens) from the inner to the outer side of the membrane. Furthermore, cytochrome b_{566} would be in direct redox equilibrium with c cytochromes at the outer side of the membrane. Thus in this case no crossover points are expected to be exerted by the $\Delta\psi$ or ΔpH component of the electrochemical proton gradient between the two types of b cytochromes and between b_{566} and c cytochromes, respectively.

These predictions have been checked by testing the effect of valinomycin and nigericin on the steady-state redox level of cytochromes. The experiment in Fig. 6 shows the effect of oligomycin and ionophores on the respiration-linked proton translocation and redox level of cytochromes in aerobic EDTA sonic particles supplemented with succinate as respiratory substrate. Addition of oligomycin enhances net proton uptake by EDTA particles. This is accom-

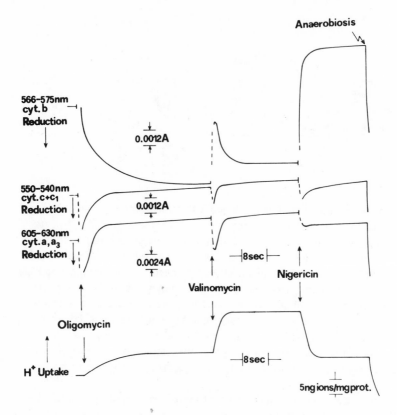

Fig. 6. Effect of oligomycin and ionophores on the aerobic steady-state redox level of cytochromes and proton translocation in EDTA particles. The redox level of cytochromes was measured with double beam, dual-wavelength spectrophotometer. The proton uptake was measured potentiometrically with a 50-100 MΩ glass electrode. The reaction mixture contained: 200 mM sucrose, 20 mM KCl, 10 mM succinate, 1 μg rotenone/mg protein, 0.2 mg/ml purified catalase, and 2 mg protein/ml EDTA particles. Respiration was activated by adding 15 μliters of 15% H_2O_2 to anaerobic particles. Where indicated 2 μg oligomycin/mg protein, 0.4 μg valinomycin/mg protein, and 0.4 μg nigericin/mg protein were added. Final volume, 1.5 ml. Final pH 7.5. Temperature 20°. For other details see ref. 3.

panied by synchronous transition of cytochrome b_{566} towards a more reduced state. Oligomycin also causes a very rapid reduction of c cytochromes and cytochrome oxidase. However, as the aerobic proton gradient increases under the influence of oligomycin, the reduction of these carriers is followed by a transition to a more oxidized state.

Thus the electrochemical proton gradient exerts a back pressure on electron flow from b cytochromes in complex III to c cytochromes and cytochrome oxidase. Collapse of the membrane potential, with the addition of valinomycin

(K^+ is present in the medium), causes rapid oxidation of cytochrome b_{566} and simultaneous reduction of c cytochromes and cytochrome oxidase. As the potential is replaced by extra ΔpH, generated by further net proton uptake, cytochrome b is again reduced up to the level presented before adding valinomycin. This secondary reduction of b cytochromes is accompanied by simultaneous reoxidation of c cytochromes and cytochrome oxidase. Further addition of nigericin abolishes the proton gradient and this is accompanied by transition of cytochrome b_{566} to a more oxidized state and of c cytochromes and cytochrome oxidase to a more reduced state.

When the redox state of b cytochromes was monitored at the wavelength pair 562-575 nm, instead of 566-575 nm, a pattern similar to that shown in Fig. 6 was observed. However, the cyclic oxidation of b cytochromes monitored at 562 nm was significantly smaller than that at 566 nm.

Fig. 7 shows the pH dependence of the valinomycin induced oxidation of b cytochromes monitored both at 562-575 and 566-575 nm. It can be seen that the oxidation of b cytochromes is, especially at the lower pH values, much higher at 566-575 than at 562-575 nm. In both cases the extent of the valinomycin-induced oxidation decreases as, by increasing the pH, the b cytochromes change from electron to hydrogen carriers.

The results of these experiments are consistent with the sequence and spatial arrangement of the b-c_1 region as envisaged by model (a). Dissipation of the membrane potential by valinomycin promotes electron flow from cytochrome b_{566} at the outer to cytochrome b_{562} at the inner site. Net reduction of cytochrome b_{562} is not observed since electrons rapidly pass further within complex III to cytochrome c_1 which, in fact, is reduced rapidly. The reason

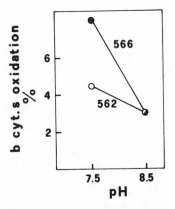

Fig. 7. pH dependence of the transient oxidation of b cytochromes induced by the addition of valinomycin to respiring EDTA particles. For experimental details see the legend to Fig. 6.

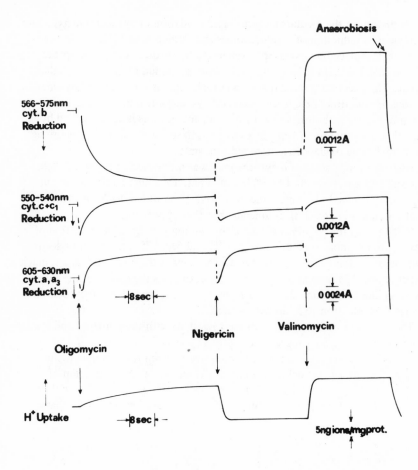

Fig. 8. Effect of oligomycin and ionophores on the aerobic steady-state redox level of cytochromes and proton translocation in EDTA particles. For experimental details see legend to Fig. 6.

oxidation rather than reduction of b cytochromes is monitored at 562-575 nm is that both b cytochromes contribute to the absorbance changes measured at these wavelengths. Furthermore, it is possible that cytochrome b_{562} is also located in part at the outer site, as is cytochrome b_{566}.

It should be remembered that when electron flow from cytochrome b to c_1 is interrupted by antimycin, valinomycin causes a marked acceleration of the aerobic oxidation of c cytochromes (3). Evidently in the uninhibited system, when the membrane potential is rapidly abolished by valinomycin, the rate at which electrons jump from b cytochromes to c_1 within complex III is higher than that for the transfer of electrons from cytochrome c to the acceptors in complex IV.

The finding that the valinomycin-induced abrupt oxidation of b cytochromes and reduction of c cytochromes is followed, as the potential is replaced by an extra ΔpH, by simultaneous reduction of b cytochromes and oxidation of c cytochromes shows the existence of a crossover point between b and c cytochromes under the specific control of transmembrane ΔpH. This crossover indicates, as predicted by model (a), the existence of a hydrogen carrier and an outward hydrogen current between b and c cytochromes.

Additional evidence in favor of this view is provided by the experiment of Fig. 8, which shows that collapse of the transmembrane ΔpH, caused by nigericin addition to actively respiring particles, is accompanied by transition of b cytochromes to a more oxidized and c cytochromes to a more reduced steady state. Nigericin also causes a larger reduction of cytochrome oxidase which is followed, however, by subsequent reoxidation up to the original level. Further addition of valinomycin gives large oxidation of b cytochromes and reduction of cytochrome oxidase. Cytochrome c shows only a small deflection toward a more oxidized state.

In conclusion, the data presented apparently favor model (a) of the redox proton pump of complex III (see Fig. 5). However, a full understanding of this process requires further characterization of the components of complex III, elucidation of the redox carrier which directly accepts electrons from ubiquinol and identification of the hypothetical hydrogen carrier of complex III.

REFERENCES

1. Papa, S., Guerrieri, F. and Lorusso, M. (1974) In *Dynamics of Energy-Transducing Membranes* (Ernster, L., Estabrook, R.W. and Slater, E.C., eds.) pp. 417-432, Elsevier, Amsterdam.
2. Papa, S., Guerrieri, F. and Lorusso, M. (1974) In *Membrane Proteins in Transport and Phosphorylation* (Azzone, G.F., Klingenberg, M.E., Quagliariello, E. and Siliprandi, N., eds.) pp. 177-186, North-Holland, Amsterdam.
3. Papa, S., Guerrieri, F. and Lorusso, M. (1974) *Biochim. Biophys. Acta* *357*, 181-192.
4. Papa, S., Lorusso, M. and Guerrieri, F. (1975) *Biochim. Biophys. Acta* *376*, 231-246.
5. *Electron Transport and Energy Conservation* (1970)(Tager, J.M., Papa, S., Quagliariello, E. and Slater, E.C., eds.) Adriatica, Bari.
6. Schneider, D.L., Kagawa, Y. and Racker, E. (1972) *J. Biol. Chem. 247*,

4074-4079.

7. Mitchell, P. and Moyle, J. (1970) In *Electron Transport and Energy Conservation* (Tager, J.M., Papa, S., Quagliariello, E. and Slater, E.C., eds.) pp. 575-587, Adriatica, Bari.

8. Hinkle, P.C. and Leung, K.H. (1974) In *Membrane Proteins in Transport and Phosphorylation* (Azzone, G.F., Klingenberg, M.E., Quagliariello, E. and Siliprandi, N., eds.), pp. 73-78, North-Holland, Amsterdam.

9. Guerrieri, F. and Nelson, B.D. (in press) *FEBS Letters.*

10. Lawford, H.G. and Garland, P.B. (1973) *Biochem. J. 130*, 1029-1044.

11. Dutton, P.L. and Wilson, D.F. (1974) *Biochim. Biophys. Acta 346*, 165-212.

12. Baum, H., Silman, H.I., Rieske, J.S. and Lipton, S.H. (1967) *J. Biol. Chem. 242*, 4876-4887.

13. Nishibayaashi-Yamashita, H., Cunningham, C. and Racker, E. (1972) *J. Biol. Chem. 247*, 698-704.

14. Das Gupta, U. and Rieske, J.S. (1973) *Biochem. Biophys. Res. Commun. 54*, 1247-1254.

15. Papa, S. (1973) In *Mechanisms in Bioenergetics* (Azzone, G.F., Ernster, L., Papa, S., Quagliariello, E. and Siliprandi, N., eds.) pp. 464-472, Academic Press, New York and London.

16. Papa, S., Guerrieri, F., Lorusso, M. and Simone, S. (1973) *Biochimie 55*, 703-716.

17. Prince, R.C. and Dutton, P.L. (1975) *Biophys. J., Abstr. 19th Ann. Mtg. Biophys. Soc.* , 278a.

18. Papa, S., Scarpa, A., Lee, C.P. and Chance, B. (1972) *Biochemistry 11*, 3091-3098.

19. Wikstrom, M.K.F. (1973) *Biochim. Biophys. Acta 301*, 155-193.

20. Mitchell, P. (1972) In *Mitochondria, Biomembranes*, Vol. 28, pp. 353-370, North-Holland, American Elsevier, Amsterdam and New York.

21. Papa, S., Guerrieri, F., Rossi Bernardi, L. and Tager, J.M. (1970) *Biochim. Biophys. Acta 197*, 100-103.

ENERGY TRANSDUCTION IN ISOLATED
CHLOROPLAST MEMBRANES

M. Avron

Biochemistry Department
Weizmann Institute of Science
Rehovot, Israel

Chemiosmotic coupling (1,2) predicts that oxidation-reduction energy is converted via proton movements into energy stored in the form of a proton-gradient and a membrane-potential. These secondary forms can, in turn, serve as the driving force for ATP formation. Fig. 1 illustrates one possible variation

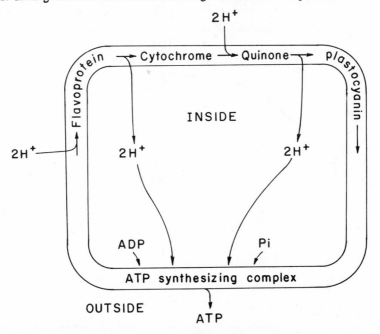

Fig. 1. A chemiosmotic model for energy coupling in chloroplasts.

of chemiosmotic coupling in isolated chloroplast membranes.

During the last few years we have concentrated our efforts in designing and utilizing techniques which enabled us to measure accurately the size of such proton gradients and membrane potentials under a variety of conditions (3-15). It was found that under steady-state conditions optimal for ATP synthesis, proton concentration gradients as large as 10,000/(ΔpH = 4) could be measured, but only insignificant membrane potentials were observed. This seemed in agreement with our earlier results (16,17) which indicated that agents which would be expected to completely abolish a membrane potential, had one existed, were ineffective in decreasing the efficiency of ATP production by isolated chloroplasts.

The elegant experiments of Jagendorf and collaborators (2) have clearly illustrated that chloroplasts can remove a large number of protons from the medium in an energy dependent manner. Such proton removal is easily observed when the pH of an appropriate chloroplast suspension is monitored during a light-dark cycle (Fig. 2). Evidence for the concentration of these protons in the inner vesicular space was provided when it was shown that the number of protons accumulated is a function of the inner buffer capacity, and can be increased as much as 20-fold by adding an appropriate internal buffer such as phenylene-diamine (Fig. 3). It was further shown that such accumulated protons, even when internally buffered by artificially added compounds, can serve

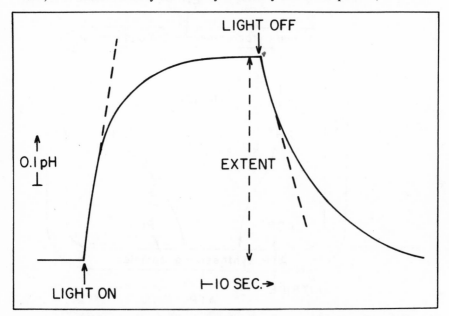

Fig. 2. Light dependent proton removal from medium as monitored by a pH electrode.

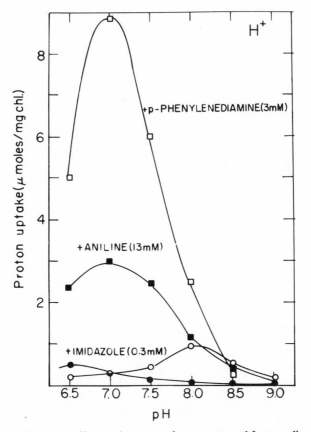

Fig. 3. Effect of internal buffers on the extent of proton removal from medium. (From Avron (13)).

as the driving force for ATP formation in a post-illumination phase (Fig. 4). Thus, it is clear that chloroplasts are capable of utilizing energy in the form of proton gradients as a driving force for ATP formation in the absence of any other independent energy source.

We next asked the question of whether chloroplasts can utilize energy stored in the form of a membrane-potential as a driving force for ATP formation. A membrane potential was produced across the chloroplast membrane by a rapid injection of a K^+-free chloroplast suspension into a medium containing KCl plus valinomycin. It was found that when such an externally produced membrane potential was superimposed on a suboptimal proton gradient, chloroplasts clearly used the membrane potential energy as a driving force for ATP formation. It was immaterial whether the suboptimal proton gradient was achieved by

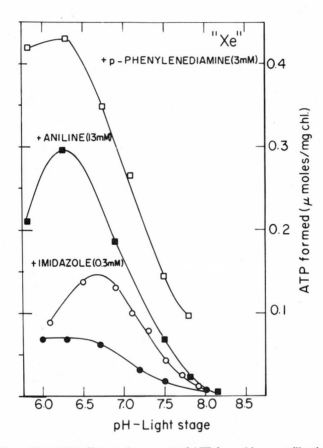

Fig. 4. Effect of internal buffers on the amount of ATP formed in a post-illumination experiment. (From Avron (13)).

lowering the light intensity, or lowering the pH of the dark stage. In either case, the superimposition of a membrane-potential resulted in a marked increase in the observed yield of ATP. It should be noted (Table I) that this increase in yield was not observed when the proton gradient was itself maximal.

We can conclude, therefore, that chloroplasts have the capability to use energy stored either in the form of proton gradients or membrane-potentials as a driving force for ATP formation. However, in the steady state only the former seems to be present to an energetically significant magnitude.

We have now turned our attention to the relation between the magnitude of this pH gradient and the ability to synthesize ATP. The technique of choice which we have introduced and employed to measure the pH gradient across the chloroplast membrane has been by following the quenching of the fluorescence

TABLE I

Membrane Potential as a Driving Force for ATP Formation
by Postillumination ATP Synthesis in Chloroplasts

pH		Light Intensity	ATP formed	
Light stage	Dark stage		Choline chloride	KCl plus valinomycin
		ergs x cm^{-2} x sec^{-1}		
6.5	8.5	3.5 x 10^5	154	160
6.5	8.5	9.0 x 10^3	9	29
6.5	7.5	3.5 x 10^5	61	144
6.5	7.5	9.0 x 10^3	3	26

From Schuldiner et al. **(6).**

of 9-aminoacridine (5). It has since been further employed and validated in studies using liposomes and chromatophores (18,19). However, since some reservations have recently been raised regarding the validity of this technique (20) it may be well to describe somewhat further its properties. As expected from theoretical considerations (5), 9-aminoacridine was found to distribute itself between two water phases separated by a hydrophobic phase, in accordance with the pH gradient pre-established across the water phases (Table II).

TABLE II

Distribution of 9-Aminoacridine in Response to pH Gradients
Between Two Water Phases Separated by a Hydrophobic Phase

pH of phase I	pH of phase II	$\log \dfrac{(H^+)^{II}}{(H^+)^I}$	$\log \dfrac{(9AA\ H^+)^{II}}{(9AA\ H^+)^I}$
7.90	9.05	-1.15	-1.24
9.05	9.05	0.00	0.00
9.00	7.90	1.10	1.24
9.80	9.10	0.70	0.75
7.95	7.05	0.90	0.60
9.00	6.90	2.10	1.90
9.00	5.95	3.05	2.50

From Pick et al. **(11)**

In chloroplasts, the quenching of the fluorescence of 9-aminoacridine can be utilized to follow in the dark an acid base transition induced by the injection of a small volume of Tris to a suspension pre-equilibrated at a lower pH (Fig. 5). When such experiments were performed over a wide variety of artificially induced pH gradients, good agreement between the induced and calculated gradients was obtained (Table III).

Finally, when the 9-aminoacridine response in this system was tested at different inner vesicular volume (i.e. different osmotic volumes), the response was again in reasonable agreement with that expected theoretically when the osmotic volume was varied by varying the quantity of chloroplasts employed, and deviated from expectation only at very high osmolarities when sorbitol was employed to control the inner vesicular volume (Table IV).

It is clear, therefore, that with proper reasonable precautions the 9-aminoacridine quenching technique can be conveniently employed to follow pH gradients across the chloroplast membrane.

Using this technique, we may ask whether the rate of ATP formation depends upon the magnitude of ΔpH, and what is the form of that dependence? Fig. 6A

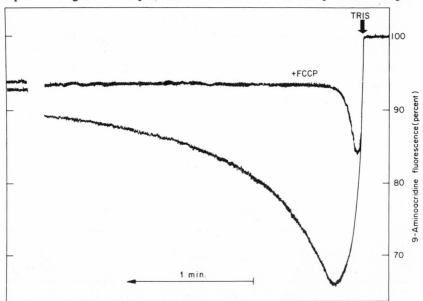

Fig. 5. Quenching of the fluorescence of 9-aminoacridine as a measure of acid-base transition in chloroplasts. Reaction medium contained, in a total volume of 2.5 ml: KCl, 10 mM; $MgCl_2$, 3 mM; sodium succinate, 3 mM; 9-aminoacridine, 2 μM; and chloroplasts containing 33 μg chlorophyll; pH, 5.35; temperature, 5° C. Where indicated, 0.5 ml of 25 mM Tris, pH 10, was injected with a syringe. Final pH, 8.30. (Pick and Avron, unpublished).

TABLE III

Correspondence Between the Magnitude of an Induced pH Gradient in Chloroplasts and the pH Gradient Calculated from the Quenching of 9-Aminoacridine

Acidic pH	Basic pH	ΔpH		
		Induced	Calculated	Induced / calculated
5.3	8.3	3.0	2.95	0.98
5.7	8.7	3.0	2.97	0.99
6.5	9.7	3.2	3.17	0.99
5.0	7.6	2.6	2.75	1.05
6.0	8.6	2.6	2.84	1.09
7.0	9.7	2.7	2.84	1.05

Details as under Fig. 5, except as indicated. (Pick and Avron, unpublished).

TABLE IV

Chloroplast concentration	Sorbitol concentration	ΔpH induced/calculated
μg/3 ml	mM	
5	0	1.03
10	0	1.06
25	0	1.10
50	0	1.05
100	0	1.00
200	0	0.96
11	0	1.01
11	20	1.01
11	30	1.00
11	50	1.03
11	100	1.06
11	300	1.20

Details as under Fig. 5, except as indicated. (Pick and Avron, unpublished).

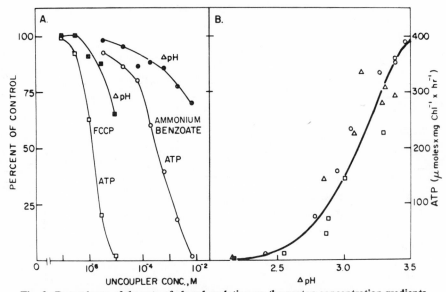

Fig. 6. Dependence of the rate of phosphorylation on the proton concentration gradients, as monitored by the addition of uncouplers. (From Pick et al. (12)).

illustrates the effect of two uncouplers on the rate of ATP formation and the magnitude of ΔpH. Fig. 6B shows this data plotted to show the rate of ATP formation as a function of ΔpH (12). It is clear that with either uncoupler tested two phases can be clearly distinguished: (a) In the range of ΔpH up to a threshold value of about 2.5 no ATP formation was observed; and (b) above this threshold value there was a sharp dependence of the rate of ATP formation on the magnitude of ΔpH. The data obtained by varying several other parameters (light intensity, electron transport inhibitors, pH of light or dark stage in postillumination experiments and ΔpH in acid base experiments) resulted in a similar dependence of ATP formation on ΔpH (11,12).

It was previously reported (21) that at very low light intensities it takes some time till the rate of phosphorylation attains its steady state rate. Fig. 7A demonstrates that at pH 7.2 and at very low light intensities the rate of phosphorylation indeed shows an intensity dependent time lag. However, no such lag can be seen in the build-up of ΔpH. In Fig. 7B the rate of phosphorylation at any time point is plotted as a function of the ΔpH at that point. Clearly, a threshold value is again observed (2.4 in this case). Thus the time lag may be a function of the time needed to build up this required threshold ΔpH value.

Finally, the question should be asked, are the values of ΔpH measured sufficient to drive ATP synthesis by the proposed chemiosmotic mechanism (1)? In order to form a molecule of ATP by the transfer of two protons, ΔpH must exceed 5.5, assuming no membrane potential and accepting the measured phos-

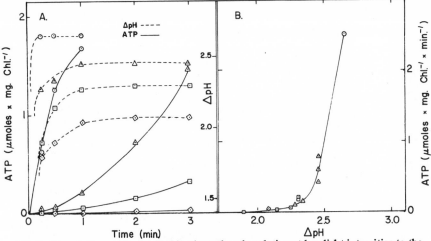

Fig. 7. The relation of the time lag in photophosphorylation at low light intensities to the magnitude of the proton concentration gradient. (From Pick et al. (11)).

phate potential in chloroplasts of 15 Kcal/mole (22,23). Experimentally, ΔpH values exceeding 4.5 have never been observed. Since 15 Kcal/mole is a maximal value, and most of the reported experiments were performed under a lower phosphate potential, we designed an experiment to test whether the ΔpH dependence of ATP formation is sensitive to changes in phosphate potential. Fig. 8 illustrates the results of such an experiment, where ΔpH was controlled by varying the light-intensity and the phosphate potential was changed by a factor of 400 by varying the ADP/ATP ratio (12). It is clear that the dependence of the rate of ATP formation on the magnitude of ΔpH was only marginally, if at all, dependent on the phosphate potential during the reaction. We have also tried to compare the actual phosphate potential maintained by chloroplasts in the steady state under different light intensities with the ΔpH maintained by the same preparations (Table V).

It can be seen that the values of the electrochemical gradient found are inconsistent with the formation of ATP by a two-proton transfer, but can be accomodated by a three proton transfer per ATP.

In conclusion, it is amply clear that the energy stored in the form of proton gradients or membrane potentials can be used by chloroplasts to drive ATP formation. Nevertheless, the ΔpH values measured which permit (or are in steady state with) ATP formation are thermodynamically inconsistent with a movement of two protons providing sufficient energy for the synthesis of an ATP molecule.

The answer to this dilemma may lie either in introducing a variation in the chemiosmotic hypothesis which will permit the coupled transfer of more pro-

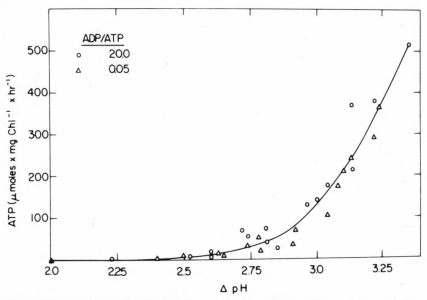

Fig. 8. The effect of varying the phosphate potential on the dependence of photophosphorylation on the proton concentration gradient. (From Pick et al. (11).

TABLE V

Comparison of Phosphate Potential and ΔpH Maintained by Chloroplasts
in the Steady State Under Different Light Intensities

Light intensity	Steady state phosphate potential	Steady state electrochemical gradient		
		Required assuming		Found
		$2 H^+/ATP$	$3 H^+/ATP$	
ergs x cm^{-2} x sec^{-1}	Kcal/mole			
2×10^5	13.4	290	193	230
4×10^4	12.8	276	184	197
2×10^4	11.9	258	172	183

Reaction mixture contained: KCl, 20 mM; phosphate, pH 8.0, 2 mM; MgCl$_2$, 4 mM;
ADP, 0.2 mM; phenazine methosulfate, 10 μM; 9-aminoacridine, 1 μM; and chloroplasts
containing 18 μg/ml chlorophyll. It was illuminated with red light for 10-20 min until a
steady state phosphate potential was measured. ΔpH was followed continuously by the
quenching of 9-aminoacridine fluorescence and ADP content was measured at 0 time and
at selected time intervals by following the number of protons taken up by the phosphorylation of a sample of the reaction mixture by pyruvate kinase and phosphoenolpyruvate.
$\Delta G^{o'}$ of 7.80 Kcal/mole was used at pH 8.0, 10 mM Mg^{++}, 25° C, 0.l ionic strength (23).
(Avron, unpublished).

tons per electron per site, and so a higher proton-to-ATP ratio, or in suggesting that the proton gradients and membrane potentials are energy storage devices in equilibrium with another, yet undefined, high-energy intermediate state, which directly drives the synthesis of ATP.

REFERENCES

1. Mitchell, P. (1968) *Chemiosmotic Coupling and Energy Transduction*, Glynn Research Bodmin.
2. Jagendorf, A.T. (1975) In *Bioenergetics of Photosynthesis* (Govinjee, ed.) pp. 413-492, Academic Press, New York.
3. Rottenberg, H., Grunwald, T. and Avron, M. (1971) *FEBS Letters 13*, 41-44.
4. Rottenberg, H., Grunwald, T. and Avron, M. (1972) *Eur. J. Biochem. 25*, 54-63.
5. Schuldiner, S., Rottenberg, H. and Avron, M. (1972) *Eur. J. Biochem. 25*, 64-70.
6. Schuldiner, S., Rottenberg, H. and Avron, M. (1972) *FEBS Letters 28*, 173-176.
7. Schuldiner, S., Rottenberg, H. and Avron, M. (1973) *Eur. J. Biochem. 39*, 455-462.
8. Schuldiner, S., Padan, E., Rottenberg, H., Gromet-Elhanan, Z. and Avron, M. (1974) *FEBS Letters 49*, 174-177.
9. Bamberger, E.S., Rottenberg, H. and Avron, M. (1973) *Eur. J. Biochem. 34*, 557-563.
10. Pick, U., Rottenberg, H. and Avron, M. (1973) *FEBS Letters 32*, 91-94.
11. Pick, U., Rottenberg, H. and Avron, M. (1974) In *Proceedings 3rd International Congress on Photosynthesis* (Avron, M., ed.) Vol. II, pp. 967-974, Elsevier, Amsterdam.
12. Pick, U., Rottenberg, H. and Avron, M. (1974) *FEBS Letters 48*, 32-36.
13. Avron, M. (1972) In *Proceedings 2nd International Congress on Photosynthesis* (Forti, G., Avron, M. and Melandri, eds.) Vol. 2, pp. 861-871, N.V. Junk, The Hague.
14. Avron, M. (1974) In *Membrane Transport in Plants* (Zimmerman, U. and Dainty, J., eds.) pp. 225-249, Springer-Verlag, Heidelberg.
15. Avron, M. (in press) In *Batelle Memorial Institute Symposium*, Seattle.
16. Karlish, S.J.D. and Avron, M. (1968) *FEBS Letters 1*, 21-24.
17. Karlish, S.J.D., Shavit, N. and Avron, M. (1969) *Eur. J. Biochem. 9*, 291-298.
18. Deamer, D.N., Prince, R.C. and Crofts, A.R. (1972) *Biochim. Biophys. Acta 274*, 323-335.
19. Melandri, B.A., Zannoni, D., Casadio, R. and Baciarini-Melandri, A. (1974)

In *Proceedings 3rd International Congress on Photosynthesis* (Avron, M., ed.) pp. 1147-1162, Elsevier, Amsterdam.

20. Fiolet, J.W.T., Bakker, E.P. and Van Dam, K. (1974) *Biochim. Biophys. Acta 368*, 432-445.

21. Sakurai, H., Nishimura, M. and Takamiya, A. (1965) *Plant Cell Physiol. 6*, 309-324.

22. Kraayenhof, K. (1969) *Biochim. Biophys. Acta 180*, 213-215.

23. Rosing, J. and Slater, E.C. (1972) *Biochim. Biophys. Acta 267*, 275-290.

V
Regulatory Functions
of Membranes

MEMBRANE STRUCTURE IN RELATION TO THE PRINCIPLE OF ENERGY TRANSDUCTION AND TO THE PRINCIPLE UNDERLYING THE CONTROL OF MEMBRANE FUNCTION

D.E. Green

Institute for Enzyme Research
University of Wisconsin
Madison, Wisconsin 53706

Membranes subserve a wide variety of functions — energy transductions, immune responses, electrical insulation, protein secretion, pinocytosis, propagation of electrical impulses, etc. This wide diversity of uses to which membranes can be put has encouraged the view that there can be no generalized module of structure for all membranes, and that each membrane has a unique structure appropriate to its specific function (1). But it must be remembered that all biological membranes have at least two functions in common — the capacity to transduce energy and the capacity to control the exercise of function. Some membranes, like myelin, may lose these capacities in their final differentiated form. Nonetheless, these capacities are present in such membranes prior to this differentiation. Given that biomembranes generally have two functionalities in common, it would necessarily follow that all membranes must share a common structural modality. Underlying all energy transductions is a principle of energy. The structure of biomembranes must be appropriate for the translation of this energy principle which will be the same regardless of which energy transduction is applicable. Thus, the structure of all biomembranes must have the common features that permit the translation of the energy principle. The same reasoning would apply to the control of the transductional functions of membranes. Whatever the principle of control, it must be compatible with, and complementary to the principle of energy and it must be compatible with the common structure of all membranes dictated by the energy principle. Thus, again we are led to the conclusion that incorporated in the structure of all biomembranes are the features which permit the translation of both the common energy and control principles. To explain the diversity of membrane function, we must then distinguish between the universal and the variable features of membrane structure. The universal features translate the energy and control principles; the variable features translate the functions which distinguish one membrane from another. But again, it follows that whatever the variable function, the structural translation of that variable function must be compatible with the structural features that translate the energy and control principles. In this way, diversity of membrane function is still compatible with invariant features of biomembrane structure.

It will be the thesis of my introductory remarks that the basic structural modality of the mitochondrial inner membrane, as well as the principle of energy transduction and the principle underlying the control of energy transduction, have finally been recognized. When these two principles have been defined and developed, then it should be possible to see how the structural modality of the mitochondrial inner membrane translates these two principles.

THE STRUCTURAL MODALITY OF THE MITOCHONDRIAL CRISTAL MEMBRANE

Under a variety of experimental conditions, Haworth and Green (2) have shown by electron microscopy that the mitochondrial cristal membrane can undergo tubularization. This involves the transition of a single flat continuous membrane sheet to a multiplicity of interconnected tubules (see Fig. 1). At high magnification, the segmented character of each tubular element is readily recognizable (Fig. 2). It is as if the membrane sheet developed tubular evaginations and these evaginations grow until essentially all the components of the membrane flow into these evaginations.

Each tubular element contains paired headpiece-stalk projections that extend the length of the tubule and show a precise periodicity, and also two paired ribbon continuum structures linked, one to another, via bilayer lipid (Fig. 3). The headpiece-stalks are projections from these ribbon structures which are predominantly protein in nature (3).

We have interpreted this tubularization of the inner membrane to mean that the components of this membrane can exist in either of two states – the two-dimensional state of the flat membrane sheet and the three-dimensional state of the tubular evaginations. The conditions used by Haworth and Green (2) strongly shift the equilibrium in the direction of the tubular state of the cristal membrane.

The critical point to be made is that tubularization is a process in which the ribbon structure of the inner membrane plays a critical role (Fig. 4). When the membrane-forming elements of the ribbons undergo a transition in geometry (we predict a shift to a more spherical geometry), the membrane is

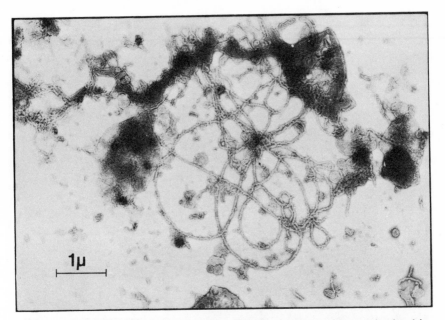

Fig. 1. Tubularization of the cristal membrane of frozen heavy beef heart mitochondria induced by a schedule of swelling and alternate warming and cooling in 10 mM Tris chloride (pH 8.0). The electron micrograph was prepared by R.A. Haworth. Negatively stained specimen.

no longer stable as a flat sheet and undergoes progressive tubularization. Thus, tubularization and a change in the curvature of the ribbon continua are two sides of the same coin. The coupling capability of the inner membrane is not compromised by the process of tubularization (2).

Lysolecithin has proved to be a powerful reagent for probing the ribbon structure of the cristal membrane (3-5); and from studies with this reagent, it can be deduced that the ribbons are composites of two tightly associated continua — the tripartite continuum and the electron transfer continuum. Lysolecithin can cleanly separate the electron transfer from the tripartite continuum (Fig. 5). But in addition to this lateral separation of the two continua, lysolecithin cleaves each of the two continua into smaller and smaller units (4,5). The tripartite continuum can be fragmented into units containing only four headpiece-stalk projections (4,5) and correspondingly, the electron transfer continuum appears to be cleaved to the stage of individual complexes (6). From various lines of evidence, we have deduced that the tripartite continuum contains the system required for coupled ATP synthesis and hydrolysis, active transport of cations, and translocation of metabolites (7). The electron transfer continuum contains the four complexes of the electron transfer chain.

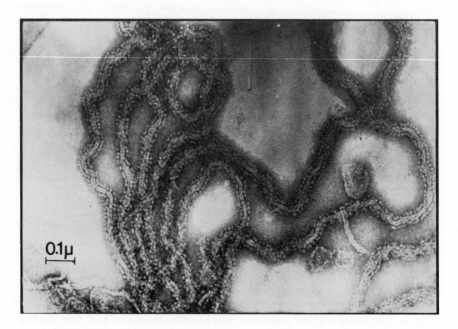

Fig. 2. High magnification view of the tubules generated by the procedure described in the legend for Fig. 1. The highly regular arrays of headpiece-stalk projections are clearly visible. The electron micrograph was prepared by R.A. Haworth. Negatively stained specimen.

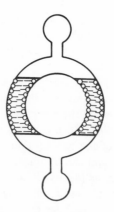

Fig. 3. Diagrammatic representation of the arrangement of the component elements that make up tubular structure. Basically, two ribbons are linked, one to another, by bilayer phospholipid.

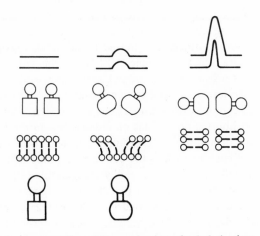

Fig. 4. Diagrammatic representation of the mechanism of tubularization.

Each of the two continua is capable of generating membranes *de novo* when supplemented with phospholipid (8,9).

It could be argued that the ribbon structure of the inner membrane may not be pre-existing in the intact membrane, but may be generated by the re-agents used to demonstrate such structures. Is there direct evidence for rib-bon structure in intact mitochondria that does not involve the use of reagents required for negative staining? If we consider tubularization of the inner

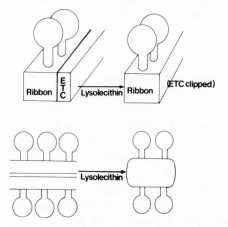

Fig. 5. Fragmentation by lysolecithin of the basic ribbon into two continua – the tri-partite continuum and the electron transfer chain. In addition to this lateral separation, lysolecithin cleaves each of the continua into their lowest common denominator.

membrane as a manifestation of a change in geometry of ribbon structure, then indeed one can point to the tubularized state of certain mitochondria (adrenal cortex, amoeba) as *prima facie* evidence of pre-existing ribbon structure. We and several others have indeed found indications of ribbon structure visualizable in the intact membrane (10-12). In the orthodox configuration, the cristae of beef heart mitochondria clearly show evidence of the ribbon continua as manifested by the extreme regularity and spacing of the head-piece-stalk projections. These projections are uniformly separated by less than 10 Å from neighboring projections on either side. In the fused state of mito-chondria in the orthodox configuration − a state in which the base-pieces are tightly apposed back to back − the continuum nature of the ribbons carrying the headpiece-stalk projections is readily recognizable (12). However, there is no evidence to support the notion that there is comparable ordered ribbon structure in the cristae of mitochondria in the aggregated configuration. Lo-cal regions of order have indeed been observed by several investigators, but the direction and extent of this order is highly local. Our present interpreta-tion is that ribbons can fluctuate from large to small domains and that there is an equilibrium governing the size of ribbon structures. In the aggregated configuration, the equilibrium may favor ribbon structures of limited size and a certain degree of randomness in the orientation of the ribbons. In the or-thodox configuration, the equilibrium appears to favor highly regular and ex-tended ribbon structures.

When mitochondria in the aggregated configuration are exposed to reagents such as silicotungstate (13), the inner membrane undergoes blanket tubulariza-tion, but the headpiece-stalk projections in such tubules are directed inwards and not outwards, as in the tubules generated from swollen mitochondria. The internal space of the silicotungstate-induced tubules is therefore the matrix space. Again, if we consider tubularization as a property derivative from ribbon structure, we may consider the silicotungstate tubularization of mitochondria in the aggregated configuration as evidence for ribbon structure.

The mode of tubularization, i.e. whether enclosing or exposing the head-piece-stalk projections, is determined by the volume of the matrix space. When the matrix space is maximally expanded, as in the orthodox configura-tion, the internal space of the tubules is the matrix space; conversely, when the matrix space is maximally contracted, as in the aggregated configuration, the internal space of the tubules is the intracristal space.

Another parameter of ribbon structure is the phenomenon of membrane fusion and defusion. Lysolecithin generates from mitochondria and submito-chondrial particles paired ribbons which are fused back to back (4,5), i.e. no space separates the two ribbon continua (Fig. 6). The headpiece-stalk projec-tions of the two ribbons extend in opposite directions − a token that the paired ribbons were derived from apposed membranes. Since such fused rib-

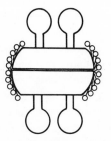

Fig. 6. The ultimate unit of lysolecithin cleavage of submitochondrial particles is a particle with no internal space containing two headpiece-stalk projections on either side of the particle. The lipid in such a fused particle has to be a monolayer.

bons contain mixtures of lysolecithin and phospholipid in addition to protein, it would necessarily follow that lipid would cover the hydrophobic flanks of the fused ribbons and that this lipid could only be in the form of a mono-layer. It is the disruption of the bilayer modality which endows fused membranes with such unusual properties (14)(fused membranes, for example, are readily perforated; moreover, the ends of two fused membranes can readily be spliced to form one continuous membrane). The capacity for undergoing the fusion-defusion cycle is a general property of all membranes (15) and there are now compelling reasons for relating this cycle to the ribbon structure of membranes.

Just as there are two modes of tubularization, so there are two modes of fusion — fusion between two membranes separated by the matrix space and fusion between two membranes separated by the intracristal space (15). It is intriguing that Mg^{++}, but not Ca^{++} can induce the first mode of fusion, whereas both Mg^{++} and Ca^{++} can induce the second kind of fusion. This is readily understandable, since Mg^{++} is known to induce contraction of the headpiece-stalks into the ribbon continuum — a precondition for this mode of fusion — whereas Ca^{++} induces the extension of the headpiece-stalk projections and thereby prevents this mode of fusion.

In summary, then, we may conclude that the structural modality of the cristal membrane is a ribbon continuum. The proteins, like the phospholipids, form continua and there is thus alternation of protein and lipid continua. The protein ribbon is in fact a composite of two continua — the tripartite and electron transfer chain continua, respectively. How this structural modality translates the transductional and control principles is a question to which we will return in a later section.

THE PRINCIPLE OF ENERGY COUPLING

The essence of the energy principle is that a negatively charged ion in a transmembrane complex moves energetically downhill and this movement is coupled to the complementary uphill movement of a positively charged ion in a tightly apposed transmembrane complex (16). The model which incorporates this principle of energy coupling is known as the paired moving charge (PMC) model. The ground rules for energy coupling according to this model (8) are summarized in the diagrams shown in Fig. 7. The initial event in energy coupling is paired charge separation followed by paired charge movement and paired charge elimination. The charges move within tracks that traverse the membrane. The track could be an electron transfer chain in the case of the electron, or the inner core of a "track" protein in the case of an ionophorous moving charge. The distance separating the two moving charges lies between 20 and 30 Å and by virtue of the constraints of the track, this distance is maintained constant during the motion of the paired charges.

The two moving ions are coupled by virtue of the principle that energy is

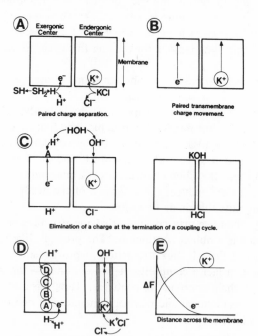

Fig. 7. The ground rules for energy coupling.

minimized when the two ions move in synchrony and in parallel. Since the negatively charged species is moving down its electrochemical gradient, there is in effect a transfer of free energy from one moving ion to the other (17). Note that coupling takes place under ground state conditions. The energy transfer reflects the accomodation of the protein system to these charge movements. The system adjusts to achieve energy minimization and this adjustment compels the positive ion to move in synchrony with the negatively charged species.

There are two molecular devices for charge separation — electron transfer chain (16) and the ionophore (18,19). Fig. 8 provides a diagrammatic representation of how the electron is separated from its proton in Complex I of the electron transfer chain. The flavin-iron sulfur link is the point in the chain where the charge separation takes place. The electron is sucked into the iron sulfur center, whereas the proton is ejected into the aqueous medium. Note that the reduction of flavin by the substrate (NADH) achieves a partial intramolecular separation of electron and proton which is preparatory to the *de facto* charge separation that takes place at the next transfer step. At the end of the coupling cycle, the charge is eliminated from the system. The electron is transferred to the terminal acceptor (ubiquinone) and this transfer is synchronized with the uptake of a H^+ from the aqueous medium on the side of the membrane opposite to the side where the coupling cycle was initiated. Thus, the electron is separated from its proton at one side of the membrane where dehydrogenation takes place and reunited with a proton (not the same one) at the other side of the membrane where hydrogenation

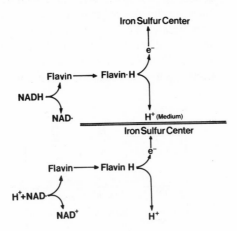

Fig. 8. Separation of the electron from its proton in Complex I of the electron transfer chain.

takes place.

Ionophores are the molecular instruments of charge separation in all coupling complexes other than those of the electron transfer chain (19). Ionophores can separate cations from anions or anions from cations (Fig. 9). We must distinguish between short-range charge separation of a single pair of charges, which excludes the possibility of energy coupling, and long-range charge separation of two pairs of charges which is a *sine qua non* for energy coupling.

The activation of anions, such as inorganic phosphate and ADP, by encapsulation of these anions within ionophores is an entirely new dimension of the catalytic role of ionophores. This activation requires both a protein and a divalent metal in addition to the ionophore (7). Composite ionophoric species such as $(\underline{Mg\text{-}O\text{-}Pi})^+$ and $(\underline{Mg\text{-}O\text{-}Pi})^-$ can be generated — species which play a key role in oxidative phosphorylation and coupled ATP hydrolysis, respectively. A neutral ionophore is required to generate $(\underline{Mg\text{-}O\text{-}Pi})^+$ and an acidic ionophore to generate $(\underline{Mg\text{-}O\text{-}Pi})^-$.

It is possible to analyze coupling in terms of unit processes in each of the two complexes participating in coupling. These processes would be charge separation, charge movement, and charge elimination (7). The components which carry out these unit processes in an electron transfer complex, in a cation-transporting complex, and in ATP synthetase are listed in Table I. Whenever moving charges are ionophore-metal ion complexes, a special set of components comes into play — ionophoroproteins, ionophoroenzymes, track proteins and charge-eliminating proteins. The ionophoroproteins are the charge-separating devices in endergonic complexes such as ATP synthetase, and the ionophoroenzymes are the charge-separating devices in exergonic complexes such as ATP hydrolase. The ionophore-metal ion complex, once generated, moves within internal tracks in specialized transmembrane proteins known as track proteins (7). These track proteins have a hollow core, probably filled with phospholipid; the ionophore-metal ion complex traverses the

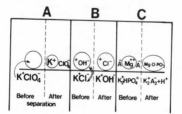

Fig. 9. **Charge separation by ionophores: A, cation from anion; B, monovalent anion from cation; C, polyvalent anion from cation.**

TABLE I

Extraction of Components of Ionophore-Dependent Coupling Complexes

Ionophores	Butanol-acetic acid-H_2O ± tryptic digestion
Ionophoroproteins	Chloroform-methanol
Ionophoroenzymes	
Track proteins	Chloroform-methanol
Cation-binding proteins	Chloroform-methanol

The roster of components in Complex III (an electron transfer complex), ATP hydrolase, ATP synthetase and cation-transporting complexes which implement the unit processes in paired charge movements.

membrane via this internal passageway. The ionophore-metal ion complex is discharged at the other end of the membrane by interaction with binding groups in a specific protein referred to as the charge-elimination protein (7). In both the charge-separating site and the charge-elimination site, there are channels in the protein through which ions can selectively enter or leave. These channels provide the link between the ionophores and the aqueous phases on both sides of the inner membrane. Thus, the ionophores are sealed within the internal passageways in each complex and communicate with the aqueous phases only via these channels in the proteins at either side of the membrane.

The integrated picture of ion flow within complexes during energy coupling is represented diagrammatically in Fig. 10. We have already considered each of the unit processes in energy coupling which is illustrated in this fig., but there are some additional details that merit some consideration. In endergonic complexes, the ionophore is tightly associated with the activating ionophoroprotein prior to charge separation. As soon as charge separation takes place and the ionophore becomes charged by uptake of a divalent metal, then the charged ionophore is ejected from the activating protein into the track and is then driven by the driving ion (electron or $(Mg\text{-}O\text{-}Pi)^-$) through the track to the charge-elimination site. After discharge of its charge, the moving ion returns to its mooring site in the activating protein.

In the ATP synthesizing complex, the discharge of the uncompensated moving ion is followed by the uptake of inorganic phosphate to generate a neutral ionophoric species. Thus, inorganic phosphate is encapsulated on the matrix side of the inner membrane and charge-separated on the intracristal

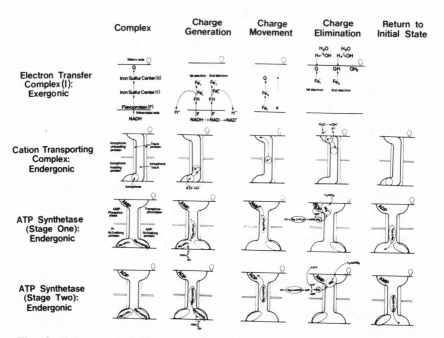

Fig. 10.   Unit processes in three of the coupling systems – an electron transfer complex, a cation-transporting complex and ATP synthetase.

side. This corresponds to the experimental evidence that during ATP synthesis, inorganic phosphate has to penetrate the inner membrane and enters the membrane from the matrix side. The same considerations which hold for Pi apply as well to ADP.

The detailed molecular mechanisms of coupled ATP synthesis and ATP hydrolysis, respectively, are depicted in Fig. 11. Note that the charge-elimination site in the synthetase is bound AMP (20), whereas bound AMP is part of the charge-separation site in the hydrolase. The charge-separating site in the synthetase comprises two ionophoroproteins – one for activation of Pi (21) and the other for activation of ADP (22). The corresponding site in the hydrolase is an ionophoroenzyme which in effect splits ATP into Pi^+ and ADP^- (Pi^+ reacts with bound AMP to generate bound ADP).

The critical question which we posed earlier, namely, how the structure of the membrane accomodates the principle of energy, can now be answered. The principle of energy requires tight apposition of vectorially aligned transmembrane complexes. The ribbon structure of the inner membrane ensures this tight apposition and close proximity. Consider how the electron transfer continuum is tightly bonded to the tripartite continuum. This requirement

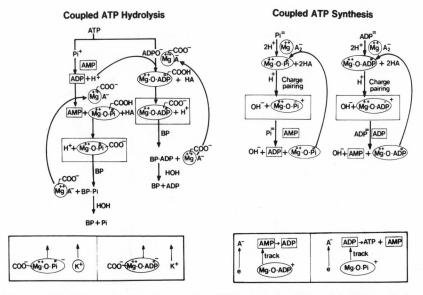

Fig. 11. Molecular mechanisms for coupling ATP hydrolysis and coupled ATP synthesis.

of tight apposition can be satisfied by a ribbon of very limited size. Why, then, does the ribbon have to be a continuum? This particular feature of membrane structure is highly relevant to the principle of control and we must first define that principle before the logic of extended ribbon structure can be rationalized satisfactorily.

THE CONTROL OF COUPLING FUNCTION

It is a remarkable fact that until very recently, there was no awareness that the mitochondrial system contained an elaborate control mechanism which regulated the coupling options. The studies on the effect of mercurials and heavy metals on mitochondrial coupling provided the first clear evidence of such a control mechanism (23,24). In the presence of mercurials such as fluorescein mercuric acetate (FMA), mitochondria lose the capacity for oxidative phosphorylation and active transport of Ca^{++} (high affinity) and acquire the capacity for active transport of K^+ and Mg^{++} (see Fig. 12). When the orthodox configuration is induced by active transport of Ca^{++} ions, a similar alteration in coupling pattern has been described (25). We may deduce from this correspondence that mercuric ions and Ca^{++} induce comparable changes in the coupling pattern of mitochondria.

During the aggregated-to-orthodox configurational transition induced by Ca^{++}, the headpiece-stalk sectors that were contracted into the membrane

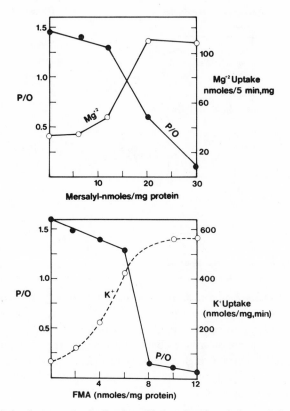

Fig. 12. Inverse relation between the decline in oxidative phosphorylation and the emergence of the capacity for active transport of K⁺ and Mg⁺⁺ (23,24).

are ejected and now project away from the membrane (26). The change in coupling pattern can be correlated perfectly with this extension or contraction of the headpiece-stalk sectors. Mercurials mimic the action of Ca^{++}, but the configuration induced by mercurials is a variant of the orthodox (27). Undoubtedly, the headpiece-stalks are ejected from the membrane, but the stalk is not fully extended. We interpret this parallelism between the change in coupling pattern and the change in the position of the headpiece-stalk sector in terms of the capacity for oxidative phosphorylation residing in the headpiece-stalk sector. When the headpiece-stalk sector is in the membrane, the capacity for oxidative phosphorylation is manifested. When the headpiece-stalk sector is extended from the membrane, this capacity is no longer manifested and capacities previously latent now become exposed.

Hunter and his colleagues have studied in depth the mechanism by which Ca^{++} triggers the change in coupling mode (28). Both Ca^{++} and Pi are absolute requirements for this trigger action. In fact, there is demonstrably a 1:1 stoichiometry of Ca^{++} and Pi. Arsenate can replace Pi at considerably lower concentrations. In oxidative phosphorylation, the reactive charge-separated species are respectively Mg-O-Pi and Mg-O-ADP, encapsulated within a neutral ionophore (7). These species with a net positive charge of 1 react successively with bound AMP and ADP to generate ATP and bound AMP. But if Mg^{++} is replaced by Ca^{++}, the active species (Ca-O-Pi) also with a net positive charge of 1, cannot react with bound AMP. Instead, it reacts with binding groups in a charge-eliminating protein – the protein which plays the charge-elimination role in the active transport of both Ca^{++} and Mg^{++}. This interaction of Ca-O-Pi with the charge-elimination protein is assumed to be the trigger action that sets off the changes that lead to the extension of the headpiece-stalk and the change in configuration of the inner membrane (Fig. 13).

Mitochondria in the aggregated configuration can carry out both oxidative phosphorylation and active transport of Mg^{++} and Pi (29). In fact, there is a competition between these two coupled processes – the competition being greatly in favor of oxidative phosphorylation. Since the same active ionophoric species (Mg-O-Pi) is involved in both active transport of Mg^{++} and oxidative phosphorylation, what controls the coupling options? The control is exercised by ADP. In the presence of ADP, oxidative phosphorylation proceeds preferentially; in the absence of ADP, only active transport can take place. All this suggests that the active ionophoric species can react either with bound AMP to generate ATP, or with the charge-elimination protein to generate a Mg-Pi complex of the binding groups. In the first option, only Pi is unloaded; in the second option, both Mg and Pi are unloaded. ADP favors the first option; Ca^{++}, even in the presence of ADP, favors the second option.

We may consider that the charge-elimination protein can exist in either of

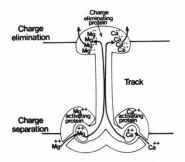

Control center: Active transport complex.

Fig. 13. The component elements in the control center. These are, respectively, ionophoroproteins for transport of Mg^{++} and Ca^{++}, the track proteins and the charge-elimination protein. The two ionophoroproteins can take up both divalent metal and inorganic phosphate.

two states – a Mg^{++} state which is compatible with oxidative phosphorylation and a Ca^{++} state which is incompatible with oxidative phosphorylation (Fig. 14). The headpiece-stalk sector is contracted into the membrane when the charge-elimination protein is in the Mg^{++} state and extended from the membrane when the charge-elimination protein is in the Ca^{++} state. Presumably, the stalk is imbedded within the charge-elimination protein and undergoes a change in state coincident with the transition from the Mg^{++} to the Ca^{++} state of the charge-elimination protein.

This is as far as the available knowledge permits us to analyze the causative events in the Ca^{++}-induced change in coupling pattern. But enough is already known to enable us to recognize the principles which underlie the control mechanism. The systems which are the determinants of the switch in coupling pattern are intrinsic to the ribbon continuum. In essence, active transport of Ca^{++} and Pi initiates a change in state of a key protein – the charge-eliminating protein. In turn, this change in state of the charge-eliminating protein induces the extension of the headpiece-stalk sector away from the membrane. The ribbon thus assumes a new configurational state and a new coupling pattern is established. The reversal of this configurational transition requires active transport of Mg^{++} + Pi. The composite ion interacts with binding groups in the same charge-elimination protein and thus induces a state of the protein which compels the contraction of the headpiece-stalk sector into the membrane and a change in configuration of the ribbon continuum.

In the operation of the control mechanism, we have dependence on the principle of energy coupling. The triggering ions are charge-separated ions encapsulated within ionophores. The triggering ions, after traversing the track protein, react with a protein concerned with charge elimination. Discharge of the triggering ions in effect alters the conformation of the charge-elimination protein and thereby induces a set of changes which eventually compels a configurational change in the ribbon and a change in the coupling pattern of the

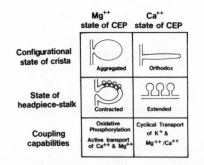

Fig. 14. Correlation between the configurational state of the cristae, the state of the headpiece-stalk, and the coupling capabilities of the mitochondrion.

coupling systems.

Whatever the uncertainties as to the details of the operation of the mitochondrial control mechanism, the basic strategy of the mechanism is readily apparent. Control is exerted not on individual proteins, but rather on proteins which are part of a continuum. By virtue of the cooperative character of ribbon structure, it is the ribbon as a unit which responds to the trigger action of the control mechanism. The mitochondrion behaves as if it were one giant protein molecule that undergoes shape changes dictated by the ion fluxes in the membrane. The continuum nature of the mitochondrial inner membrane is therefore fundamental to the tactic of the control of energy coupling.

SUMMARIZING COMMENT

It has been our working hypothesis that the structure of the membrane translates both the principle of energy and the principle of the control of the coupling options. The ribbon structure of the membrane provides the tight apposition of complexes and the vectorial alignment of complexes essential for energy coupling. Moreover, it provides the cooperative feature which underlies the control strategy. How general is the ribbon structure, the energy principle and the principle of control? Are these characteristic of a few specialized membranes like mitochondria, or do they apply across the board to all membranes? It is our prediction that these are universal attributes of all membranes. To be specific, we are predicting that all membranes have ribbon structure; that energy coupling invariantly involves paired charge separation, paired charge movement and paired charge elimination; and finally, that control is exercised at the level of ribbon continua and not of individual proteins and enzymes.

REFERENCES

1. Bretscher, M.S. (1974) In *Perspectives in Membrane Biology* (Estrado-O, S. and Gitler, S., eds.) pp. 3-24, Academic Press, New York.
2. Haworth, R.A. and Green, D.E. Manuscript in preparation.
3. Sadler, M.H., Hunter, D.R. and Haworth, R.A. (1974) *Biochem. Biophys. Res. Commun. 59*, 804-812.
4. Komai, H., Hunter, D.R. and Takahashi, Y. (1973) *Biochem. Biophys. Res. Commun. 53*, 82-89.
5. Hunter, D.R., Komai, H. and Haworth, R.A. (1974) *Biochem. Biophys. Res. Commun. 56*, 647-653.
6. Hatefi, Y. (1974) *Ann. N.Y. Acad. Sci. 227*, 504-520.

7.   Green, D.E. and Blondin, G.A. (in press) *Chem. Eng. News, Supplement.*
8.   Tzagoloff, A., MacLennan, D.H., McConnell, D.G. and Green, D.E. (1967) *J. Biol. Chem. 242*, 2051-2061.
9.   Haworth, R.A. and Komai, H. Unpublished studies.
10. R.A. Haworth. Unpublished studies.
11. Sjostrand, F. (1968) In *Regulatory Functions of Biological Membranes 11* (Jarnefelt, J., ed.) pp. 1-20, BBA Library.
12. Crane, F.L., Stiles, J.W., Prezbindowski, K.S., Rujecka, F.J. and Sun, F.F. (1968) In *Regulatory Functions of Biological Membranes 11* (Jarnefelt, J., ed.) pp. 21-56, BBA Library.
13. Smoly, J. and Wakabayashi, T. Unpublished studies.
14. Green, D.E. (1972) *Ann. N.Y. Acad. Sci. 195*, 150-172.
15. Wakabayashi, T. and Green, D.E. Unpublished studies.
16. Green, D.E. and Reible, S. (1974) *Proc. Nat. Acad. Sci. (U.S.A.) 71*, 4850-4854.
17. Kemeny, G. (1974) *Proc. Nat. Acad. Sci. (U.S.A.) 71*, 3064-3067.
18. Green, D.E. and Reible, S. (1975) *Proc. Nat. Acad. Sci. (U.S.A.) 72*, 253-257.
19. Green, D.E., Blondin, G., Kessler, R. and Southard, J.H. (1975) *Proc. Nat. Acad. Sci. (U.S.A.) 72*, 896-900.
20. Roy, H. and Moudrianakis, E. (1971) *Proc. Nat. Acad. Sci. (U.S.A.) 68*, 464-468.
21. Kessler, R. Unpublished studies.
22. Tyson, C. and VandeZande, H. Unpublished studies.
23. Southard, J., Nitisewojo, P. and Green, D.E. (1974) *Fed. Proc. 33*, 2147-2153.
24. Southard, J.H. and Green, D.E. (1974) *Biochem. Biophys. Res. Commun. 61*, 1310-1316.
25. Southard, J.H. Unpublished studies.
26. Fernandez-Moran, H., Oda, T., Blair, P.V. and Green, D.E. (1964) *J. Cell Biol. 22*, 63-100.
27. Lee, M.J., Harris, R.A., Wakabayashi, T. and Green, D.E. (1971) *Bioenergetics 2*, 13-31.
28. Hunter, D.R., Haworth, R.A. and Southard, J.H. Unpublished studies.
29. Brierley, G.P., Murer, E., Bachmann, E. and Green, D.E. (1963) *J. Biol. Chem. 238*, 3482-3489.

ALDOSTERONE–MEDIATED CHANGES IN LIPID METABOLISM IN RELATIONSHIP TO Na+ TRANSPORT IN THE AMPHIBIAN URINARY BLADDER

Howard Rasmussen, David B.P. Goodman and Eric Lien

Departments of Biochemistry and Pediatrics
University of Pennsylvania
Philadelphia, Pennsylvania 19174

INTRODUCTION

A unique feature of naturally-occurring phospholipids is the asymmetric distribution of their constituent fatty acids. Long-chain polyunsaturated fatty acids are found predominantly in the 2- or β- position of the α-glycerol phosphate backbone and these fatty acids turn over more rapidly than those in the 1- or α- position of the phospholipids or than the glycerol backbone. The importance of this structural and metabolic uniqueness is not known. However, a question of great current interest, which is discussed in this symposium (1), concerns the possible relationship of membrane lipid composition to the distribution and function of membrane proteins. This question may be directly related to membrane fatty acid composition, since the fatty acid composition of membrane phospholipids is a major determinant of membrane fluidity.

Much evidence has been adduced in support of the concept that a change in fluidity can alter the function of specific membrane proteins (2-9). Much of this work has relied upon studying the changes in fluidity and membrane protein activity as a function of changes in temperature.

However, more recent studies (9-10) have been concerned with considering the effects of changing membrane lipid composition upon altered cell function. Some of the results of these studies imply that changes in membrane fatty acid composition may have specific effects upon protein function in addition to any gross effects they may exert upon membrane fluidity. Nonetheless, few studies have approached the problem from a physiologic point of view in the sense of asking whether change in physiologic state can lead to a change in membrane function which can be attributed to an altered membrane lipid environment.

This report deals with studies (11-16) concerned with the question of whether certain hormones may alter membrane function by altering the fatty acid composition of membrane phospholipids. Such a potential might explain the evolutionary value of the rapid turnover of the fatty acids in membrane phospholipids. Such a potential might explain the evolutionary value of the rapid turnover of the fatty acids in membrane phospholipids. This turnover may serve another function also brought to focus by our studies and those of others (17-19): this latter function being the ability to replace polyunsaturated fatty acids after they have undergone oxidation by atmospheric oxygen, thereby allowing the membrane to maintain a proper lipid environment for its functional proteins.

THE SYSTEM UNDER STUDY

Before describing our studies, it is necessary to review the basic biology of the system under study: the isolated amphibian urinary bladder (20-24). This tissue is capable of sustained net transcellular (mucosal to serosal) Na^+ transport *in vitro* which can be monitored by measuring the short-circuit current by the technique of Ussing and Zerahn (25). Two hormones alter tissue function: the peptide hormone, arginine vasotocin, and its homologs, and the steroid hormone, aldosterone. Addition of vasotocin, oxytocin, or vasopressin leads to an immediate increase in transcellular Na^+ transport, an increase in bulk water flow during an imposed osmotic gradient (27). These effects occur within 1-3 min after hormone addition, are not altered by inhibitors of protein and RNA synthesis, and are thought to be mediated in part by changes in cAMP (27) and Ca^{2+} within the cell.

In contrast, aldosterone has no direct effect upon H_2O permeability or H_2O flow although it increases the magnitude of the effects of vasopressin on both Na^+ transport and bulk water flow (28,29). It also increases the sensitivity of this tissue to inhibition by ouabain (28,29). On the other hand, aldosterone stimulates Na^+ transport (Fig. 1) after a lag of 60-90 min. Its effect is blocked by inhibitors of protein and RNA synthesis (30,31). Based on these facts, the reported nuclear localization of ^{3}H-labeled aldosterone (30) and the model of steroid hormone action in other tissues (32), it has been proposed that this steroid hormone acts by turning on one or more specific genes, which direct the synthesis of specific mRNA's and thus specific proteins involved in the Na^+ transport process (22).

In spite of the different modes of action of the two hormones on Na^+ transport, the effects of both on Na^+ transport are inhibited either by the mucosal addition of amiloride (33) or the serosal addition of ouabain, implying that the pathway of Na^+ transport influenced by the two hormones is the same but that they act in different ways on this pathway.

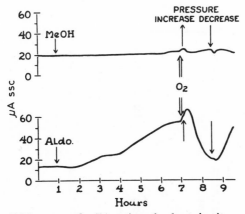

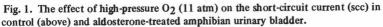

Fig. 1. The effect of high-pressure O_2 (11 atm) on the short-circuit current (scc) in control (above) and aldosterone-treated amphibian urinary bladder.

RESPONSE TO HYPERBARIC OXYGEN

Our initial studies dealt with the biochemical basis of oxygen poisoning. We selected the toad bladder for these studies because of its physiologic and biochemical properties, its ready availability, and because it was possible to monitor continuously a physiologic function, i.e., transcellular Na$^+$ transport by the Ussing technique. When bladders were exposed to high pressures of oxygen (HPO), even 6-10 atmospheres, there was little change in rate of trans-cellular Na$^+$ transport (Fig. 1). However, when bladders were pretreated with aldosterone and then exposed to HPO, a dramatic effect was observed (Fig. 1). After a short initial rise in short-circuit current, there was a rapid and profound inhibition. If the bladder was then returned to room air, short-curcuit current rapidly returned to high values, and often overshot in the sense of going higher than it had been before HPO exposure.

Bladders pretreated with vasopressin did not show this behavior, thus ruling out the trivial explanation that simply an increase in the rate of Na$^+$ transport led to this marked change in sensitivity to HPO.

Using the pattern of change in Na$^+$ transport after HPO exposure of an aldosterone-treated bladder as a guide, the changes in energy metabolism were measured before HPO exposure and 60-75 min after exposure (Fig. 2). At a time when Na$^+$ transport was reduced by 60%, the tissue ATP concentration was higher than normal. There was a significant increase in the rates of both pyruvate and glucose decarboxylation, and particularly of [1-^{14}C] glucose de-carboxylation, and the NADP/NADPH ratio (Fig. 2) and NAD/NADH (not shown) ratios were increased. These results indicated that the cause of the decrease in Na$^+$ transport was not an inhibition of energy metabolism or a lack

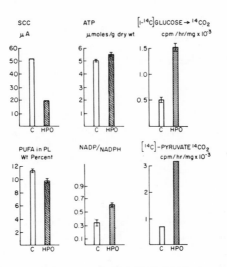

Fig. 2. The metabolic effects of HPO upon the aldosterone-treated amphibian urinary bladder. The control values are from tissues treated with aldosterone for 7.5 hr; the experimental from similar tissues treated with aldosterone for 6 hr and then exposure to 11 atm of O_2 for 90 min.

of ATP for Na^+ pump activity of oxygen action.

From these data and additional studies (13), a sequence of events was postulated: increased O_2 pressure → increased membrane phospholipid peroxidation → increased utilization of reducing equivalents → increased NADP/NADPH and NAD/NADH ratios → increased glycolytic, hexose monophosphate shunt, and citric acid cycle flux. Whether the cell became 'poisoned' by O_2 depended upon a number of factors, particularly the ability to generate reducing equivalents and the content of oxygen-sensitive (presumably lipid) components of the cell membrane.

This model predicted that following exposure of an aldosterone-treated bladder to HPO, there would be a fall in the weight percentage of long-chain polyunsaturated fatty acids in membrane phospholipids and a rise in their concentration upon reversal of Na^+ transport inhibition, i.e., the return of the poisoned bladder to room air. This prediction was verified (Fig. 2).

ALDOSTERONE: EFFECTS UPON LIPID METABOLISM

The dramatic interaction of aldosterone and HPO in altering the function of this tissue led us to reexamine the then-current models of the mode of action of this hormone. Our data raised the distinct possibility that changes in lipid metabolism and membrane lipid turnover might be involved in aldosterone action.

Within 20-30 min after aldosterone addition there is an increased incorporation of label from [U-^{14}C] glucose, [2-^{14}C] pyruvate and [1-^{14}C] oleate, palmitate, acetate and stearate into total lipid and into phospholipid. The hormone does not preferentially increase the labeling of any particular phospholipid class, but it does cause preferential changes in the specific activities of phospholipid fatty acids when acetate served as substrate (Fig. 3). There was also a specific marked effect upon incorporation of label from [1-^{14}C] oleate into all other lipid classes. Aldosterone treatment also increased the rate of decarboxylation of linolenate but not other labeled fatty acids in the first 30 min, and it stimulated the decarboxylation of [1-^{14}C] glucose. In addition, when the hormone was added to bladders in which the phospholipid fatty acids had been labeled by preincubation with [^{14}C] acetate the rate of removal of labeled fatty acids from the phospholipids was increased, i.e., aldosterone stimulated the deacylation of membrane phospholipids. It is noteworthy that all of these effects upon lipid turnover were seen within 20-30 min, long before there was any change in aldosterone-induced Na⁺ transport.

When lipid metabolism was studied at progressively longer times after aldosterone addition, there were progressive changes.

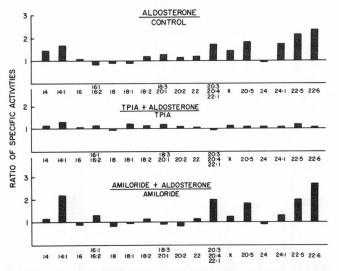

Fig. 3. The relative specific radioactivities of phospholipid fatty acids obtained from amphibian urinary bladders which had been pulsed for 30 min with [^{14}C] acetate. Top: aldosterone vs control tissue. Middle: both control and experimental tissues were preincubated with 2-methyl-2-[p-(1,2,3,4-tetrahydro-1-napthyl)-phenoxy] -propionic acid (TPIA) and before the experimental were exposed to aldosterone and then both control and experimental pulse-labeled with [^{14}C] acetate. Bottom: aldosterone vs control in amiloride-pretreated tissues.

For example, if after 4 hr of exposure to aldosterone the bladders were exposed to labeled [^{14}C] oleate for 30 min, there was a marked stimulation of label into a large number of fatty acids. However, if [9-10-^{3}H] oleate was used as precursor, there was a less general effect, but a marked effect upon conversion of oleate to 18:2; 18:3; 20:1; and 22:6 as well as an inhibition into 20:2 and 22:0.

When bladders were preincubated with [2-^{14}C] pyruvate, and relative specific activities of labeled acids in aldosterone-treated vs control tissues were examined at 90 and 360 min after hormone addition, there were dramatic changes. There was a marked increase in the labeling of 16:1; 16:2; 18; and 18:1 at 90 min. At 360 min, a time when the maximal rise in Na$^+$ transport was seen, there was a marked increase in the specific activities of 18:1; 20:2; 20:4; 22:1; 24:1; 22:5ω6; and 22:5ω3; 22:6 with lesser changes in 16:1; 16:2; 18; 18:2; 20; and 18:3 with a fall of 20:5.

These data suggested that aldosterone caused an increase in the incorporation of specific long-chain polyunsaturated fatty acids into membrane phospholipids. This conclusion was substantiated by measurements of the weight percentage of fatty acid in membrane phospholipids in control and aldosterone-treated tissues. Aldosterone treatment caused a fall in the weight percentage of 16 and an increase in 24:1; 22:5ω6; 22:5ω3; and 22:6 (Fig. 4). Thus, the long-term-labeling studies and the measurement of weight percentage of fatty acids both indicated a specific effect of aldosterone upon the fatty acid composition of membrane phospholipids. Furthermore, studies of phospholipid preparations obtained from bladders incubated with [2-^{14}C] pyruvate with or without aldosterone showed that phospholipase A caused a much greater release of labeled fatty acid from the aldosterone-treated tissue than from control tissue.

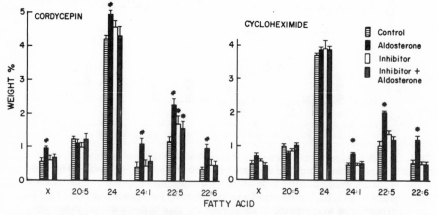

Fig. 4. The effect of inhibitors of RNA and protein synthesis on aldosterone-induced changes in phospholipid fatty acids.

These results indicate that aldosterone stimulated an increase in the acylation and deacylation of fatty acids in the 2- position of membrane fatty acids and an alteration in the ratio of different fatty acids in this position: specifically, an increase in the content of a number of long-chain polyunsaturated fatty acids. These changes in membrane fatty acid composition could account for the fact that aldosterone sensitizes this tissue to the toxic effects of O_2 (Fig. 1); that aldosterone sensitizes this tissue to the inhibitory effects of Ouabain; and that aldosterone pretreatment enhances the effects of vasopressin upon Na^+ transport and bulk water flow in this tissue. However, the most important question is whether or not these changes in membrane lipid structure can account for the ability of aldosterone to stimulate transcellular Na^+ transport.

We considered three possible explanations for the relationship between the changes in Na^+ transport and the changes in membrane lipid structure: (1) the change in lipid structure was a consequence rather than a determinant of the changes in Na^+ transport; (2) the primary effect of aldosterone was to control the synthesis of one or more species of new membrane proteins which, when incorporated into the membrane, required and determined a change in membrane lipid structure; or (3) the effect of the hormone upon lipid structure was the primary event which, by altering the catalytic activity of the membrane proteins, enhanced transcellular Na^+ transport.

EFFECTS OF TRANSPORT AND METABOLIC INHIBITORS

In order to test these alternative hypotheses, we have examined the effect of various metabolic and Na^+ transport inhibitors upon the metabolic and physiologic aspects of aldosterone action. The inhibitors examined were: (1) cordycepin, an inhibitor of mRNA synthesis (34); (2) cycloheximide, an inhibitor of protein synthesis (35); (3) amiloride, an inhibitor of the mucosal entry of Na^+ (33); and (4) 2-methyl-2-[p-(1,2,3,4-tetrahydro-1-napthyl)-phenoxy]-propionic acid (TPIA), a supposed inhibitor of acetyl-CoA carboxylase (36,37).

Previous studies by others had shown that both cordycepin and cycloheximide blocked the effects of aldosterone on Na^+ transport without blocking the effects of vasopressin on either Na^+ transport or H_2O flow (30,31,35). Pretreatment of the tissues for 2 hr with cordycepin blocked the ability of aldosterone to stimulate acetate incorporation into lipid; blocked the ability of aldosterone to increase the weight percentage of long-chain polyunsaturated fatty acids in the membrane phospholipids (Fig. 4) and altered the ability of aldosterone to cause higher specific activities of many long-chain polyunsaturated fatty acids (16). Cordycepin by itself produced some changes in labeling pattern after [^{14}C] acetate incubation but did not change the composition of membrane phospholipid fatty acids as measured by weight percentage. Cyclo-heximide also blocked the effect of aldosterone on incorporation of labeled precursor into phospholipid acids and on the fatty acid composition of membrane fatty acids (Fig. 4).

When amiloride was added to the mucosal medium of an isolated bladder, it caused an immediate and complete inhibition of transcellular Na^+ transport and an inhibition of the effects of both aldosterone and vasopressin on Na^+ transport. However, amiloride had no effect upon the aldosterone-induced changes in lipid metabolism. Aldosterone caused an increase in the specific activities of membrane fatty acids in the presence of amiloride (Fig. 3) and an increase in the weight percentage of long-chain polyunsaturated fatty acids in membrane phospholipids (Fig. 5). Furthermore, if amiloride was washed out of the muco-sal bath after a period of 4-6 hr of aldosterone treatment, at a time when aldo-sterone had induced a modification of membrane lipid structure, there was an immediate increase in Na^+ transport, indicating that aldosterone had modified the Na^+ transport system. These data mean that the changes in lipid metabol-ism and membrane lipid structure are not secondary to an aldosterone-induced increase in Na^+ transport.

When the presumptive acetyl-CoA carboxylase inhibitor, TPIA, was added to the mucosal-bathing solution at a concentration of 2 mM, it altered neither the basal rate of transcellular Na^+ transport nor the effect of vasopressin on Na^+ transport (Fig. 6). However, it blocked the effect of aldosterone on Na^+ trans-port. This inhibition of aldosterone action was reversible. When TPIA was

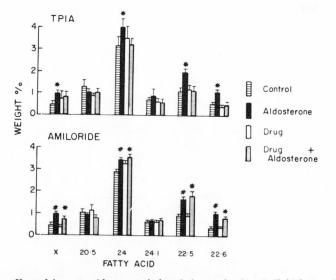

Fig. 5. The effect of drugs on aldosterone-induced changes in phospholipid fatty acids.

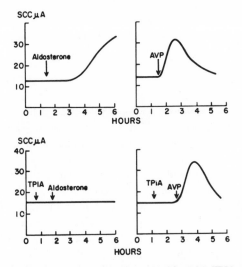

Fig. 6. The effect of pretreatment of the urinary bladder with TPIA upon its response, as measured by short-circuit current, to aldosterone and arginine vasopressin (AVP).

washed out of the mucosal medium there was a rise in Na$^+$ transport in aldosterone-treated bladders but only after a lag period of 90-120 min (Fig. 6), i.e., in a similar or slightly longer time than the usual lag period after aldosterone addition. Thus, in contrast to the situation with amiloride, where removal of an inhibitor in an aldosterone-treated tissue led to an immediate expression of aldosterone action, there was no apparent expression of aldosterone action in the TPIA-treated tissue, so that only upon removal of inhibitor did the hormone begin to act, and it required the usual lag period in order to exert its physiologic effect.

These results implied that TPIA was specifically inhibiting the biochemical effects of aldosterone directly responsible for its ability to increase transcellular Na$^+$ transport. Because of the previous evidence that inhibition of RNA and protein synthesis blocked aldosterone action, we examined the effect of TPIA on the rates of incorporation of appropriate precursors into RNA and protein. Neither protein nor RNA synthesis were blocked by TPIA either in the presence or absence of aldosterone, indicating that this agent did not have a general toxic effect on metabolism. This conclusion was also supported by the observation that vasopressin induced a normal response in TPIA-treated tissues.

In contrast to its lack of effect upon protein and RNA synthesis, TPIA had striking effects upon lipid synthesis and turnover, and upon the aldosterone-mediated changes in lipid metabolism. When TPIA was present, the rate of incorporation of [^{14}C] acetate into lipid was only 70% of control values and addition of aldosterone (which usually stimulated acetate incorporation) had no effect. Even more striking, when TPIA-treated bladders were incubated for 30 min with $[^{14}C]$ acetate and then the relative specific activities of the phospholipid fatty acids of TPIA-treated and control tissues compared, the results shown in Fig. 7 were obtained. The relative specific activities of both C16 and C18 saturated acids were higher in the TPIA-treated than control tissue in spite of the fact that total incorporation was decreased. Conversely, the relative specific activities of all unsaturated acids and the longer-chain saturated fatty acids were all reduced in TPIA compared to control tissues. Thus, TPIA had a significantly greater effect on inhibition of chain elongation and desaturation than on total fatty acid synthesis.

When bladders were pretreated with TPIA and then treated with or without aldosterone and given a 30 min pulse of ^{14}C-acetate, the results shown in Fig. 3 were obtained. Aldosterone in the absence of TPIA pretreatment caused a significant increase in the rate of acetate incorporation into nearly all the longer-chain polyunsaturated fatty acids. In contrast, exposure of TPIA-pretreated tissue to aldosterone led to no increase in the specific activities of any of these acids. Thus, TPIA blocked the specific effects of aldosterone on acetate incorporation into membrane phospholipid fatty acids. Pretreatment of the tissue with TPIA also blocked the effect of aldosterone upon membrane lipid fatty acid composition (Fig. 5). As noted above, after 6 hr of aldosterone exposure there was a significant increase in the weight percentage of certain long-chain polyunsaturated fatty acids in membrane phospholipids. Pretreatment with TPIA alone had little effect upon the relative weight percentage of the various fatty acids. However, pretreatment with TPIA totally blocked the effect of aldosterone upon membrane fatty acid composition.

DISCUSSION AND CONCLUSION

These results show that aldosterone-induced changes in lipid metabolism and membrane lipid structure are of fundamental importance in the action of this

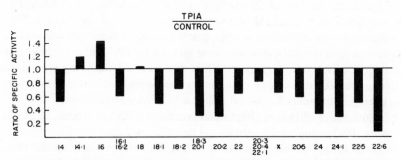

Fig. 7. The effect of TPIA upon the relative specific activities of phospholipid fatty acids after a 30 min pulse of $[^{14}C]$ acetate.

hormone upon transcellular Na$^+$ transport in the isolated amphibian urinary bladder. Although not yet complete enough to prove a direct relationship between a change in membrane lipid fluidity and a change in Na$^+$ transport, they do raise at least four important questions: (1) the biological significance of membrane phospholipid fatty acid turnover; (2) the relationship of this process to oxygen poisoning; (3) the mechanism by which any agent — in this case, aldosterone — brings about an integrated change in this deacylation-reacylation cycle and how selection is made of the particular mix of fatty acids which is maintained in the 2- position of membrane phospholipids; and (4) the possible alternative explanation for the relationship between the change in membrane lipid structure and change in transcellular Na$^+$ transport.

When one reexamines the modulatory effect of aldosterone upon the response of this tissue to hyperbaric oxygen in the light of the evident relationship between aldosterone action and membrane lipid structure and turnover, a logical conclusion is that the change in lipid metabolism and structure is responsible for the heightened sensitivity of hormone-treated tissue to the toxic effects of O_2. This heightened sensitivity may be due to at least two factors: (1) the fact that the most readily oxidizable fatty acids are the long-chain polyunsaturated acids and their membrane content increases after aldosterone action; and (2) there are increases in lipid (fatty acid) biosynthesis, chain elongation and desaturation, all of which require reducing equivalents, thereby increasing the demand for NADPH production. The latter would affect the capacity of the cell to respond to another stimulus, increased ambient oxygen concentration, that also places an additional demand for NADPH synthesis.

Perhaps the most interesting data, because of the questions raised, concerns the response of either hormone-treated or non-hormone-treated tissue to increased ambient oxygen concentrations above the normal values but below the toxic concentrations. When one raises the partial pressure of O_2 from 1/5 to 1 atmosphere in control tissue, there is a slight increase in rate of [1-^{14}C] glucose decarboxylation, and as one raises the O_2 pressure to 6 atmospheres a progressive increase in the rate of [1-^{14}C] decarboxylation is seen without any inhibition of Na$^+$ transport. In contrast, in aldosterone-treated tissue an increase from 1/5 to 1 atmosphere leads to a marked increase in decarboxylation (to rates equivalent to those seen in control tissues exposed to 6 atmospheres of O_2) without inhibition of Na$^+$ transport, but an increase in O_2 pressure of 6 atmospheres leads to little further increase in [1-^{14}C] glucose decarboxylation but a marked inhibition of Na$^+$ transport. Thus, in both types of tissue there is a significant range of ambient O_2 concentrations over which a rise in the partial pressure of O_2 leads to an increased demand for the generation of reducing equivalents without, however, being toxic in the sense of interfering with the physiologic function of the tissue.

There are at least two possible ways in which this O_2-dependent increase in

NADPH synthesis may be related to lipid metabolism and membrane turnover. Increased NADPH may be required for the direct reduction of partially-oxidized membrane fatty acids via the glutathione requiring lipid peroxide reductase (38-40), or the increased need for NADPH may reflect an increase in rate of fatty acid synthesis, elongation and desaturation to make new fatty acids to replace oxidized fatty acids in the membrane, i.e., an increase in the partial pressure of O_2 may increase the rate of turnover of the deacylation-reacylation cycle of unsaturated fatty acids in membrane phospholipids. At present, we do not have the data needed to decide between these two alternatives, but such work is in progress. However, if the second alternative is the correct one, it would provide an evolutionary explanation for the existence of this energetically costly and apparently wasteful cycle of deacylation and reacylation of the unsaturated fatty acids in the 2- position of membrane phospholipids: It is a necessary adaptation to an oxygen-rich environment, this adaptation being necessary to maintain these lipids in their reduced and therefore fluid state.

If this interpretation is valid, it raises the possibility that defects in this complex metabolic system may lead to changes in membrane structure that alter cell function sufficiently to lead to cell dysfunction and human disease.

The second question raised by these observations is: When the deacylation-reacylation cycle of membrane phospholipids is stimulated, what determines the mix of unsaturated fatty acids that gets inserted into the 2- position of membrane phospholipids? Obviously there are two quite different possibilities: (1) the selection is made by the enzymes involved in the reacylation step; or (2) this enzyme has a relatively broad specificity and is relatively unselective with the selection being determined by the relative amounts of the various long-chain polyunsaturated acids which get made under a particular physiologic circumstance. At present, we have no evidence which allows us to choose between these alternatives, but experimental testing of the two hypotheses is feasible.

The final point of discussion concerns the relationship of these hormone-induced lipid changes to the hormone-induced change in Na^+ transport. The first point to be made is that aldosterone has multiple effects upon lipid metabolism (Fig. 8). It increases the rates of both deacylation and acylation of phospholipid fatty acids, increases the rate of chain elongation and desaturation, the rate of decarboxylation of certain fatty acids, and the rate of de novo fatty acid synthesis. Some of the effects upon lipid metabolism are immediate, and it seems unlikely that they are brought about by an aldosterone-mediated gene activation. On the other hand, the final expression of aldosterone action clearly depends upon maintenance of RNA and protein synthesis. An attractive hypothesis is that one of the effects of aldosterone is to regulate the synthesis and thereby the concentration of some key enzyme or carrier in lipid metabolism.

Regardless of how protein turnover and concentration are involved, the

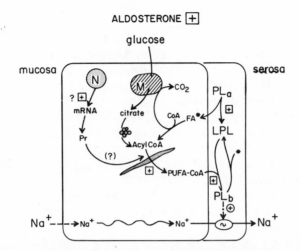

Fig. 8. A model of the effects of aldosterone upon lipid metabolism and membrane lipid turnover in the amphibian urinary bladder. The hormone increases $\mp$ the rate of deacylation and reacylation of the 2- position in membrane phospholipids; it stimulates the microsomal chain elongation and desaturation system; it stimulates the rate of fatty acid oxidation and probably of the de novo synthesis of fatty acids; and it may induce the synthesis of one or more proteins involved in lipid metabolism in this tissue.

evidence strongly suggests a direct relationship between membrane lipid changes and changes in transcellular Na^+ transport. The most compelling argument in favor of this conclusion is the data obtained with the inhibitor, TPIA. In this tissue, this compound exerted a relatively specific inhibition of chain elongation and/or desaturation of fatty acids. By doing so, it completely but reversibly inhibited the effect of this hormone on transcellular Na^+ transport.

Attractive as this model of hormone action may be, we still lack the information to define, in any biochemical or molecular sense, how the changes in lipid structure determine the changes in ion transport. However, our data do raise the possibility that some hormones can alter specific tissue functions by altering the lipid structure of membranes, thereby altering either the binding of proteins to the membrane or the catalytic activities of membrane-associated proteins.

ACKNOWLEDGEMENTS

This work was supported by grants from the Office of Naval Research (NR 202-005) and the U.S. Public Health Service (AM 09650 and CA 14345).

REFERENCES

1. Metcalfe, J.C., Warren, G.B. and Houslay, M.D. This volume.
2. Tsukagoshi, N. and Fox, C.F. (1973) *Biochemistry 12*, 2822-2829.
3. Smith, M.W. and Kemp, P. (1969) *Biochem. J. 114*, 659-661.
4. deGier, J., Marderšloot, J.G. and Van Deenen, L.L.M. (1968) *Biochim. Biophys. Acta 150*, 666-675.
5. Grisham, C.M. and Barnett, R.E. (1973) *Biochemistry 12*, 2635-2637.
6. Kimelberg, H.K. and Parahadjopoulos, D. (1974) *J. Biol. Chem. 249*, 1071-1080.
7. Linden, C.O., Wright, K.C., McConnell, H.M. and Fox, C.F. (1973) *Proc. Nat. Acad. Sci. (U.S.A.) 70*, 2271-2275.
8. Overuth, P. and Träuble, H. (1973) *Biochemistry 12*, 2625-2631.
9. Wisniesky, B.J., Williams, R.E. and Fox, C.F. (1973) *Proc. Nat. Acad. Sci. (U.S.A.) 70*, 3669-3673.
10. Horwitz, A.F., Halten, M.E. and Burger, M.M. (1974) *Proc. Nat. Acad. Sci. (U.S.A.) 71*, 3115-3119.
11. Allen, J.E. and Rasmussen, H. (1971) *Int. J. Clin. Pharmacol. 5*, 26-33.
12. Goodman, D.B.P., Allen, J.E. and Rasmussen, H. (1971) *Biochemistry 10*, 3825-3831.
13. Allen, J.E., Goodman, D.B.P., Besarab, A. and Rasmussen, H. (1973) *Biochim. Biophys. Acta 320*, 708-728.
14. Goodman, D.B.P., Wong, M. and Rasmussen, H. (1975) *Biochemistry 14*, 2803-2809.
15. Lien, E.L., Goodman, D.B.P. and Rasmussen, H. (in press) *Biochemistry*.
16. Lien, E.L., Goodman, D.B.P. and Rasmussen, H. Submitted for publication.
17. O'Malley, B., Mengel, C.E., Meriwether, W.D. and Zirkle, L.G., Jr. (1966) *Biochemistry 5*, 40-45.
18. Schauenstein, E. (1967) *J. Lipid Res. 8*, 417-428.
19. Taylor, D.W. (1958) *J. Physiol. 140*, 37-47.
20. Leaf, A., Anderson, J. and Page L.B. (1958) *J. Gen. Physiol. 41*, 657-668.
21. Sharp, G.W.G. and Leaf, A. (1966) *Physiol. Rev. 46*, 593-633.
22. Edelman, I.S. and Fimognari, G.S. (1968) *Rec. Prog. Hormone Res. 24*, 1-44.
23. Crabbé, J. (1963) *J. Clin. Invest. 40*, 2103-2110.
24. Leaf, A. (1967) *Am. J. Med. 42*, 745-756.
25. Ussing, H.H. and Zerahn, K. (1951) *Acta Physiol. Scand. 23*, 110-127.
26. Maffly, R.H., Hays, R.M., Lamdin, E. and Leaf, A. (1960) *J. Clin. Invest. 39*, 630-641.
27. Orloff, J. and Handler, J. (1967) *Am. J. Med. 42*, 575-768.
28. Handler, J.S., Preston, A.S. and Orloff, J. (1969) *J. Clin. Invest. 48*, 823-

833.

29. Goodman, D.B.P., Allen, J.E. and Rasmussen, H. (1969) *Proc. Nat. Acad. Sci. (U.S.A.) 64*, 330-337.
30. Edelman, I.S., Bogoroch, R. and Porter, G.A. (1963) *Proc. Nat. Acad. Sci. (U.S.A.) 50*, 1169-1177.
31. Chu, L.L. and Edelman, I.S. (1972) *J. Membrane Biol. 10*, 291-310.
32. Jensen, E.V. and DeSombre, E.R. (1972) *Annu. Rev. Biochem. 41*, 203-230.
33. Bentley, P.J. (1968) *J. Physiol. 195*, 317-330.
34. Siev, M., Weinberg, R. and Penman, S. (1969) *J. Cell Biol. 41*, 510-520.
35. Lahav, M., Dietz, T. and Edelman, I.S. (1973) *Endocrinology 92*, 1685-1699.
36. Maragoudakis, M.E. (1969) *J. Biol. Chem. 244*, 5005-5013.
37. Maragoudakis, M.E. (1971) *J. Biol. Chem. 246*, 4046-4052.
38. Christophersen, B.O. (1969) *Biochim. Biophys. Acta 176*, 463-470.
39. O'Brien, P.J. (1969) *Can. J. Biochem. 47*, 485-492.
40. O'Brien, P.J. and Little, C. (1969) *Can. J. Biochem. 47*, 493-499.

CHOLERA TOXIN, MEMBRANE GLYCOLIPIDS AND THE MECHANISM OF ACTION OF PEPTIDE HORMONES

Pedro Cuatrecasas, Vann Bennett, Susan Craig,
Edward O'Keefe and Naji Sahyoun

Department of Pharmacology and Experimental Therapeutics
and
Department of Medicine
The Johns Hopkins University School of Medicine
Baltimore, Maryland 21205

Cholera toxin (choleragen) stimulates ubiquitously the membrane bound enzyme, adenylate cyclase (1). Although in clinical cholera the intestine appears to be the only tissue affected, virtually all other tissues which have been examined *in vitro* have been shown to be sensitive to the actions of this toxin (Fig. 1). In all these tissues the toxin simulates the specific biological actions expected for cyclic AMP in that particular tissue. To illustrate the extraordinary specificity and potency of this toxin, it has been demonstrated (2) that exposure for a very short time to concentrations of toxin as low as 10^{-14} M, followed by washing of the cells, results in a potent inhibition of serum- or epidermal growth factor-stimulated DNA synthesis in human skin fibroblasts (Fig. 2). Similarly, it has been shown (3) that a brief exposure of melanoma cells in tissue culture to extraordinarily low concentrations of the toxin results in induction of the enzymes responsible for the biosynthesis of melanin. These cells are thus stimulated to differentiate, to change their morphology and to synthesize and deposit pigment in a way similar to the effects of the natural hormone, melanocyte stimulating hormone.

Such biological actions show that there must be considerable selectivity and specificity in the cell membrane components responsible for the recognition of and reaction with cholera toxin. This problem has been studied in great detail in our laboratory with ^{125}I-labeled derivatives of choleragen (4-8). The binding of this labeled toxin, which has been shown to be fully

EFFECTS OF CHOLERA TOXIN

All tissues - ↑ adenylate cyclase, cAMP

Intestinal epithelium - ↑ secretion salts and water

Fat cells - ↑ lipolysis

Liver and platelets - ↑ glycogenolysis

Adrenal cells - ↑ steroidogenesis, changed morphology

Melanocytes - ↑ tyrosinase, pigmentation, changed
  morphology, differentiation

Fibroblasts - ↓ DNA synthesis, ↓ growth

Transformed cells - ↑ contact inhibition, ↓ growth

Skin - ↑ permeability, ↑ hair growth

Pregnant mice - abortion

Intravenous - hyperglycemia, hyponatremia

Lymphocytes - ↓ mitogenesis B and T (PHA, ConA, LPS),
  ↓ cytolysis and cytotoxicity, ↓ histamine
  release by AG and IgE, ↓ plaque formation

Fig. 1. Various metabolic effects of cholera toxin in isolated tissues and cells.

active biologically, to a variety of cells and membranes has been studied. In isolated adipose tissue cells of the rat, a good correlation exists between the binding and biological activities of the toxin. However, in most tissues, with higher concentrations of the toxin, considerable binding can be seen at concentrations beyond which the biological effects are saturated (2,3,9,10). In the fat cell the relatively unusual situation exists that saturation of binding parallels closely saturation of the biological effects. As will be discussed later, there are few excess receptors in isolated fat cells compared to other tissues.

Various studies (4,5,9,11-15) have demonstrated that the principal component in the cell membrane which is responsible for the very high affinity binding of choleragen is a specialized glycolipid, ganglioside G_{M1} (Fig. 3). The specificity for G_{M1} is exquisite, since very minor changes, for example as occur with ganglioside G_{M2} or G_{D1A}, result in virtually complete loss of binding activity. Addition of low concentrations of gangliosides to the

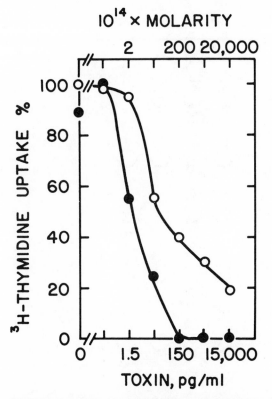

Fig. 2. Inhibition by cholera toxin of DNA synthesis in human fibroblasts (HF) stimulated by serum or epidermal growth factor (EGF). HF cells at confluency (100 μg of protein) were re-fed with fresh growth medium containing 0.1% bovine albumin and either 10% fetal calf serum or 10 ng/ml of EGF in the presence of increasing amounts of cholera toxin. [^{3}H]Thymidine incorporation, measured at 23 hr, is expressed as a percent of the maximum incorporation. Maximum thymidine incorporation was 12,900 $\pm$ 155 cpm for serum and 20,900 $\pm$ 1862 cpm for EGF. Serum, O–O, EGF, ●–●. (Data from ref. 2.)

medium can inhibit the binding as well as the biological effects of the toxin. In fat cells such studies (6,7) again show good correlation between inhibition of binding and biological activity, and provide greater confidence that at low concentrations the binding observed is a relatively accurate measure of biologically functional and significant receptors. Further evidence for this comes from studies with a structural analog of choleragen, choleragenoid, which can inhibit binding but has no intrinsic biological activity itself (7). This molecule is a convenient competitive antagonist of toxin action. With this analog it is possible to show parallel loss of binding as well as of biological activity. Choleragenoid is a subunit ("binding") of the toxin molecule which contains

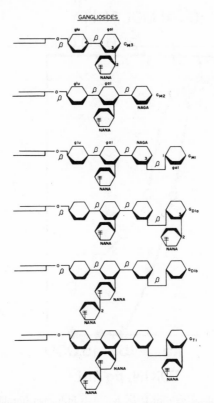

Fig. 3. Structure of various gangliosides tested (5) for inhibition of cholera toxin binding and action. Ganglioside G_{M1} appears to be the specific receptor for this toxin. The ceramide portion of the molecule is indicated by the parallel lines.

all of the ganglioside binding activity, but it lacks the subunit ("active") which is necessary for biological activation (8)(Fig. 4).

It has been shown (4,5) that preincubating cells or membranes with ganglioside G_{M1}, followed by thorough washing of these membranes or cells, results in a substantial elevation in the capacity to bind cholera toxin (Fig. 5). These studies suggest that the ganglioside is spontaneously being incorporated in the membrane in a way which can subsequently bind choleragen. The binding properties of the increased and newly appearing binding sites appear to be very similar to those which occur normally. More importantly, it has been shown that the prior incubation of cells with ganglioside can enhance the biological effects of cholera toxin (5,16). Thus, there is an increased sensitivity to the toxin, which can be correlated with an increase in binding sites. It is important that these effects can be demonstrated only in cells

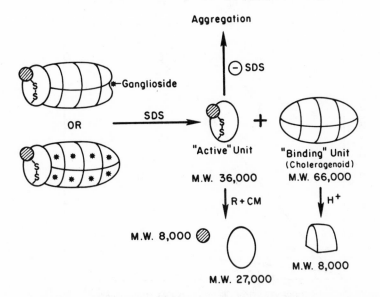

Fig. 4. Schematic representation of the structure of cholera toxin. The "binding" subunit (choleragenoid) can be dissociated into 8000 MW components by heating (60°, 30 min) in SDS or by exposure to 0.1 N HCl. Reduction and carboxymethylation (R and CM) of the "active" subunit results in the release of an 8000 MW component, but these two polypeptides are not joined by a disulfide bond since heating for 50 min at 100° in 1% SDS in the absence of reducing agents also results in dissociation. Of the two proposed structures of the native toxin the bottom one is more likely since it is now known (19) that the toxin (and choleragenoid) is multivalent with respect to ganglioside binding. (Data from ref. 8 and unpublished.)

which have low and limited numbers of binding sites to begin with, as occurs in fat cells, or special means must be devised (16) to decrease the existing binding sites experimentally before incorporating the exogenous, new receptor sites. In addition to biological effects such as lipolysis (5), it has been possible to demonstrate (16) increased sensitivity toward stimulation of adenylate cyclase. Such studies have also been performed recently by suppressing the existing binding sites by first adding choleragen. In this way, subsequently inserted gangliosides can be examined under conditions where the normal binding and activity are suppressed. The regaining of sensitivity to cholera toxin stimulation of adenylate cyclase can thus be demonstrated. These studies show again an important correlation between binding to gangliosides and activation of biological responses, and they establish the true receptor nature of this glycolipid. In effect, such studies represent the artificial reconstitution of a hormone-type of receptor-adenylate cyclase system. Such a

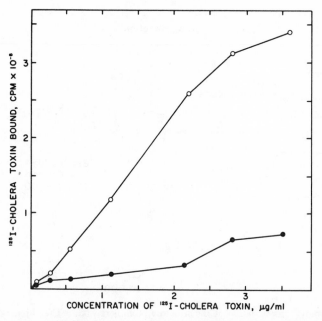

Fig. 5. Effect of increasing the concentration of cholera toxin on the binding of the toxin to ganglioside-treated liver membranes. Liver membrane suspensions (3 mg of protein/ml) were incubated for 50 min at 24° in Krebs-Ringer–bicarbonate buffer, 0.1% (w/v) albumin, in the absence (●) and presence (○) of 0.1 mg/ml of bovine brain gangliosides. The suspensions were then diluted ten-fold with the same buffer (without gangliosides) and centrifuged for 20 min at 30,000 g. The pellets were resuspended in the same buffer, and 0.2 ml fractions containing 14 μg of membrane protein were incubated for 15 min at 24° with varying concentrations of ^{125}I-labeled cholera toxin (2 μCi/μg). (Data from ref. 5.)

system is amenable to experimental manipulation and therefore may be of considerable value in studying the fundamental properties and mechanisms of regulation of this important enzyme.

The processes which transpire after the toxin binds ganglioside, after the toxin-receptor complex forms, and the activation of adenylate cyclase are very complex (6,10,17,18). Characteristically, there is a lag period of about 30 min between binding and biological activation. The nature of this lag period has been studied in considerable detail in intact fat cells (6,10) as well as in erythrocyte ghosts (10) where adenylate cyclase has been studied. The lag period is a highly temperature-dependent process which does not depend on protein or RNA synthesis, on metabolic or energy-requiring processes, on microtubular or microfilament-dependent processes, or on prostaglandin biosynthesis.

On the basis of many studies it has been suggested (6,7,10,17,18) that choleragen binds first to gangliosides, forming an initially inactive complex. The same complex is formed with choleragenoid (7). With native cholera toxin, a temperature and time dependent transition occurs within the three-dimensional structure of the membrane which leads to a relocation of the toxin molecules such that eventually there is a direct association between toxin and adenylate cyclase molecules. In this way, there is a resultant, virtually permanent activation of adenylate cyclase. Choleragenoid, by virtue of lacking the subunit which is required for biological activity, can only undergo the initial binding reaction. Some evidence for this hypothesis is illustrated by the following observations. It has been demonstrated (6) that the binding, particularly the extent which does not dissociate, increases with time and temperature of preincubation, suggesting marked changes in the nature of the membrane-bound toxin material.

Extensive studies (10) on the kinetics of activation of adenylate cyclase in toad erythrocytes and in isolated fat cell membranes have shown that there are three distinct phases in the activation process. First, there is a lag period, which is absolute. This is followed by an exponential phase of increasing activation, and a final very slow phase which may continue for some hours. The latter phase has not been studied in detail but may represent the biosynthesis of new enzyme molecules which are activated soon after their incorporation into the membrane (3). In toad erythrocyte ghosts (10) the major effects on adenylate cyclase are an increase in basal activity, a greatly enhanced sensitivity to hormonal (catecholamine) stimulation, and a paradoxical inhibition of the fluoride-stimulated response (Fig. 6). All three of these effects occur simultaneously during the lag, exponential and slow phases of activation. Studies of the temperature dependence of activation demonstrate (10) a relatively sharp transition which varies with the nature of the species from which the cells are obtained. It is interesting that in erythrocytes from toads, turkeys and rats, the transition temperature seems to reflect the ambient temperature of the species (Fig. 7). This is consistent with the postulate that some fluidity or lateral mobility of the toxin molecules within the plane of the membrane is important in the process of activation. One of the more interesting properties of the toxin-activated adenylate cyclase is the increased sensitivity to a variety of hormones such as isoproterenol, glucagon, ACTH and vasoactive intestinal polypeptide (10). This is reflected as an apparent increase in the affinity of the hormone for activation of the enzyme (Fig. 8).

Effects of the kinetics of activation and their dependence on the concentration of toxin illustrate some important principles which are pertinent to the mechanism of action (10). In the first place, the lag period is clearly absolute, and is independent of the concentration of the toxin. In other words, it is zero order with respect to toxin concentration. This is quite an

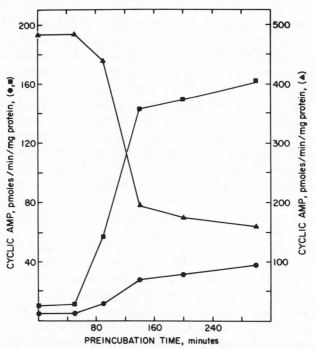

Fig. 6. Effect of increasing the time of incubation of cholera toxin with intact toad erythrocytes on the adenylate cyclase activity of plasma membranes assayed with 20 μM (——)-epinephrine (■), 20 mM sodium fluoride (▲), or with no additions (●). The cells were thoroughly washed with amphibian-Ringer, pH 7.5, suspended in this buffer, and divided into six 30 ml portions, each containing about 2×10^8 cells. The cell samples were incubated in a 30° water bath and cholera toxin (5.6×10^{-11} moles) was added at various times. One sample (zero incubation time) received no toxin. After 5 hr, the cells were rapidly chilled, and the plasma membrane adenylate cyclase activity determined. All values were determined in triplicate. (Data from ref. 10.)

extraordinary situation, with few if any parallels in the biochemical literature. It clearly distinguishes the mechanism of toxin activation from that of diphtheria toxin, which is known to act intracellularly. Furthermore, the exponential phase of activation is related to toxin concentration in a classical Langmuir absorption isotherm pattern (10). This can be studied properly only by correlating not the concentration but the number of molecules bound (as the concentration of the toxin is increased) with the exponential velocity of activation. Such data reveal linear reciprocal plots which suggest that the activation is of a bimolecular type (Fig. 9). Such data indicate (in toad erythrocytes) that half-maximal activation occurs with about 2000 molecules per cell (10). These results argue very strongly against all catalytic mechanisms of toxin action. For example, one toxin molecule cannot activate many

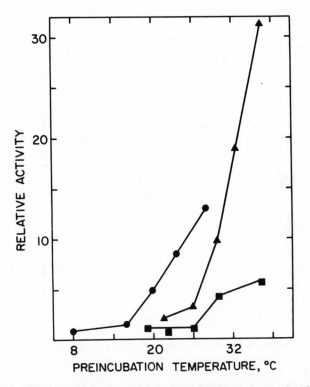

Fig. 7. Temperature dependence of the activation by cholera toxin of the adenylate cyclase activity of toad erythrocytes (●), turkey erythrocytes (■), and rat adipocytes (▲). Adenylate cyclase activity was measured at 33°. The values were determined in triplicate, and are expressed as the enzyme activity relative to control values (cells incubated under identical conditions but in the absence of cholera toxin). (Data from ref. 10.)

molecules of adenylate cyclase, by whatever mechanism. The data is most consistent with a bimolecular type of reaction which, as suggested earlier, may be one that involves direct interaction between the toxin and the enzyme molecules. Other studies which extend these observations further demonstrate that if the enzyme is submaximally stimulated by relatively low concentrations of the toxin, and if after the lag period more toxin is added, the second burst of activation occurs after another lag phase (10). This also argues against a catalytic mechanism of activation. Unlike the action of diphtheria toxin, one molecule of toxin cannot continue to have increasing effects with greater time of incubation, as would be expected with a catalytic mechanism. The exclusion of catalytic processes by such data makes it unlikely that the toxin is working, for example, by activating another enzyme

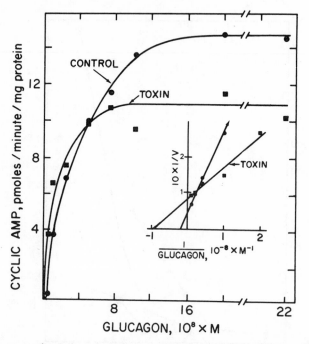

Fig. 8. Glucagon stimulation of adenylate cyclase of control (●) and cholera toxin-treated (■) rat liver membranes, showing a change in the apparent affinity for the hormone. Male rats (130 g) were injected intracardially with 0.4 ml of Krebs-Ringer – bicarbonate containing 100 μg of cholera toxin. Twelve hr later the livers were homogenized and plasma membranes were purified. Cyclase activity is expressed as stimulated activity (stimulated minus basal). The basal activities of the control and toxin-treated membranes were 2.2 pmole/min/mg and 25.7 pmoles/min/mg, respectively. Similar effects were reproduced in at least five different consecutive experiments. (Data from ref. 17.)

which subsequently acts on adenylate cyclase, or by the toxin itself having some intrinsic enzymatic property such as phospholipase activity.

Recent studies (18) have shown that cholera toxin can stimulate the activity of adenylate cyclase in isolated and purified plasma membrane preparations. These experiments have proven difficult mainly because of the lability of adenylate cyclase, which makes it difficult to incubate membranes at sufficiently high temperatures for the length of time required for the lag period. However, by manipulating the experimental conditions such that there is protection and stabilization of the enzyme, it has been possible to incubate the membranes and the toxin at the required temperatures and thus to demonstrate activation which shows the characteristic lag phase (18). Such studies exclude the involvement of intracellular, cytoplasmic events in the

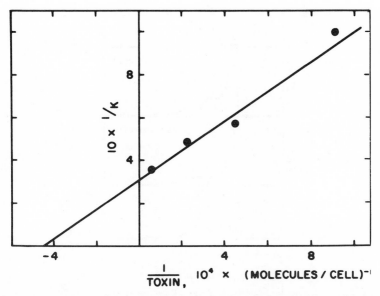

Fig. 9. The influence of increasing the concentration of cholera toxin on the rate of activation of toad erythrocyte adenylate cyclase. Washed erythrocytes were suspended in amphibian-Ringer and the time course of activation was examined at various fixed concentrations of cholera toxin. Adenylate cyclase activity was measured in the presence of 20 μM (—)-epinephrine. The values obtained from plots of the log of enzyme activity vs. time are described as a double-reciprocal plot of the slope of the semi-log plots determined separately vs. the number of molecules of cholera toxin bound per cell. (Data from ref. 10.)

activation process.

Considerable evidence has now accumulated which points directly to the interaction which subsequently occurs between cholera toxin and adenylate cyclase (17,18). If the enzyme is activated with radioactively labeled toxin, and the membranes are subsequently isolated and solubilized with detergents, the gel filtration pattern shows a peak of radioactivity that corresponds with adenylate cyclase (17). This is not seen, or is very minimal, if the toxin and the cells have not been preincubated under conditions that lead to activation (Fig. 10). In addition, affinity columns which have been prepared containing the "active" subunit of cholera toxin bound to agarose show absorption of adenylate cyclase (17). Perhaps most important, however, is the observation that antibodies against the active subunit of the toxin can specifically immunoprecipitate adenylate cyclase only under conditions where the enzyme has been activated by the toxin (17,18). Under appropriate experimental

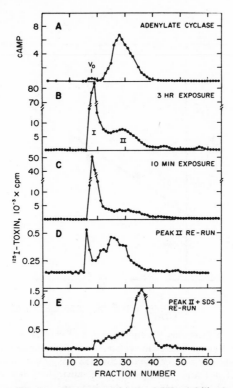

Fig. 10. Sepharose 6B chromatography of Lubrol PX-solubilized adenylate cyclase from fat cells incubated with ^{125}I-labeled cholera toxin. Cells were incubated at 37° for 3 hr. ^{125}I-Labeled cholera toxin (2.4 μg, 9.4 μCi/μg) was added either at the beginning of the incubation or for the final 10 min. Samples (0.75 ml) of the 250,000 x g membrane detergent extracts were applied to Sepharose 6B columns (1 x 57 cm; flow rate 10 ml/hr; 0.6 ml fractions) and cyclase activity (panel A) was determined. Panels A and B refer to cells exposed to toxin for 3 hr, while panel C is for the 10-min exposure to ^{125}I-labeled toxin. The peak of cyclase in panel A (fractions 28-30) was rechromatographed with (panel E) or without (panel D) heating the sample in 1% sodium dodecyl sulfate (60 min at 40°). (Data from ref. 17.)

conditions it has been possible to immunoprecipitate as much as 20% of the total adenylate cyclase activity of membrane-solubilized, activated cyclase (18).

On the basis of these studies, the model depicted in Fig. 11 has been proposed (10,17,18). During the lag period there is a translocation or some rearrangement of the toxin-ganglioside complex within the membrane that results in the breaking down of the toxin molecule with the resultant entry of the "active" subunit of the toxin into the substance of the membrane. One of the unique properties of the "active" subunit of the toxin is its

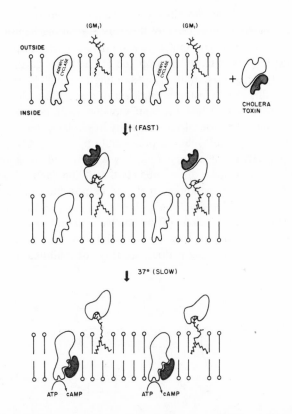

Fig. 11. Postulated (17,20,22) mechanism of action of cholera toxin. The toxin binds initially to ganglioside G_{M1} receptors in the cell membrane to form an inactive toxin-receptor complex. This complex is not associated initially with adenylate cyclase, but after a special time and temperature transition which involves movement within the structure of the membrane, the complex associates with and perturbs the enzyme. One possible mechanism for this transition, suggested in this figure, involves a dissociation of the toxin subunits with translocation of one of these which is highly water insoluble to the inner region of the membrane where the catalytic unit of the cyclase is localized.

water insolubility and tendency to aggregate. It is easily dissolved, however, with nonionic detergents, showing the hydrophobic nature of this protein. Studies using radioactively-labeled toxin have shown that with increasing time of incubation there is a relative enrichment of "active" subunit in the membrane (18), and in studies in tissue culture it has been shown that the toxin or a portion of it remains associated with the membrane for as long as weeks in tissue culture. Teleologically, the native toxin molecule may be viewed as a means that the bacterium has devised for specifically delivering a membrane-

type of water-insoluble protein across an aqueous environment and into the
membrane matrix of host cells by this very special mechanism (Fig. 12).

Since the cholera toxin molecule is multivalent with respect to its binding
of ganglioside (19), it has been proposed (18,19) that during the lag period
there may be progressive multivalent binding to gangliosides, and that at
some point this results in the breakdown of the toxin molecule with release
of the "active" subunit into the matrix of the membrane where it will dis-
solve and remain because of its unique hydrophobic properties. Once in the
membrane, the "active" subunit, presumably by lateral diffusion and special
intrinsic properties, will associate directly with the adenylate cyclase molecule. Such a mechanism, then, predicts that if the "active" subunit could be
incorporated into the membrane it should be possible to achieve activation
without having the "binding" unit. Indeed, it has recently been shown (18)
that *very* high concentrations of the "active" subunit can activate directly
various membrane preparations as well as intact cells. In this case the activation occurs without a lag period, and it is not inhibited by gangliosides or

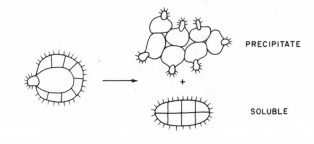

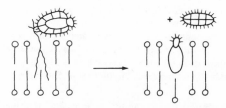

Fig. 12. Diagrammatic representation of the native cholera toxin molecule showing how
this molecule can dissociate into a water-soluble component (choleragenoid or "binding"
subunit) on a highly insoluble, hydrophobic ("active") subunit which aggregates. It is
proposed that in the process of activation the toxin molecule dissociates on the cell surface, possibly as a result of multivalent binding to surface gangliosides, with resultant
dissolution of the "active" subunit into the membrane. Once dissolved in the membrane,
this subunit can migrate laterally along the plane of the membrane and associate with
adenylate cyclase (Fig. 11).

choleragenoid (18). The "binding" subunit and gangliosides are thus not *essential* components for cyclase activation, but they very greatly facilitate the process.

In other studies (2,3,9,10,17) it has been shown that certain cells contain an enormous excess of binding activity. For example, in erythrocytes 100,000 sites may be present, whereas only 2000 to 4000 are required for maximal activation. These 100,000 sites must be unassociated with adenylate cyclase and they must be functionally equivalent with respect to their capacity for exchange, redistribution and ability to lead to a biological response (17). For example, it is possible to block one-half or even 80% of the potential binding sites with choleragenoid without affecting the biological activity of subsequently added cholera toxin (16,17). In other words, of all the potential binding sites it is only necessary to have a given number, say 2000 (*any* 2000), which presumably equals the number of cyclase molecules. Which ones of the total potential binding sites are occupied does not appear to matter, suggesting that they are potentially equivalent with respect to this special function.

The property of redistribution and of lateral mobility in the membrane has been demonstrated by studies using fluorescein-labeled cholera toxin (19). It has been shown that lymphocytes are labeled diffusely at 4°, that at this temperature patching and clusters occur with time and that at higher temperatures capping occurs. As in studies of immunoglobulin redistribution and capping, cholera toxin cap formation is energy-dependent and is inhibited by drugs which disrupt microfilaments and microtubules. It is a very temperature-dependent process. These studies show the multivalency of toxin molecule as well as its potential for redistribution on the surface of the cell. Presumably, these studies also show the lateral mobility and probable fluidity of ganglioside molecules within the plane of the membrane.

On the basis of these studies, a mobile receptor theory of hormonal receptor-adenylate cyclase interaction has been proposed to explain the mechanism by which hormone receptors activate adenylate cyclase (17,20-22). It has been suggested that hormone receptors are normally relatively free to diffuse laterally along the membrane, and that modulation of enzyme activity depends on direct encounters with receptors. Productive collisions would be favored when the receptors are complexed with hormones. This mechanism is in effect a two-step process (Fig. 13). Several unique features follow from such a scheme. For example, hormones may dissociate at different rates from free receptors than from the receptor-cyclase complexes, thus permitting changes in hormone binding properties without directly perturbing the receptor binding site. Since enzyme activation would involve two separate and sequential steps, the extent or apparent affinity of hormonal activation

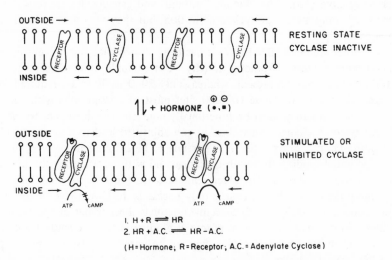

Fig. 13. General two-step, mobile receptor hypothesis for the mechanism of modulation of adenylate cyclase activity of cell membranes by hormone receptors. The central feature is that the receptors and the enzyme are discrete and separate structures which acquire specificity and affinity for complex formation only after the receptor has been occupied by the hormone. These structures can combine after binding of the hormone because of the fluidity of cell membranes. The hormone binding sites of the receptor are on the external face, exposed to the acqueous medium, and the catalytic site of the enzyme is facing inward, toward the cytoplasm of the cell. This sequential, two-step hypothesis for the action of membrane hormone receptors involves the discrete steps of: (a) binding of hormone (H) to the receptor (R), followed by (b) binding of the hormone-receptor complex (H-R) to another separate membrane structure, such as adenylate cyclase (A.C.), whose activity is subsequently modified.

may not be reflected in a simple way by the binding properties of the hormones.

These considerations could help to explain why cholera toxin increases the apparent affinity of various hormones for binding and activation (10,17). Similar effects of the toxin on the apparent affinity of various hormones for activation in a variety of cells suggests a generalized mechanism. The ability of GTP to enhance hormone stimulation and decrease the apparent hormone affinity simultaneously may also be rationalized in this framework. A novel feature of the mobile receptor model is the concept of functionally important and dynamic collisions and associations between membrane proteins. This has been postulated previously on the basis of estimates of the collision frequency of rhodopsin molecules in rod outer segment membranes (23).

REFERENCES

1. Finkelstein, R.A. (1973) *Crit. Rev. Microbiol. 2*, 553-623.
2. Hollenberg, M.D. and Cuatrecasas, P. (1973) *Proc. Nat. Acad. Sci. (U.S.A.) 70*, 2964-2968.
3. O'Keefe, E. and Cuatrecasas, P. (1974) *Proc. Nat. Acad. Sci. (U.S.A.) 71*, 2500-2504.
4. Cuatrecasas, P. (1973) *Biochemistry 12*, 3547-3558.
5. Cuatrecasas, P. (1973) *Biochemistry 12*, 3558-3566.
6. Cuatrecasas, P. (1973) *Biochemistry 12*, 3567-3576.
7. Cuatrecasas, P. (1973) *Biochemistry 12*, 3577-3581.
8. Cuatrecasas, P., Parikh, I. and Hollenberg, M.D. (1973) *Biochemistry 12*, 4253-4264.
9. Hollenberg, M.D., Fishman, P.H., Bennett, V. and Cuatrecasas, P. (1974) *Proc. Nat. Acad. Sci. (U.S.A.) 71*, 4224-4228.
10. Bennett, V. and Cuatrecasas, P. (in press) *J. Membrane Biol.*
11. Holmgren, J., Lonnroth, I. and Svennerholm, L. (1973) *Inf. Immunol. 8*, 208-214.
12. Holmgren, J., Lonnroth, I. and Svennerholm, L. (1973) *Scand. J. Infect. Dis. 5*, 77-78.
13. King, C.A. and van Heyningen, W.E. (1973) *J. Inf. Dis. 127*, 639-647.
14. Pierce, N.F. (1973) *J. Exp. Med. 137*, 1009-1023.
15. van Heyningen, S. (1974) *Science 183*, 656-657.
16. O'Keefe, E. and Cuatrecasas, P. Manuscript in preparation.
17. Bennett, V., O'Keefe, E. and Cuatrecasas, P. (1975) *Proc. Nat. Acad. Sci. (U.S.A.) 72*, 33-37.
18. Sahyoun, N. and Cuatrecasas, P. Manuscript in preparation.
19. Craig, S. and Cuatrecasas, P. Manuscript in preparation.
20. Cuatrecasas, P. (1974) *Annu. Rev. Biochem. 43*, 169-214.
21. Cuatrecasas, P. (1974) *Biochem. Pharmacol. 23*, 2353-2361.
22. Cuatrecasas, P. and Bennett, V. (1974) In *Perspectives in Membrane Biology* (Estrada-O, S. and Gitler, C., eds.) pp. 439-453, Academic Press, New York.
23. Poo, M. and Cone, R.A. (1974) *Nature 247*, 438-441.

MECHANISM OF CARRIER TRANSPORT AND THE ADP, ATP CARRIER

M. Klingenberg, P. Riccio, H. Aquila, B.B. Buchanan and K. Grebe

Institut für Physiologische Chemie und Physikalische Biochemie
Universität München, BRD

INTRODUCTION

A central feature of biomembranes is their ability to facilitate the transport of specific substances through a basically impermeable lipid barrier. This catalysis requires a much higher degree of specialization and information than the basic function of membranes to separate cellular compartments. It is therefore understandable that highly specific proteins, "carriers," are responsible for transport catalysis and that relatively nonspecific lipids constitute the barrier. In accordance, research progress has been much faster in elucidating the chemistry and function of lipids in the membrane compared to elucidating the more complex and difficult to handle membrane proteins.

A participation of membrane proteins in transport is especially logical for the transport of metabolites which are selected with high specificity. It can be expected that the relatively complicated metabolite molecules require a carrier characterized by a much higher information content than the transport of small cations which in some cases can be selectively catalyzed by carriers of low molecular weight. It is obvious that carrier proteins are intrinsic membrane proteins *par excellence*. Their very function depends on the existence of an intact membrane in contrast to membrane bound enzymes where the enzymatic reaction *a priori* does not require association to membranes.

The most powerful carriers in respiring eukaryotic cells can be considered to be the catalysts for the transport of ADP, ATP and P_i through the inner mitochondrial membrane (Fig. 1). The high activity of these transport processes must supply the ATP for a multitude of metabolic reactions in the cytosol and ADP and P_i for the production of ATP inside the inner mitochondrial mem-

Abbreviations: ATR, atractylate; BKA, bongkrekate; CAT, carboxy atractylate; BHM, beef heart mitochondria; NEM, N-ethylmaleimide.

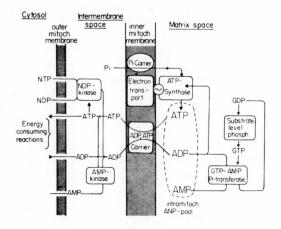

Fig. 1. The role of the ADP,ATP carrier in mitochondrial phosphate transfer reactions. The function of the intramitochondrial ANP pool as an intermediate in the synthesis of the extramitochondrial ATP. Localization of transport systems on the impermeable inner mitochondrial membrane and diffusion through the permeant outer mitochondrial membrane. Exclusion of ANP by the inner mitochondrial membrane and localization of AMP re-utilizing phosphate transferases in the perimitochondrial and intramitochondrial space.

brane. During the past twelve years many facets of ADP,ATP transport have been investigated, so that our knowledge of that process has reached an advanced state (1). Recently a protein provisionally identified as the carrier has been isolated and partially characterized (2,3). The following discussion will deal with general problems of membrane carrier function and will focus on the ADP,ATP carrier.

THE ADP,ATP CARRIER AS A SINGLE–SITE, REORIENTING MECHANISM

The classical mobile carrier models (4) must be considered as describing a carrier mechanism in analogy to an enzyme mechanism which is used for deriving laws of enzyme kinetics (Fig. 2). Accordingly, the mobile carrier model has been very useful for describing transport kinetics, such as dependency on substrate concentration inside and outside the membrane, as was discussed in detail for glucose transport in erythrocytes and for other transport systems (4,5). Of particular interest is the exchange mechanism described by the mobile carrier model which was central to the formulation of carrier concept and which is also now a feature of ADP,ATP transport presently discussed.

The mobile carrier model encompasses, in fact, both mobile and fixed carrier mechanisms (Fig. 3). In the mobile mechanism the site to which the substrate binds (carrier site) may translocate to the other side of the membrane by

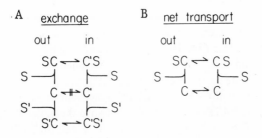

Fig. 2. Scheme for the carrier catalysis. A. Circulating (mobile carrier) in counter-exchange. B. Carrier in net transport.

rotational or translational movement of the protein in the membrane; in the fixed mechanism the protein remains fixed and where the substrate is squeezed across the carrier, the substrate site is either open to the outside or inside. In both cases, in line with the formalism of Fig. 2, the carrier exists in two states as its binding site is directed to the inside or outside. In the obligatory exchange mechanism, in which transport into the mitochondria is coupled to transport out of the mitochondria, it is further postulated that an unloaded binding site cannot be translocated by rotational or "gated pore" mechanism. The carrier becomes mobilized only when bound to the substrate. For an exchange mechanism, a seemingly attractive formulation is a double site carrier where binding sites are available on the inside and outside. Only when substrates bind simultaneously to both sides does the carrier become activated and channel the substrate to the other side. In this case the carrier can be visualized to be a dimer in which the subunits activate each other for the translocational process. This

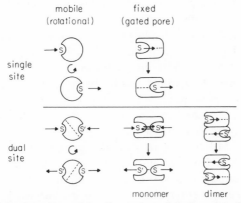

Fig. 3. Mobile and fixed carrier models. Comparison of single and dual site carriers, operating in counter exchange.

formulation has been discussed by us previously for the exchange of the ADP, ATP carrier (6).

Dimeric or even tetrameric carriers with multiple sites have been proposed also for the glucose carrier (7). However, the reasoning in this case was based only on kinetic "anomalies" which later were explained more convincingly (8), by retaining a single site carrier model, on the basis of microscopic unstirred layers in front of the binding sites. A dual site mechanism has also been proposed for the $Na^+ K^+$ ATPase (9). In this instance the molecular evidence, however, speaks in favor of a single site for both K^+ and Na^+, although evidence actually has been furnished here for a dimer form of the carrier. The two subunits in this model, however, act in parallel rather than in series (10).

EXPERIMENTAL EVIDENCE FOR THE SINGLE SITE REORIENTING CARRIER

The experimental demonstration of a reorientation mechanism involving a single site has been possible in the case of the ADP,ATP carrier because of the existence of two different highly specific inhibitors for this carrier: atractylate (ATR) or its homologue, carboxy–atractylate (CAT) and bongkrekate(BKA). These inhibitors differ in that ATR and CAT are membrane impermeant whereas BKA is permeant. ATR and CAT can therefore reach the carrier only on the outside, whereas BKA can also reach the carrier when it faces the inside. In extensive studies on the binding of these ligands, it was possible to define the carrier sites, their affinity to ADP and the other ligands and, quite unexpectedly, also to gain an insight into the dynamics of carrier function.

A number of surprising results arose which cannot be explained by a simple interaction of ligands with protein. The problems could be more elegantly explained by taking into account the possibility that binding involved the switching of the carrier sites across the membrane. Carrier sites located inside should show particular "odd" binding effects because they require a translocation step before they are in equilibrium with the ligand added on the outside. If one chose to interpret these results simply as "allosteric" or "conformational" effects of the carrier, he would miss the opportunity to exploit the results for a more complete description of the transport mechanism. Although only a superficial account of these studies can be given, I hope to make clear that the results converge to support the reorienting carrier mechanism.

The binding of ADP to the carrier was the first stage in our efforts to define the carrier sites (11). Only that portion of bound ADP which can be removed by ATR can be considered as ADP bound specifically to the carrier. These values are obtained by a differential procedure and then analyzed in mass action plots (Fig. 4A), which show two apparent types of binding sites, 80% with low

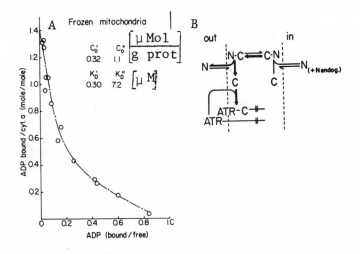

Fig. 4. Binding of ADP to the carrier in BHM differentiated by sensitivity to ATR. A. Mass action plot of the specific binding (difference of total ADP to ADP + ATR), giving the amount of high and low affinity sites and K_d. The curve is fitted with these constants to the measured values. B. Scheme for the removal of ADP (N) from inner and outer carrier sites by ATR as an impermeant ligand.

and 20% with high affinity. In the usual evaluation of ligand binding with protein, one would arrive at the conclusion that a smaller amount of carrier has high and a larger amount has low affinity sites.

A much more interesting, and for further research more valuable, interpretation of these results is obtained by making use of the reorientating carrier mechanism (Fig. 4b). On binding of ADP, the previously empty external carrier sites which were immobile are activated and switched to the inside. After equilibration, the added labelled ADP then binds both to the inner and outer sides of the membrane. ATR as an impermeant and immobilizing ligand can be expected to bind carrier sites only to the outside. By fixing the carrier in the ATR–C form, ADP–C complexes from the inside are pulled to the outside so that eventually all carrier accumulates in the ATR–C complex outside. Thus the total ADP released by ATR comes in part from the outer and in part from the inner sides.

It is then quite easy to explain the existence of the two different affinities. The smaller portion of high affinity sites corresponds to those inside since these sites are always saturated with the endogenous nucleotides even at "non-saturating" external concentrations. Their high affinity therefore is not real but is only a result of their location inside in the internal pool. Low affinity sites, being located outside, are in a direct equilibrium with external ADP and exhibit true "K_d's."

A much more striking application of the reorienting carrier mechanisms, which results actually in its first experimental demonstration on the molecular level, comes from the interaction of BKA with bound ADP (12,13,14). Since BKA, similar to ATR, is an inhibitor of ADP transport, it can be expected to compete also with the ADP binding to the carrier. Surprisingly, the apparent affinity of ADP to the membranes was increased by BKA to such an extent that the bound ADP appeared to be even resistant to ATR (Fig. 5). This major effect was accompanied by a number of peculiar side effects such as dependence on the sequence of addition of BKA and ADP, unusually strong dependence on time, on temperature, and on pH. These phenomena resulted in a very detailed study of the influence of BKA on ADP binding (12,13).

At first the application of a reorienting carrier mechanism encountered great difficulties. The new finding actually led us to accept the notion for some time that the carrier has different binding sites for BKA and for ADP (15). With BKA at the regulatory site the binding of ADP is increased, whereas, conversely, ATR at this site decreases the affinity for ADP (C = carrier).

$$ADP \rightarrow C \leftarrow BKA \qquad \text{or} \qquad ADP \leftarrow C \leftarrow ATR$$

The assumption of a ternary complex, ADP–C–BKA, could also originally explain the finding that the BKA induced fixation of ADP is smaller when BKA is added before rather than after ADP. This difference in ADP binding was attributed to the amount of inner carrier sites present before the addition of BKA, as explained by Fig. 6. In the case of adding first ^{14}C-ADP and then BKA (Fig. 6A), both the external and internal sites equilibrate with the ^{14}C-ADP so that BKA as a permeant ligand can then form the immobilized BKA $-$[14]C–ADP complex (BA·C·N*) both outside and inside; if BKA is added

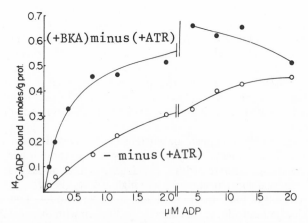

Fig. 5. Influence of BKA on the ADP binding (BHM). The plot demonstrates the increase of affinity for ADP under the influence of BKA (see ref. 12). The difference of the binding (+BKA) minus (+ATR) equals the specific binding portions to the carrier either under the influence of BKA or without BKA.

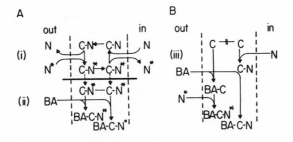

Fig. 6. Differentiation between binding of BKA to inner and outer sites, based on the difference of ^{14}C-ADP (N*) binding when added before BKA (case A) or after BKA (case B). Contribution of endogenous ^{12}C-ADP (N) and exogenous ^{14}C-ADP (N*). Assumption of a ternary complex BKA-carrier-ADP (= BA-C-N)(16).

first (Fig. 6B), it fixes the endogenous ^{12}C–ADP to the internal sites so that on the following addition of ^{14}C–ADP only the external sites are free to bind ^{14}C–ADP.

This explanation also complements the finding of two types of ADP binding sites measured as the ^{14}C–ADP released by the impermeant ATR (Fig. 4A). There were, however, a number of reasons to assume that all of the BKA which functions to increase the affinity of ADP, and not only a small portion as in Fig. 6, has to be inside and that BKA binds exclusively from the inner membrane side to the carrier. Evidence to this point includes the extremely high pH dependence, temperature dependence and also some inconsistencies concerning the interaction between endogenous ADP and BKA in the previous scheme. Although space does not allow a detailed discussion of the arguments here, let it be said that the new explanation is attractive because of its simplicity and of certain surprising consequences for the carrier mechanism.

Firstly, it is now assumed that BKA forms only a binary complex with the carrier, C–BKA, instead of a ternary complex. Possibly there is the same binding site for all ligands, the substrates, ADP, as well as the inhibitors, ATR, CAT and BKA. From this site ADP is displaced by ATR and BKA:

The reaction sequence of binding increase of ADP by BKA is the following (Fig. 7): at non-saturating concentrations of ADP, some carrier sites form the ADP–C complex and are activated for switching to the inside. BKA which has

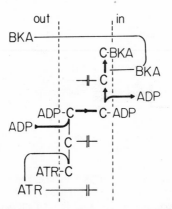

Fig. 7. Scheme based on the circulating (mobile) carrier model. Reorientating carrier mechanism assuming a binary complex of BKA and carrier where BKA replaces ADP similarly to ATR. It is assumed that BKA can penetrate the membrane and ATR is impermeant.

penetrated to the inside displaces ADP and forms the impermeant, tight C–BKA complex. Eventually all outer carrier sites will bind ADP and then become trapped inside. Thus, as much ^{14}C–ADP is trapped inside as there originally were carrier sites outside. This would correspond exactly to the apparent increase of ADP binding to the carrier sites.

The tightness and complete inaccessibility of the trapped ADP to ATR is explained by blockage of the carrier. This agrees with the finding that the effect of BKA is so strong that it can even remove ^{35}S–ATR which first has trapped carrier sites outside. This is shown by using ^{35}S–ATR and ^{3}H–ADP (Fig. 8). With increasing concentration of BKA, ^{35}S–ATR is removed and, conversely, ^{3}H–ADP is bound (15). This experiment can also be arranged to demonstrate that ADP is required for the removal of ^{35}S–ATR by BKA. The synergistic effect of ADP and BKA, as seen also in other instances, and the opposing effects of ADP and ATR are explained easily by the BKA-induced

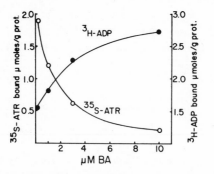

Fig. 8. The displacement of ^{35}S–ATR by BKA (+^{3}H–ADP). Simultaneous measurement of ^{35}S–ATR and ^{3}H-ADP binding to BHM (15).

switching mechanism of Fig. 7. Only in the presence of ADP can the carrier sites switch to the inside and become available to bind BKA. Why BKA preferentially binds to the carrier in the inside will be further elucidated below.

THE CONFORMATIONAL CHANGE OF THE MEMBRANE

Unexpected support for the mobile carrier mechanism came from a "macroscopic" change in the mitochondrial structure which was closely linked to the reorientation of the carrier. In beef heart mitochondria the configuration of the cristae membranes is changed in a characteristic manner on the binding of various ligands to the ADP,ATP carrier (17,19). As in other swelling-shrinkage phenomena of mitochondria, energy transfer in oxidative phosphorylation is excluded. At first it seemed somewhat improbable that binding of ADP or ATP to the ADP carrier should induce such prominent changes in the membrane configuration. Strong evidence that the binding of ligands to the carrier is involved in a change in mitochondrial structure, came from the tendency that BKA (12) behaves similar to ADP and opposite to ATR. As inhibitors of nucleotide transport both ATR and BKA should prevent any changes due only to the uptake of ADP and its utilization in the matrix. The ability to record turbidity changes in a photometer made these changes a valuable tool to follow the kinetics of the binding and reorientation processes induced by the ligands.

It could be concluded from a detailed comparison of ADP-binding and absorbance changes that the configurational changes reflect the carrier orientation, independently of which ligand is bound. The mitochondria are in a relaxed form when the sites are turned outside and in a contracted form when the sites face inside. On this basis full agreement between the reorientation mechanism and the turbidity changes is found (Fig. 9): Before addition of ADP, the carrier sites are largely outside since they are trapped in the unloaded state. With ADP, the sites distribute inside and outside and a partial contraction is observed. This effect can be either reversed by ATR, which traps all sites on the outside, or further increased by BKA, which traps all the carrier sites on the inside.

The rate of contraction induced by BKA has an unusually high pH and temperature dependence. These measurements established that a diffusion of the undissociated $BKA-H_3$ form is a rate-determining step for the contraction and, therefore, also for the effect on the ADP binding. It corroborated impressively that BKA fixes the ADP and simultaneously induces the contraction only by binding from the inside to the carrier.

A mechanism for configurational changes of the mitochondrial cristae has been suggested by reducing them to an elementary concept (14,18): one mem-

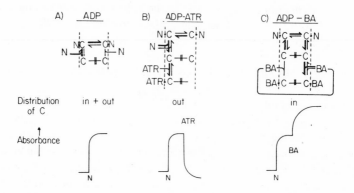
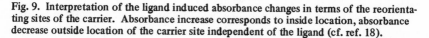

Fig. 9. Interpretation of the ligand induced absorbance changes in terms of the reorientating sites of the carrier. Absorbance increase corresponds to inside location, absorbance decrease outside location of the carrier site independent of the ligand (cf. ref. 18).

brane surface area is enlarged as compared to the other so that the membranes become curved (Fig. 10). The direction happens to coincide with the orientation of the carriers: when the carrier sites accumulate with BKA on the matrix side, this membrane surface becomes convex and when sites accumulate on the outside with ATR, the surface becomes concave. Possibly it is not the movement of the carrier protein *per se* but, rather, an electrical surface charge excess related to the carrier's orientation which causes the spreading of the membrane surface areas.

In conclusion, combined studies on ADP binding and on membrane configuration change lead to a first demonstration of the reorientation of a membrane carrier. Whereas, when binding substrate, carrier sites are partially distributed to the inside and outside, we have the unique possibility to bring with the impermeant ATR the carrier exclusively to the outside, and with the lipophilic

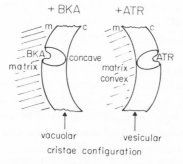

Fig. 10. Correlation of carrier orientation to the morphological change of the inner mitochondrial membrane, in terms of a transition from a convex (as seen from the matrix) to a concave form.

permeant BKA exclusively to the inside location. Whereas with the substrates alone it would remain unclear whether the carrier has single, double or more sites operating in the counter-exchange, as illustrated in Figs. 2 and 3, the possibility of fixing the carrier exclusively on one or the other site is evidence that it has only a single site which can reorient itself only by the translocation process (with the help of ADP) to the other site.

THE ASYMMETRY OF THE CARRIER

Because they face different cellular compartments, it is to be expected that biomembranes are asymmetrical. With the reorientating carrier mechanism the question thus arises as to what degree the carrier site changes to face the outside or inside. This question has also great consequences for the mechanism of carrier action. The most striking indication for asymmetry is again furnished by experiments with the two inhibitors, ATR and BKA. In sonic particles (in which the inner mitochondrial membrane becomes inverted) the inner face is accessible for the use of impermeant ligands such as ATR. Here the transport shows a difference to mitochondria in that the exchange with ADP is insensitive to ATR but fully inhibited by BKA (2,21). Such exchange is inhibited by BKA independent of pH and without any time delay (20), thus indicating that BKA binds to the carrier site directly on the outside. Only if ATR is bound before sonication is nucleotide transport inhibited in the sonic particles. Direct binding studies of ADP also reflect the same difference from intact mitochondria in that ADP is removed by BKA, although only to a small extent, but more than by ATR (M. Klingenberg, unpublished).

In view of these results it is useful to define the carrier in terms of its localization on the inner membrane surface in two states: C_m-state, facing the matrix (m-) side and C_c-state facing the cytosol (c-) side. The asymmetry of the carrier in the two states is revealed by its affinity for the two different ligands (2):

in c-state: medium affinity for ADP and ATP, high for ATR and CAT, low for BKA.

in m-state: medium affinity for ADP and ATP, very low for ATR and CAT, high for BKA.

It appears that the asymmetry of the carrier is a result of a difference in the state of conformation of the binding site when it faces the m- or c- side. The application toward the carrier mechanism will be discussed below.

An asymmetry of the carrier toward the substrates ADP and ATP is established in a somewhat different manner. In kinetic studies of translocation, asymmetry is reflected with regard to the specificity of ATP vs. ADP. It is clear that for a purely catalytic process such difference in specificity would violate the second law of thermodynamics and therefore can be expected to exist only at the expense of free energy. Accordingly, only in energized mitochondria is there observed a strong preference in transport for ADP as compared to ATP for influx and, in contrast, a preference for ATP over ADP for efflux (22). In the uncoupled state this difference is abolished. It is clear that this asymmetry results in an "active" type of exchange, in which there is a preferential accumulation of ATP outside and ADP inside (22,23).

In this connection it can be expected that the affinities for ADP and ATP are different on the c- and m- sides of the membrane. Although measurements are not reliable for the m- side in sonic particles, the ratio of affinities for ADP to ATP is reflected in transport specificity. Under energized conditions a difference in the affinity ratio appears to develop such that: affinities-$(ATP/ADP)_c <$ affinities-$(ATP/ADP)_m$. Under de-energized conditions ("ground state") the ratios may be more equal (22,24).

Another indication of the difference of the carrier conformations on both sides of the membrane comes from the influence of the alkylating reagent, NEM, which inhibits the carrier only when activated by ADP (25,26). In further research the effect of ADP was simply explained by its ability to bring the carrier more to the C_m- state (1). Thus ATR completely prevented this alkylation, whereas BKA promoted it. Concomitant with the NEM alkylation there was a contraction of the membrane similar to that observed with BKA. Obviously, only the C_m- state exposes a group available for alkylation by NEM.

MECHANISM OF ASYMMETRIC CARRIER ACTION

Two conclusions can be derived from these studies on the molecular mechanism of carrier action: (i) The carrier orients itself alternatively in two states that are highly asymmetric. (ii) The asymmetry is of such magnitude that ligands showing high affinity for the carrier are largely excluded from binding when the carrier is in the opposite orientation state.

While the reorientation mechanism allows for either a mobile or fixed gated pore, the strong asymmetry of the carrier binding site, as discussed above, is in favor of the "gated pore" (Fig. 11). There is a notable difference as to how a substrate is bound on each side of the carrier in the two mechanisms. In a rotational model, the substrate can be visualized to bind and dissociate from both sides of the membrane in the same molecular orientation. In the "gated pore," the substrate can be visualized to be bound with opposite moieties of the molecules to the carrier on both sides of the membrane. When channeled through the "gated pore," it leaves head first and reversely is bound tail first.

This marked asymmetry of the conformation of the substrate binding sites might be reflected in the difference toward the binding of ATR and BKA (Fig. 11). Common binding sites for the ligands are obviously the common three anionic groups. As illustrated in Fig. 12, another surface conformation

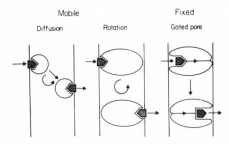

Fig. 11. Comparison of substrate orientation to a mobile (diffusing or rotating) and fixed, gated pore carrier.

of the binding site is sensed with BKA compared to ATR. The two molecules may reflect, therefore, the conformational difference of the ADP binding site in the asymmetric "gated pore" process.

ISOLATION OF THE CARRIER PROTEIN

Most biochemists would hold the view that it is unwise to speak about a mechanism before one has investigated the pertinent purified protein. While this attitude is valid for soluble enzymes, it may be disadvantageous for the isolation of a carrier protein where, in fact, the opposite situation exists, i.e., some knowledge of the carrier mechanism is prerequisite to the isolation of carrier protein in a functional state. Otherwise, the attempts to isolate the functional carrier protein may be largely abortive or result in inactive protein.

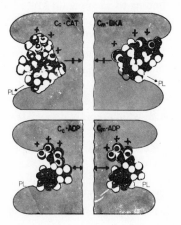

Fig. 12. The hypothetical arrangement of inhibitors and substrate at the carrier binding site, according to an asymmetric binding site configuration, using space filling models of the ligands. The (+) charges at the carrier surfaces attract the (−) charges of the ligands, as indicated. PL, phospholipid.

In the case of the ADP,ATP carrier, knowledge of the functional states of the carrier not only facilitated isolation but also handling of the isolated solubilized protein. A breakthrough was achieved after it was understood that only the fixation of the carrier in the intact membrane in a certain state by a high affinity ligand would permit the solubilization and isolation of functional carrier protein (20,27). The carrier is first tagged with ^{35}S-CAT in the mitochondria and can then be identified in the soluble extract as long as the complex remains intact. Among the many detergents tested, only Triton X-100 proved to be successful in solubilizing and at the same time maintaining the carrier-CAT complex intact. If the parent mitochondria were solubilized with Triton X-100 prior to addition of CAT no binding was observed. Obviously CAT added before the solubilization protects the carrier protein against denaturation by Triton X-100. Addition of cholate or other ionic detergents to the solubilized carrier-CAT complex causes the rapid dissociation of CAT.

For the solubilization, high concentrations of Triton X-100 (about 5%) and of salt are necessary, indicating the high lipophilic character of the protein. It belongs to one of the least extractable proteins of the mitochondrion.

The purification and isolation of the protein which is based on hydroxylapatite chromatography is remarkably simple. The pass-through of the column to which the crude extract of mitochondria is applied consists of more than 60% of the carrier protein complex. On polyacrylamide gel electrophoresis the progress of the purification can be assayed by following the appearance of a protein with a MW of 29,000 (Fig. 13). In the final purification state a single band with MW of 29,000 is obtained. At this stage about 18 μmol CAT/g protein are found, corresponding to an apparent MW of 55,000. It appears, therefore, quite possible that the solubilized CAT binding protein is actually a dimer comprised of two subunits of MW 29,000.

The identity of the CAT binding protein with the carrier, i.e., with proteins which also ligand with ADP and BKA, is indicated by the specific displacement of CAT on addition of ADP or BKA (Fig. 14). No ADP is bound in this process, thus indicating that after displacement of the CAT an intermediary labile carrier-ADP complex is formed which rapidly dissociates and results in an inactive protein.

Another striking line of evidence for CAT conferring a particular conformation to the carrier comes from behavior of the protein on the hydroxylapatite column. As shown in Fig. 15, the pass-through of this column contains a large amount of the 29,000 MW protein when the parent mitochondria were first loaded with CAT, whereas in the unloaded state or with ADP, most of this protein is adsorbed on the column. Obviously CAT conveys a non-adsorbing property to the carrier. It should also be mentioned that on binding BKA, the carrier appears to be more easily extractable.

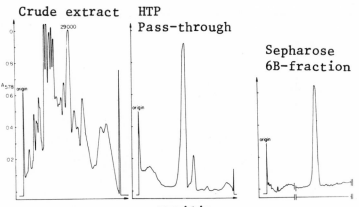

Fig. 13. Sodium dodecylsulfate polyacrylamide gel electrophoresis (SPAGE) at various stages of the purification of the CAT binding protein (P. Riccio and H. Aquila, unpublished) from BHM. HTP, hydroxylapatite. The single band of the purified CAT-binding protein corresponds to MW 29,000.

Particularly convincing evidence for a specific conformation of the CAT protein complex comes from an investigation of its antigenic properties. With the purified CAT binding protein as an antigen, an antibody can be generated with a relatively high antigenic titer in the purified γ-globulin fraction of rabbit serum (Fig. 16). This antibody is highly specific for the CAT binding protein, as is demonstrated in Fig. 17. Only with an extract of mitochondria

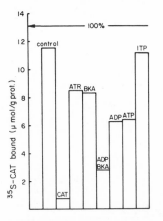

Fig. 14. Demonstration of the specificity of ligand interaction with the CAT-binding protein as determined by measuring displacement of bound ^{35}S-CAT. ^{35}S-CAT protein (of about 70% purity) in Triton X-100 solution was added to the indicated ligands and subjected to equilibrium dialysis at room temperature for 5 hrs (3).

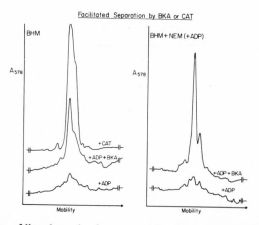

Fig. 15. Influence of ligands on the chromatography of the carrier protein (MW 29,000). SDS-polyacrylamide electrophoresis of the "pass through" of hydroxylapatite column, loaded with Triton X-100 extracts of BHM, under the addition of various ligands to the ADP,ATP carrier.

which has been preloaded with CAT is an antigenic reaction observed. Other experiments show that there is no cross reaction with hydroxylapatite fraction of rat liver mitochondria or with the isolated purified CAT protein from *Neurospora crassa*; such studies also demonstrated that CAT and cardiolipin (DPG), which is present in the isolated carrier, can be excluded as haptenes.

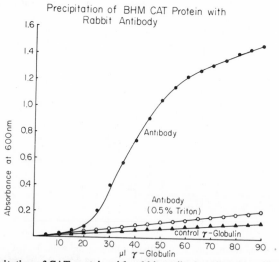

Fig. 16. Precipitation of CAT protein with rabbit antibody. Titration of turbidity increase on addition of γ-globulin fraction.

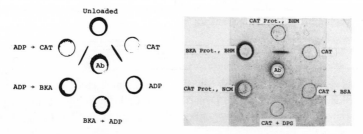

Fig. 17. Specificity of CAT-protein antibody in Ouchterlony double diffusion in agar plates stained with Coomassie Blue. Left plate: Center well contained anti-CAT protein γ-globulin. The outer wells contained hydroxylapatite column pass-through fractions of the Triton X-100 extract of beef heart mitochondria which were loaded as indicated with CAT, ADP, BKA → ADP, ADP → BKA, ADP → CAT. Right plate: The center well contained anti-CAT protein γ-globulin. The outer wells contained CAT-protein, or corresponding amounts of CAT, BSA, DPG and CAT-protein from NCM (Neurospora crassa) or a BKA-protein fragment from BHM.

The highly specific antibodies appear to react exclusively with the protein in the conformation conferred by binding CAT.

The specific differential labeling of the mitochondria with NEM described above had shown that the incorporation of ^{14}C-NEM (which is stimulated by ADP and prevented by CAT) is largely concentrated in a protein fraction also with a MW around 30,000 (20). Therefore, it appears that NEM is incorporated into the same protein to which CAT is bound. It is not possible to use the advantages of CAT binding for purification of the NEM-protein since the binding of CAT and NEM are mutually exclusive. However, with BKA an easier separation of the NEM-labelled protein is possible (H. Aquila and W. Eiermann, unpublished), as shown in Fig. 16, and a corresponding enrichment has been achieved.

CONCLUSIONS

The characterization of the ADP,ATP carrier in the intact inner mitochondrial membrane provides an advanced insight into the carrier mechanism. This characterization is facilitated considerably by: (i) the high concentration of this carrier in the membrane and the resultant possibility to identify its binding sites in relatively high amounts; and (ii) the existence of highly specific inhibitors such as ATR and BKA which in a certain sense show "opposite" properties in binding to the carrier.

The same factors also proved to be advantageous for isolating and characterizing a nearly functional isolated carrier ligand complex. Its ease of isolation and availability of large amounts, make the ADP,ATP carrier a highly suitable target for a number of chemical and biophysical studies in characterizing the structure and function of a transport protein.

ACKNOWLEDGEMENTS

This work was supported by Sonderforschungsbereich 51 (DFG). One of us (B.B.B.), on leave from the Department of Cell Physiology, University of California, Berkeley, gratefully acknowledges support of a J.S. Guggenheim Fellowship.

REFERENCES

1. Klingenberg, M. (in press) In *Membrane-Bound Enzymes*, Plenum Publishing Corp., New York.
2. Klingenberg, M., Riccio, P., Aquila, H., Schmiedt, B., Grebe, K. and Topitsch, P. (1974) In *Membrane Proteins in Transport and Phosphorylation* (Azzone, G.F. *et al.*, eds.) pp. 229–243, North-Holland, Amsterdam.
3. Riccio, P., Aquila, H. and Klingenberg, M. (in press) *FEBS Letters*.
4. Wilbrandt, W. and Rosenberg, T. (1961) *Pharmacol. Rev. 13*, 109–149.
5. Eilam, Y. and Stein, W. (1974) In *Methods in Membrane Biology* (Korn, E., ed.) pp. 283–352, Plenum Publishing Corp., New York.
6. Klingenberg, M. (1967) In *Mitochondrial Structure and Compartmentation* (Quagliariello, E. *et al.*, eds.) pp. 317–319, Adriatica, Bari.
7. Lieb, W.R. and Stein, W.D. (1972) *Biochim. Biophys. Acta 265*, 187–207.
8. Regen, D.M. and Tarpley, H.L. (1974) *Biochim. Biophys. Acta 339*, 218–228.
9. Skou, J.C. (1965) *Physiol. Rev. 45*, 596–625.
10. Kyte, J. (1974) *J. Biol. Chem. 249*, 3652–3660.
11. Weidemann, M.J., Erdelt, H. and Klingenberg, M. (1970) *Eur. J. Biochem. 16*, 313–335.
12. Erdelt, H., Weidemann, M.J., Buchholz, M. and Klingenberg, M. (1972) *Eur. J. Biochem. 30*, 107–122.
13. Klingenberg, M. and Buchholz, M. (1973) *Eur. J. Biochem. 38*, 346–358.
14. Klingenberg, M., Scherer, B., Stengel–Rutkowski, L., Buchholz, M. and Grebe, K. (1973) In *Mechanisms in Bioenergetics* (Azzone, G.F. *et al.*, eds.) pp. 257–284, Academic Press, New York/London.
15. Klingenberg, M., Buchholz, M., Erdelt, H., Falkner, G., Grebe, K., Kadner, H., Scherer, B., Stengel–Rutkowski, L. and Weidemann, M.J. (1971) In

Biochemistry and Biophysics of Mitochondrial Membranes (Azzone, G.F. *et al.*, eds.) pp. 465–486, Academic Press, New York/London.

16. Weidemann, M.J., Erdelt, H. and Klingenberg, M. (1970) *Biochem. Biophys. Res. Commun. 39*, 363–370.

17. Klingenberg, M., Grebe, K. and Scherer, B. (1971) *FEBS Letters 16*, 253–256.

18. Scherer, B. and Klingenberg, M. (1974) *Biochemistry 13*, 161–170.

19. Stoner, C.D. and Sirak, H.D. (1973) *J. Cell Biol. 56*, 51–64.

20. Klingenberg, M. (1974) In *Dynamics of Energy-Transducing Membranes* (Ernster, L. *et al.*, eds.) pp. 511–528, Elsevier Scientific Publishing Co., Amsterdam.

21. Shertzer, H.G. and Racker, E. (1974) *J. Biol. Chem. 249*, 1320–1321.

22. Klingenberg, M. (1975) In *Energy Transformation in Biological Systems, Ciba Foundation Symposium 31*, pp. 105–124, Associated Scientific Publishers, Amsterdam.

23. Klingenberg, M. (1972) In *Mitochondria/Biomembranes*, Proceedings of the 8th FEBS Meeting, pp. 147–162, Elsevier, Amsterdam.

24. Souverijn, J.H.M., Huisman, L.A., Rosing, J. and Kemp, A., Jr. (1973) *Biochim. Biophys. Acta 305*, 185–198.

25. Leblanc, P. and Clauser, H. (1972) *FEBS Letters 23*, 107–113.

26. Vignais, P.V. and Vignais, P.M. (1972) *FEBS Letters 26*, 27–31.

THE REGULATION OF MUSCLE FUNCTION BY THE
TROPOMYOSIN–TROPONIN SYSTEM

H.E. Huxley

MRC Laboratory of Molecular Biology
Cambridge CB2 2QH, England

To understand how muscular contraction is switched on and off by electrical changes in the external membrane of a muscle fibre we have, basically, to understand four different aspects of the process:
(1)   How is the signal for contraction transmitted from the external membrane to the interior of the fibre?
(2)   What is the nature of the signal that finally acts on the contractile material?
(3)   What is the molecular nature of the tension-generating process which is switched on by the signal?
(4)   On what element or elements of the mechanism does the local signal act, and how does it produce its effect?
Earlier work has already answered the first two of these questions fairly comprehensively, and I will summarise the conclusions only very briefly.

THE SIGNAL FOR CONTRACTION

The depolarization of the external membrane is, in skeletal muscle fibres, transmitted inwards as an electrical signal along invaginations of that membrane, which form the so-called transverse tubules, and which extend throughout the interior of the fibre. It is still not completely clear whether the depolarization is normally transmitted by passive spread or as a regenerative action potential, but in either case the transmission time will be only of the order of a millisecond. The transverse tubules form specialised junctions with another component of the sarcoplasmic reticulum, the longitudinal system, which has the form of flattened membrane-bound sacs lying in between the myofibrils and sometimes forming an almost complete sheath around them. In amphibian muscle (e.g. frog sartorius) the junctions occur at the level of the Z-lines, one transverse tubule lying between the longitudinal elements associated with two adjoining sarcomeres, and the junction is known as a triad. In vertebrate muscle, the junctions are located near the A-I boundaries. The reticulum shows up well in the electron microscope and has been described by a number of authors (1-3).

Calcium is sequestered in the longitudinal elements of the reticulum, and in a resting muscle the concentration of free calcium in contact with the myofibrils is kept down to a value in the region of 10^{-7} M by the very efficient calcium pump associated with the reticulum. Following stimulation, calcium is rapidly released from the reticulum so that the concentration of free calcium ions in contact with the myofibrils rises to the region of 10^{-5} M. Since the calcium has to diffuse a distance of only a micron or less, the rise in calcium concentration could follow the release with a delay of only about a millisecond. How the signal for the release of calcium is transmitted from the transverse tubules to the longitudinal reticulum (where the calcium is stored) is not known at present, and is a difficult problem to investigate directly because of the small size of the structures involved. This calcium is the chemical signal for contraction and it switches on the contractile activity of the myofibrils when it reaches them. Thus in principle the signal for contraction can be transmitted from the outside membrane to the whole of the contractile material within a few milliseconds. This is obviously advantageous to an animal which may want to start to move rather quickly. Mechanisms which depended on the diffusion inwards of some substance from the outside membrane would be much too slow to account for the known rates at which muscles can become fully activated, as A.V. Hill pointed out many years ago (4,5). However, the actual delay before a muscle becomes fully active is in many instances a good deal longer than the delay which would be introduced by inward electrical transmission and by interfibrillar diffusion. In a frog sartorius muscle at $0°$ C, for example, these events should still only introduce a delay of a few milliseconds, yet the capacity of the muscle to sustain full tension takes about 50 milliseconds to develop. The steps in the process responsible for the additional delay may include the rate at which calcium is released from the longitudinal reticulum after the signal has arrived, the rate at which calcium binds to the contractile machinery, and the rate at which the individual contractile elements, having bound calcium, can attain a tension-generating configuration. Little is known about these rates at present and they could be a very interesting subject for future research.

When stimulation of the muscle ceases, calcium is taken up by the reticulum again, the free calcium concentration falls to its original very low level, and the contractile machinery is switched off.

THE NATURE OF THE CONTRACTILE PROCESS

Given that the signal received by the actomyosin structure is a change in free calcium concentration, how does the control system operate? The level of our understanding of the control system is a function of the level of our understanding of the contractile mechanism itself. While we might be able to understand a great deal about the contractile mechanism without appreciating how it was switched on and off — indeed this was the case up to a few years ago — we cannot hope to understand much about the control process unless we have a fairly coherent general picture of the mechanism in which it has to intervene. Next, therefore, I will summarise the relevant details of our present picture of the contractile process itself.

All the available evidence is strongly in favour of a sliding filament mechanism in which the proteins actin and myosin are organised into separate but partially overlapping arrays of filaments (Fig. 1) which slide past each other and increase the extent of overlap during shortening of the muscle. This basic model was originally proposed in 1954 independently by A.F. Huxley and R. Niedergenke (6), and by myself and the late Jean Hanson (7). The sliding force between the actin and myosin filaments is believed to be generated by cross-bridges projecting outwards from the myosin filaments and attaching in a cyclical fashion to actin, as suggested by Hanson and Huxley in 1955 (8), splitting ATP as they do so and thereby releasing the energy for contraction. These cross-bridges represent the enzymatically active parts of the myosin molecule — the S_1 subunits — of which there are two per myosin molecule, both attached at the same end of a 2-chain α-helical rod portion, a considerable part of which is bonded side by

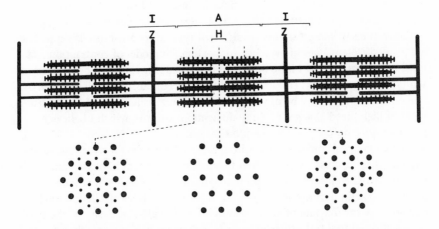

Fig. 1. Diagrammatic representation of construction of striated muscle from overlapping arrays of thick (myosin) and thin (actin-tropomyosin-troponin) filaments. Sliding force between filaments is generated by repetitive cyclic movement of cross-bridges. Attachment of cross-bridges to thin filaments is blocked by regulatory systems when muscle is switched off.

side with other myosin rods to form the backbone of the thick filaments. Present evidence indicates that the active ends of the myosin molecules can hinge out sideways so as to allow the 'head' or S_1 subunits to attach to the actin filaments alongside whose side spacing from the myosin filament backbone varies somewhat depending on muscle length.

In a resting muscle, the cross-bridges are not attached to actin, the filaments can slide past each other readily under an external force, and the muscle is plastic and readily extensible. When a muscle contracts, it is supposed that any particular cross-bridge will first attach to actin in one configuration, probably so that it is approximately perpendicular to the filament axis and then, while still attached, will undergo some configurational change so that its effective angle of attachment alters, i.e. it 'swings' or 'tilts' in such a direction as to pull the actin filament along in the direction of the centre of the A-band. When this movement is complete – the extent of movement probably being 50-100 Å – the cross-bridge can be detached from actin by the binding to it of another molecule of ATP (whose first effect is to dissociate actin and myosin). The ATP is then split by the myosin head, while still uncombined with actin (as indicated by the work of Lymn and Taylor (9)) but the reaction products remain attached to the enzyme, with the complex probably in a "strained" state, until the enzyme once more attaches to actin and releases the stored energy in the form of mechanical work. The combined effect of all the cross-bridges undergoing these asynchronous cycles of attachment, pulling and detachment is to produce a steady sliding force which will continue as long as the muscle is active.

The existence and nature of this cross-bridge cycle is clearly very relevant to considerations of the possibility of different types of control mechanism. There are many different lines of evidence supporting the model of contraction and cross-bridge action that I have described (reviewed, for example, both by myself (10,11) and by A.F. Huxley (12)), but in this discussion I will mention only one of them, which may be of some current interest as it involved the measurement of high speed changes in X-ray diffraction patterns, which I believe has not been done before, even in the physical sciences.

The lateral position of the cross-bridges, i.e. the parameter which measures whether they lie close to the myosin filament backbone, as in resting muscle, or whether they lie closer to the actin filaments, as they will do when they are attached to actin during contraction, has a profound effect on the low angle equatorial X-ray diagram of striated muscle. The relative intensity of the two principal equatorial reflections changes in a characteristic and readily-interpretable way when cross-bridge attachment takes place, and indeed provides the main evidence for the model in which such lateral movement can occur.

If this change in pattern is indeed a consequence of cross-bridge attachment, and if the sliding force is also produced by attached cross-bridges, then the very rapid development of the active state of a muscle following stimulation should be accompanied by an equally rapid change in X-ray pattern. Recent technical developments which I will describe elsewhere have now made it possible to record the X-ray diagrams sufficiently rapidly to investigate this question. Studying single twitches or short tetani of frog sartorius muscle, Dr. J.C. Haselgrove and I have found that the expected changes in the equatorial X-ray diagram do indeed occur with great rapidity after stimulation. At 10° C, for example, the change is 50% complete within about 20 milliseconds, a time when the externally-measured tension (which in a normal isometric contraction is delayed behind 'active state' by the necessity to stretch the series elastic elements before the internal activity can become manifest as tension) has hardly begun to rise.

Similarly, profound changes occur in the axial part of the X-ray diagram during contraction. In a resting muscle, there is a well developed system of layer lines with a 429 Å repeat and a very strong meridional reflection on the third order at 143 Å. There is very strong evidence that this pattern arises from a regular helical arrangement of the cross-bridges on the myosin filaments. In a contracting muscle this whole pattern becomes very much weaker than at rest, indicating that the cross-bridges are much less regularly arranged (13,14) as would be expected if they were undergoing asynchronous longitudinal movements (and possibly lateral ones too) during their tension-generating cycles of attachment to actin. Again, the change in pattern should develop very rapidly after stimulation if tension is indeed generated in the manner we have supposed. Recent measurements which we have made show that this is indeed the case. The change in the pattern begins (at 10° C in frog sartorius muscle) about 10-15 milliseconds after stimulation and is half complete by about 20 milliseconds, again well ahead of the externally measured tension but again according well with the expected temporal characteristics of the active state.

THE MECHANISM OF CALCIUM REGULATION

We believe therefore that we are on fairly secure ground in supposing that the development of the active state in a muscle is characterised by the rapid attachment to actin of the myosin cross-bridges. It is apparent, therefore, that the calcium switch must control whether or not this attachment takes place. In low concentrations of calcium attachment is prevented; in higher calcium concentrations (around $10^{-6} - 10^{-5}$ M) it can take place. Moreover, one can see very readily that there are strong reasons why any control mechanism would have to interrupt the cycle at this particular point. While a muscle's first function is to produce movement by applying a force, it also has to allow that movement to be reversed, either by other muscles or by an external force, at some subsequent time. Thus when a muscle is not actively contracting, it must be kept in a state where it offers a minimum of resistance to passive length changes.

If the cross-bridges between the actin and myosin filaments were attached, but not cycling, the muscle would be in a rigid, inextensible state — indeed this is believed to correspond to the condition of rigor. Thus, it is vital that the switch operates so that when the muscle is switched off, bridges are allowed to detach but are prevented from reattaching again. Furthermore, if attachment to actin for one part of the cycle is required for the continued splitting of molecules of ATP to take place, then the energy producing reactions will automatically be switched off at the same time that the mechanical manifestations of contractile activity cease.

How, then, does calcium effect this transition between rest and activity? In this brief survey, I will not attempt to give a full account of the development of the various lines of evidence that have led to our present views (see, for example, excellent reviews by Ebashi and Endo (15) and by Weber and Murray (16)). Instead, I will summarise the basic features of what seems to me the simplest plausible model with particular reference to the structural evidence.

Control by calcium of the contractile activity of the actin-myosin system in vertebrate striated muscle requires the presence of the regulatory proteins, troponin and tropomyosin. These proteins form part of the structure of the thin filaments (17,18,19). Troponin is a protein complex consisting of three different subunits (one copy of each); one of the subunits binds calcium very tightly, but all three together are required for regulation. Troponin is located along the length of the thin filaments at intervals of approximately 385 Å and the stoichiometry is such that there is one troponin complex and one tropomyosin molecule for every seven actin monomers (20). This stoichiometry is particularly significant if it is recalled that the actin filaments consist of two chains of actin monomers twisted round each other with a helical repeat of 360-380 Å and with a subunit repeat along each chain of 54.6 Å, so that there are approximately seven actin subunits along each chain within each helical repeat, and therefore probably exactly seven subunits per chain for each (approximately) 385 Å troponin repeat.

A particularly striking feature of the mechanism is that troponin is able to control the activity of all the actin monomers in the intervening region, keeping them all switched off in a resting muscle and switched on, probably all of them, during contraction (21). This feature of the mechanism is borne out by *in vivo* studies of the stoichiometry of the regulation process (21,22,23). Troponin appears to be a relatively globular molecule, but tropomyosin is a two-chain coiled-coil α-helical structure (24,25) present in sufficient amounts to provide two continuous strands running along the length of the actin filaments and hence able to make contact with each actin monomer (see Fig. 2). Since the presence of tropomyosin is necessary for the regulation mechanism to operate, it is natural to suppose that the influence of the troponin is transmitted to the actin monomers via the tropomyosin strands. It is therefore of interest to see whether

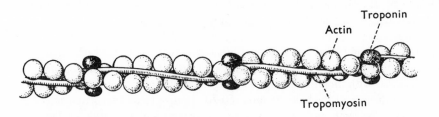

Fig. 2. Diagram showing probable arrangement of tropomyosin, troponin and actin in the thin filaments of muscle (15,17).

structural information about the thin filament complex can provide any clues as to how this mechanism operates.

STRUCTURAL EVIDENCE ABOUT REGULATION

Hanson and Lowy (17) suggested that tropomyosin might lie in the long-pitch grooves of the actin double helix and this idea has been confirmed and extended by later work. Moore, Huxley and DeRosier (26) applying the 3-D reconstruction techniques to electron micrographs of negatively stained specimens, found that preparations of purified actin showed simply a helical arrangement of relatively simple globular units, but that preparations of thin filaments prepared directly from muscle showed additional material asymmetrically located in the long-pitch grooves of the structure. The protein composition of these filaments was not investigated (since they were not available in purified form), but it was noted that the additional material might correspond to the tropomyosin-troponin which one would expect to be present.

O'Brien *et al.* (27) reported that there was usually a characteristic difference in the optical diffraction pattern given by micrographs of paracrystals of pure actin on the one hand and of unpurified actin, or actin mixed with impure tropomyosin, on the other. This difference took the form of an increase of relative intensity of the reflection on the second layer line of the actin pattern (regarding the ~365 Å layer line as the first) in the presence of the additional components, and they pointed out that it was consistent with the presence of additional material within the grooves of the structure. The problem was taken some stages further by Spudich *et al.* (28) using preparations of highly purified actin and of the same actin mixed with the purified regulatory protein complex, giving a system shown to have full calcium sensitivity in an actomyosin ATPase assay. We confirmed the result of Moore *et al.* (26) that pure actin had the form of relatively simple globular units arranged in the two-chain helix, and also the strong indications from O'Brien *et al.* (27) that it was the presence of the regulatory proteins which gave rise to the strengthening of the second and third layer line reflections. Moreover, we were able to 'reconstruct' the complex filaments and show that a fairly continuous strand of material ran along in each of the two grooves in the actin helix, situated asymmetrically, so that if the actin structure is thought of as two strings of monomers twisted around each other, one regulatory strand is more closely associated with one string of monomers, and the other with the other string. The angular position of each strand, measured from the axis of the helix, is about 60-70° relative to the associated actin monomer at the same level, as can be seen from the end-on projection of the structure shown in Fig. 3. In these reconstructions the strong 385 Å axial periodicity believed to arise from the troponin component was not included, so that the resultant structure would be expected to show the location of the tropomyosin only. Thus the tropomyosin appears to run along-side the chains of actin monomers in a position very suggestive of possible interaction. However, it must be realised that the preparations were not made in a calcium-free medium, so that even if the "switched-off" configuration was preserved by negative staining, there is no reason to suppose that the reconstructed tropomyosin configuration in these experiments corresponds to the inhibitory form. Indeed, we will see later it more probably corresponds to the "switched-on" form.

Another 3-D reconstruction result that is relevant here concerns the mode of attachment of the myosin S_1 subunits to the actin monomers, studied on so-called "decorated" actin (i.e. actin mixed with S_1 in absence of ATP) which

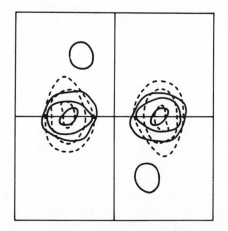

Fig. 3. Arrangement of actin and tropomyosin as seen in end-on helical projection of 3-D reconstruction of composite filament (copied directly from Plate X(a) in Spudich et al. (28). Each tropomyosin strand seems to be associated with its 'own' helix of actin monomers. In this type of view, the position of the tropomyosin peak lies at an azimuthal angle of about 60° with respect to the actin. However, a more reliable estimate of position is given by the difference Fourier (loc. cit Plate X(b)) in which the angle is 70°.

probably corresponds to the "rigor" configuration of the cross-bridges and to their probable position at the end of their working stroke. Moore *et al.* (26) found that the S_1 subunit, besides being tilted and skewed in a characteristic manner, was attached to the actin monomers in a somewhat tangential fashion, so that the end of the S_1 subunit and the contact area between it and the actin extended round into the groove in the actin structure. This can be seen in Fig. 4 showing an end-on projection of a short length of the structure.

It must be recalled that the formation of "rigor links" (i.e. in absence of ATP) between myosin and actin is not inhibited by the regulatory complex even in the absence of calcium, and therefore attachment of the type seen in Fig. 4 would not be inhibited by the troponin-tropomyosin complex.

Nevertheless the projection of myosin into the groove is very suggestive and it seems at least a working possibility that when myosin attaches to actin in the presence of ATP (or its split products), the end of the molecule projects a little bit further into the groove and that the attachment is vulnerable to the influence of tropomyosin. For the tropomyosin to just make contact with the myosin head in the rigor attachment, it would have to lie at an angle of about 60-70° to the respective actin monomer, measured from the helix axis. Fig. 4 shows the position of the rigor links, together with the outline of the probable positions of tropomyosin given by the difference Fourier of Spudich *et al.* (28) (activated state) and from the X-ray data on resting muscle (relaxed state), as discussed below.

The electron microscope evidence has been taken a stage further by the work of Wakabayashi, Huxley, Amos and Klug (29). These workers were concerned to see whether a difference in the position of tropomyosin could be detected in the electron microscope images of thin filaments in the active and in the inhibited state. Since control of the calcium level could not be assured during the negative staining process, and since much more satisfactory paracrystalline arrays of filaments (needed for good three-dimensional image reconstruction) were formed in the presence of calcium than in its absence, the inhibited state of the filaments was produced by including only troponin-I (inhibitory subunit) and troponin-T (tropomyosin combining subunit), together with tropomyosin, in the regulatory complex with which actin was combined. In these circumstances, the filaments are in the 'switched-off' state (i.e. unable to interact with myosin) whether calcium is present or not, but it is reasonable to suppose that the mechanism of inhibition is the same as that involved in the absence of calcium, when the calcium-binding subunit (troponin-C) is present. The structure of the inhibited filaments was compared with that of actin plus tropomyosin in the absence of troponin, when the filaments are always in the 'switched-on' state. It was found that a very marked difference in structure did indeed occur between the two different forms. In the active filaments, the tropomyosin

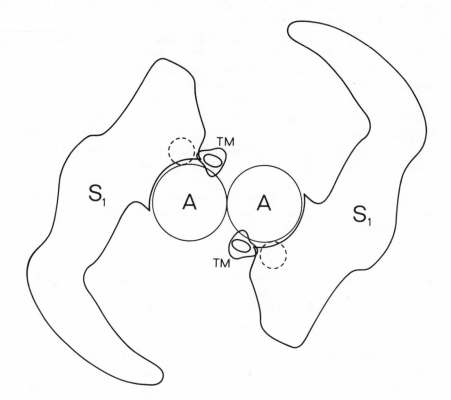

Fig. 4. Composite end-on view of actin-tropomyosin-S_1 structure. The shape of the S_1 subunits and the position at which they attach to the actin is copied directly from Plate IX of Moore et al. (26). The actin structure (24 Å diameter spheres placed at a radius of 24 Å) is that found by Haselgrove (36) to give the best agreement (for a simple model) with the strong features of the X-ray diagram. The two tropomyosin (TM) positions correspond to: solid contours, activated state taken directly from difference Fourier (Plate X(b)) of Spudich et al. (28); dotted contours, relaxed state, based on radial position of reflection on third layer line of relaxed muscle. (This assumes that the "sense" of the azimuthal position of tropomyosin with respect to the attachment site of S_1 has been chosen correctly, which has not yet been proved.) The possible way in which tropomyosin could block the attachment of a cross-bridge is very evident.

strands made relatively loose contact with the actin monomers, and though still asymmetrically disposed in the long pitch grooves between the two 'strings' of actin monomers, lay relatively closer to the centre of those grooves. In the inhibited filaments, the tropomyosin strands made very close contact with the actin monomers, in a position which could plausibly block the attachment of myosin, and further away from the centre of the long pitch helical grooves. The results therefore provide further strong support for the idea that regulation is associated with a significant lateral movement of the tropomyosin strands, and may be effected by a relatively simple and straightforward mechanism by which tropomyosin physically blocks the attachment of myosin when the thin filaments are in the inhibited state and myosin is carrying ATP (or its split products); and that the actin structure itself undergoes little or no change.

X-RAY EVIDENCE

This model is strongly supported by X-ray diffraction evidence (30,31,32,33, 34,35). The pitch and subunit repeat of the actin filaments remain virtually constant in an actively contracting muscle, and the main pattern of actin reflections do not alter their relative intensities either, indicating that the major part of the actin filament structure remains unchanged too. As was pointed out at the time, this does not rule out repetitive cyclical changes taking place in a small part of the filament structure at any given time; but since calcium will be tightly bound to the troponin of the thin filaments for the whole period of activity, any changes produced by it would be maintained virtually continuously throughout contraction; and hence they cannot be of a kind which affects the helical parameters of the thin filaments. This view is strengthened by the observation (30) that even in rigor (when the level of free calcium is normally high) the subunit repeat in the actin filaments is unchanged, and any change in the pitch of the actin helix is either relatively small or absent. Experience with other systems (e.g. haemoglobin) indicates that even relatively small internal changes in structure inside a molecule or subunit are liable to produce appreciable changes in the way the subunits pack together, and I therefore believe that the most likely interpretation of the constancy of the actin structure is a virtually complete constancy in the internal structure of the actin monomers themselves when they are switched on or off. This strongly favours the possibility that regulation is affected by a steric blocking mechanism involving tropomyosin movement (31,32,33).

Evidence for tropomyosin movement comes from X-ray diffraction observations of some of the relatively weak parts of the diagram which arises from the thin filaments in muscle. Whilst, as I have already mentioned, the main parts of that diagram remain unaltered during contraction, there are reflections, rather far away from the meridian, on the second and third layer-lines (specifically at axial spacings of approximately 190 Å and 131 Å and radial spacings of about 0.021 Å^{-1} and 0.026 Å^{-1}, respectively) in which very significant changes can be seen. There is a marked decrease in intensity of the third layer line reflection in a contracting muscle and a marked increase in intensity on the second layer line. There are strong technical grounds for believing that these parts of the diagram arise in very large part from the tropomyosin component of the thin filaments, and that the changes in them arise from a change in the position of the tropomyosin. One can say in a very general way that the change from a relatively strong third layer line to a relatively strong second layer line, unaccompanied by changes on higher layer lines, indicates an overall lateral movement by a helically wound tropomyosin strand from a position nearer to one-third of the way between the two actin strands to one nearer half-way, i.e. in the direction required by the model for regulation which we have been discussing.

These arguments can in fact be made rather detailed and rigorous (36,37, 38) and it can be shown that the changes in the X-ray diagram upon contraction can be very well accounted for by a movement of tropomyosin, away from a position which might block myosin attachment, to one lying 10-15 Å nearer the centre of the long pitch groove, a similar movement to that indicated by the 3-D reconstruction results.

There are several attractive features of this type of mechanism and I will mention three of them very briefly. First of all, there is the basic one, that if the tropomyosin strand moves as a whole and possibly as a relatively rigid structure (it is a two-chain coiled-coil molecule) it is easy to see how one tropomyosin molecule could regulate the seven actin monomers over which it extends. Secondly, in such a model, what happens at one myosin-binding site on actin (for example, attachment of a myosin head not loaded with ATP) can influence what happens on adjoining ones, by causing a displacement of tropomyosin. This could account for some of the very remarkable cooperative effects in the interaction, as discussed by Bremel and Weber (23). Thirdly, the somewhat non-specific nature of the interaction that would be required between tropomyosin and the successive actin monomers would accord very well with the known features of the amino acid sequence of tropomyosin (39), in which there is not a precise sequence of amino acid residue, which repeats at positions about 55 Å apart along the coiled-coil structure, but merely a tendency for similar types of residue to recur with this period. One would imagine that a control mechanism mediated by a structural change *within* the actin monomers and transmitted along by tropomyosin would call for a very precise and exact molecular interaction between the two proteins at each successive actin monomer repeat; and yet this is not found. Additionally, the fact that hybrid systems, involving types of actin which are not associated with troponin *in vivo* (e.g. molluscan actin), can be regulated by vertebrate troponin-tropomyosin again suggests that an absolutely exact and specific pattern of interaction between each actin monomer and tropomyosin is not a necessary requirement for regulation.

CONTROL OF TROPOMYOSIN MOVEMENT

Finally, we must discuss whether there are any clues as to how the required movement of tropomyosin might be brought about. Actin and tropomyosin combine together quite strongly even in the absence of troponin, but in the complex so formed, actin retains its ability to activate myosin ATPase — indeed in some circumstances (16) this ability may be enhanced. Thus tropomyosin is, in these circumstances, held along the actin helix in such a way that a high proportion, if not all, of the myosin-binding sites are available. The intact troponin complex will combine not only with actin-tropomyosin together but with actin and with tropomyosin separately. Thus actin and tropomyosin must, in the region where troponin attaches, be held in a specific configuration relative to each other by bonds additional to those between actin and tropomyosin directly. Accordingly, one can reasonably envisage a model in which, in the absence of calcium, tropomyosin is held in the blocking position, and where, in the presence of calcium, some structural change within the troponin complex occurs which allows or causes tropomyosin to move towards the centre of the long-pitch grooves in the actin double-helix and away from the position which sterically blocks myosin attachment (37).

Very strong positive evidence in favour of such a model has been found by Hitchcock, Huxley and Szent-Gyorgyi (40), who studied the binding to actin-tropomyosin of a combination of troponin C plus troponin I, i.e. a system in which the third part of the troponin complex, troponin T, which binds strongly to tropomyosin, was omitted. In the absence of calcium, troponin [I + C] binds strongly to actin-tropomyosin filaments (though it does not, when present in amounts equivalent to the normal stoichiometry, inhibit their ability to activate myosin ATPase). In the presence of calcium, however, at calcium concentrations (10^{-5} M) just sufficient to switch off inhibition in a fully constituted system, troponin [I + C] detaches from actin-tropomyosin filaments and can be separated from them by ultracentrifugation. This finding suggests a so-called 'two-site model' (40) in which the troponin complex binds to actin-tropomyosin via two separate sites on troponin. One of these, on the troponin-T subunit, binds permanently (i.e. in both the presence and absence of calcium) to tropomyosin, so that the intact troponin complex always remains attached to the thin filaments. The other binding site is calcium sensitive and attaches to actin-tropomyosin in the absence of calcium only. The formation of *this* link holds tropomyosin in the blocking position providing the troponin-T subunit is present. When calcium is present, the link is broken, and tropomyosin can move to a non-blocking position. Again, the evidence is consistent with a rather simple steric model.

Margossian and Cohen (41) also support the same type of model, and have observed an increase in the strength of the binding between troponin-C and troponin-T in the presence of calcium which could reflect a change in the state of the troponin complex associated with tropomyosin movement. Thus, in general, there are several interesting pieces of evidence which fit plausibly into a model for tropomyosin movement mediated by troponin. However, the real details of such a model have still to be worked out, and the model can only be accepted as a useful working hypothesis until that is done.

I should also mention that not all muscles are regulated by the troponin-tropomyosin system, though as far as we know they are all regulated by changes in the concentration of free calcium ions over similar ranges of values. Thus molluscan for example (42) is regulated by calcium sites on the myosin component and lacks troponin, and certain other species have both actin-linked and myosin-linked regulatory systems present in muscles at the same time (43).

In the earlier part of my discussion of this topic I mentioned the desirability of being able to account for the rate constants observed by mechanical measurements during the activation and relaxation of muscle in terms of rate constants of the underlying processes of calcium release, interaction with contractile proteins and retrieval by the sarcoplasmic reticulum, which could be investigated by various physical and chemical means. Clearly, the rate constants involved in the troponin interactions – and in the interaction in the myosin-linked systems too – must also be included in this category.

REFERENCES

1. Porter, K.R. and Palade, G.E. (1957) *J. Biophys. Biochem. Cytol. 3*, 269-300.
2. Andersson-Cedergren, E. (1959) *J. Ultrastruct. Res. Suppl. 1*.
3. Franzini-Armstrong, C. and Porter, K.R. (1974) *Nature 202*, 355.
4. Hill, A.V. (1948) *Proc. Roy. Soc. B135*, 446-453.
5. Hill, A.V. (1949) *Proc. Roy. Soc. B136*, 399-420.
6. Huxley, A.F. and Niedergerke, R. (1954) *Nature 173*, 971-973.
7. Huxley, H.E. and Hanson, J. (1954) *Nature 173*, 973-976.
8. Hanson, J. and Huxley, H.E. (1955) *Symp. Soc. Exptl. Biol. 9*, 228-264.
9. Lymn, R.W. and Taylor, E.W. (1970) *Biochemistry 9*, 2975-2983.
10. Huxley, H.E. (1969) *Science 164*, 1356-1366.
11. Huxley, H.E. (1971) *Proc. Roy. Soc. B178*, 131-149.
12. Huxley, A.F. (1974) *J. Physiol. 243*, 1-43.
13. Elliott, G.F., Lowy, J. and Millman, B.M. (1967) *J. Mol. Biol. 25*, 31-45.
14. Huxley, H.E. and Brown, W. (1967) *J. Mol. Biol. 30*, 383-434.
15. Ebashi, S. and Endo, M. (1968) *Prog. Biophys. Mol. Biol. 5*, 123-183.
16. Weber, A. and Murray, J.M. (1973) *Physiol. Rev. 53*, 612-673.
17. Hanson, J. and Lowy, J. (1963) *J. Mol. Biol. 6*, 46-60.

18. Pepe, F.A. (1966) *J. Cell Biol. 28*, 505-525.
19. Ohtsuki, I., Masaki, T., Nonamura, Y. and Ebashi, S. (1967) *J. Biochem. 61*, 817.
20. Potter, J.D. and Gergely, J. (1974) *Biochemistry 13*, 2697-2704.
21. Weber, A. and Bremel, D.R. (1971) In *Contractility of Muscle Cells and Related Processes* (Podolsky, R.J., ed.) p. 37, Prentice-Hall, Englewood Cliffs.
22. Spudich, J.A. and Watt, S. (1971) *J. Biol. Chem. 247*, 4866-4871.
23. Bremel, D.R. and Weber, A. (1972) *Nature 238*, 97-101.
24. Cohen, C. and Szent-Gyorgyi, A.G. (1957) *J. Am. Chem. Soc. 79*, 248.
25. Caspar, D.L.D., Cohen, C. and Longley, W. (1969) *J. Mol. Biol. 41*, 87-107.
26. Moore, P.B., Huxley, H.E. and DeRosier, D.J. (1970) *J. Mol. Biol. 50*, 279-295.
27. O'Brien, E.J., Bennett, P.M. and Hanson, J. (1971) *Phil. Trans. Roy. Soc. London B261*, 201-208.
28. Spudich, J.A., Huxley, H.E. and Finch, J.T. (1972) *J. Mol. Biol. 72*, 619-632.
29. Wakabayashi, T., Huxley, H.E., Amos, L. and Klug, A. (1975) *J. Mol. Biol. 93*, 477-497.
30. Huxley, H.E. and Brown, W. (1967) *J. Mol. Biol. 30*, 383-434.
31. Huxley, H.E. (1970) *Abstr., 8th Int. Cong. Biochem. (Interlaken)*.
32. Huxley, H.E. (1971) *Abstr., Am. Biophys. Soc.*, 235(a).
33. Huxley, H.E. (1971) *Biochem. J. 125*, 85P.
34. Vibert, P.J., Lowy, J., Haselgrove, J.C. and Poulsen, F.R. (1971) *Abstr., 1st Eur. Biophys. Cong. (Vienna)*.
35. Vibert, P.J., Haselgrove, J.C., Lowy, J. and Poulsen, F.R. (1972) *Nature New Biol. 236*, 182-183.
36. Haselgrove, J.C. (1972) *Cold Spring Harbor Symp. Quant. Biol. 37*, 341-352.
37. Huxley, H.E. (1972) *Cold Spring Harbor Symp. Quant. Biol. 37*, 361-376.
38. Parry, D.A.D. and Squire, T.M. (1973) *J. Mol. Biol. 75*, 33-58.
39. Sodek, J., Hodges, R.S. and Smillie, L.B. (1972) *Proc. Nat. Acad. Sci. (U.S.A.) 69*, 3800-3804.
40. Hitchcock, S.E., Huxley, H.E. and Szent-Gyorgyi, A.G. (1973) *J. Mol. Biol. 80*, 825-836.
41. Margossian, S.S. and Cohen, C. (1973) *J. Mol. Biol. 81*, 409-413.
42. Kendrick-Jones, J., Lehman, W. and Szent-Gyorgyi, A.G. (1970) *J. Mol. Biol. 54*, 313-326.
43. Lehman, W., Kendrick-Jones, J. and Szent-Gyorgyi, A.G. (1972) *Cold Spring Harbor Symp. Quant. Biol. 37*, 319-330.

VI
Excitable Membranes

ORGANISATION OF THE SODIUM CHANNELS
IN EXCITABLE MEMBRANES

Richard D. Keynes

Physiological Laboratory
Cambridge, England

The energy for conduction of the nervous impulse comes directly from the pre-existing concentration gradients of Na^+ and K^+ ions, and thus indirectly from the pumping mechanism that builds up these gradients. In speaking about the sodium channels in nerve membranes, I wish first to emphasize that I am concerned solely with those responsible for electrical excitability, and not with the sodium pump. There is now no doubt whatever that the uphill movements of Na^+ mediated by the operation of the sodium pump, that is by Na, K-ATPase, are entirely independent of the downhill movements through the excitability channel, if only because there are a whole variety of ways of putting one of these pathways completely out of action without any effect on the other. Let us, therefore, take the ionic concentration gradients for granted, and consider some new evidence on the way in which the sodium channels perform their essential function of opening and closing appropriately under the influence of the membrane potential.

The technique on which we depend for observing both the changes in ionic conductance and, as I shall explain, the movements of the charged gating particles that control the channels, is voltage-clamping, as first applied by Hodgkin, Huxley and Katz (1). This consists basically in a feed-back arrangement that enables the membrane to be held at a given potential and then driven abruptly to a new level or sequence of levels while the resulting changes in membrane current are measured. On depolarization from the resting potential of −60 mV to, say, +20 mV, there is an early inward current carried by Na^+, followed by a delayed outward current carried by K^+. After separating the contributions of the two ions, Hodgkin and Huxley (2) found that the time course of the change in sodium conductance, g_{Na}, consisted in a phase of activation which was well fitted by the expression

$$g_{Na} = \bar{g}_{Na}(1 - e^{-t/\tau_m})^3 \tag{1}$$

followed by an inexorable shutting off which they termed 'inactivation,' and which followed a single exponential

$$g_{Na} = \bar{g}_{Na}e^{-t/\tau_h} \tag{2}$$

If the membrane was repolarized at the peak of the conductance change, the shutting off was a rapid exponential with a time constant $\frac{1}{3}\tau_m$. The voltage-dependent time constants τ_m and τ_h were defined by the equations

$$g_{Na} = \bar{g}_{Na}m^3h \tag{3}$$

$$\frac{dm}{dt} = \alpha_m(1 - m) - \beta_m m \tag{4}$$

$$\frac{dh}{dt} = \alpha_h(1 - h) - \beta_h h \tag{5}$$

$$\tau_m = \frac{1}{\alpha_m + \beta_m} \tag{6}$$

$$\tau_h = \frac{1}{\alpha_h + \beta_h} \tag{7}$$

where $\bar{g}_{Na}$ is a constant representing the peak sodium conductance, and m and h are dimensionless quantities that vary between 0 and 1. The α's and β's are forward and backward rate constants that depend in a defined manner on potential.

As Hodgkin and Huxley (2) were careful to point out, all we have so far is simply an accurate mathematical description of the time course of the conductance change, which unfortunately can be fitted by an embarrassingly large number of molecular models, because there are many quite different physical processes that can be described equally well by two first order differential equations like (4) and (5). Is it possible somehow to get at the control mechanism itself?

Since the voltage dependence of the rate constants α and β implies that the gate must be charged, its movements when the electric field alters must result in the passage of a small surge of current that necessarily precedes the flow of ionic current. Early attempts to detect such a current were unsuccessful, because it was swamped by the hundred times larger surge of sodium current; but we now have at our disposal a highly effective blocking agent, the Japanese puffer-fish poison, tetrodotoxin (TTX), which at a concentration of well under 1 μM completely stops up the sodium channels and enables us to circumvent the difficulty. The sodium 'gating current,' as it has come to be called, was first recorded at Woods Hole in 1972 by Armstrong and Bezanilla (3), and shortly afterwards by Rojas and myself (4) at Plymouth. The records that I shall show were mainly made by Professor Rojas.

Suppose we take a squid axon in which the ionic channels have been completely blocked, sodium by substituting Tris buffer in the external medium and applying TTX, and potassium by perfusing with caesium. When we apply a large voltage-clamp pulse, a surge of capacity current is seen as the potential across the 1 μF/cm^2 membrane capacity changes with a time constant of the order of 15 sec. Recorded with low current gain and a fast sweep speed, the capacity transient looks much the same whether the pulse is negative or positive. When, however, the tails of the records are examined with high current amplification and a slower sweep, as in the upper part of Fig. 1, they are seen to be asymmetrical. For the hyperpolarizing pulse, the capacity transient is over in about 50 μsec, but for the depolarizing pulse, that is for a change in potential that would open the gate, it is evident that the transient has an additional slow component added to it. If the symmetrical capacity transient is eliminated by adding the two records together, it is seen that the asymmetrical component rises sharply at the start of the pulse and then declines exponentially. On a semilogarithmic plot it lies on a very good straight line. At the end of the pulse, the asymmetrical displacement current flows back again with a rather shorter time constant.

Processing the records in this manner is rather laborious, and we normally use a signal averager both to get rid of the capacity transient and to improve the signal-to-noise ratio. We thus obtain families of gating current records for pulses of different sizes like that shown in Fig. 2. Here it may be seen that as the potential during the pulse becomes more positive, the amplitude of the gating current increases and its 'on' time constant shortens, though it remains roughly constant for the return to the holding potential at 'off.' In this axon, as was often the case, there was some rectification of the residual leakage current during the pulse that resulted in the addition of a rectangular pedestal to the exponentially declining gating current.

I must next quickly summarize the reasons why you should believe me when I assert that this asymmetrical displacement current is caused by the movement of mobile charges or dipoles that form an integral part of the mem-

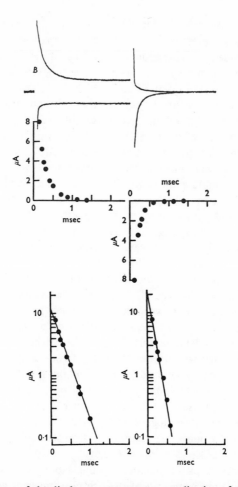

Fig. 1. Asymmetry of the displacement current on application of equal and opposite voltage-clamp pulses to a squid axon perfused with 300 mM-CsF and bathed in Na- and K-free saline containing 1000 nM-tetrodotoxin. The top traces are single-sweep records of the membrane current for ±120 mV pulses, applied at a holding potential of -100mV. The difference between them is plotted beneath on linear and logarithmic scales. (Fig. 4B from Keynes and Rojas (5).)

brane. First, the total amount of charge transferred one way at the start of the pulse should exactly balance the amount transferred back at the finish. At least for short pulses this holds good, since a plot of the area under the exponential at 'on' against the area at 'off' has a slope of 45° (5). For very long pulses it looks as though there are departures from equality, perhaps because some of the mobile charges undergo a conformational change during

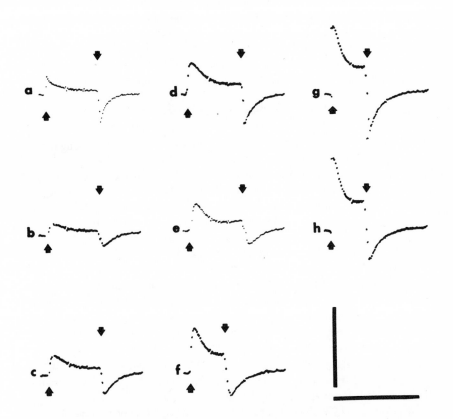

Fig. 2. A family of gating current records obtained by summation with a signal averager of the membrane currents for 60 positive and 60 negative pulses which started and finished at the arrows. The axon was perfused with 55 mM-CsF, and bathed in Na- and K-free saline containing 300 nM-saxitoxin. Pulse amplitude a-h, 40-110 mV; holding potential, -70 mV; vertical bar, 5.56 μA; horizontal bar, 2500 μsec; membrane area, 0.06 cm^2; temperature, 7° C. (Fig. 6 from Keynes and Rojas (5).)

prolonged depolarization that increases their relaxation time. This is one of the points on which the experimental evidence is still far from complete. Second, the charge transfer reaches a definite saturation level when sufficiently large pulses are applied (5). Third, the number of charges, and hence the total charge transferred during a large pulse, is unaffected by changing the temperature, although the relaxation time constants have Q_{10}'s of 3 (5).

A quantitative examination of the charge distribution as a function of potential shows that it conforms rather well with a Boltzmann distribution of particles with two alternative energy states whose effective valency, calculated from the slope at the midpoint of the S-shaped steady-state distribution curve, is 1.3 (5). Changes in the fixed-charge surface potential brought about either by reducing the internal ionic strength or by raising the external calcium concentration, shift the charge distribution curve in a positive direction along the voltage axis by just about the expected amount without affecting its slope (5).

In order to decide whether these mobile charges are indeed the hypothetical sodium gating particles, we have made a quantitative comparison between the kinetic and steady-state properties of the two systems (6). The gating current relaxation time constant τ_{Ig}, plotted against membrane potential, lies on a symmetrical bell-shaped curve whose maximum falls at about $-40\,mV$ with an absolute magnitude of just under 500 μsec at $6°$ C. The sodium conductance time constant τ_m, defined and measured according to equations (1) to (6), lies on a virtually identical curve with a maximum of the same size at $-36\,mV$. The agreement between the positions and slopes of the steady-state charge distribution curves on the one hand, and of the curve relating the cube root of the sodium conductance to potential on the other, is nearly as good. There is thus good reason not only for identifying the asymmetrical displacement current with the operation of the sodium gating mechanism, but also for concluding that there are precisely three gating particles for each sodium channel, all of which have to be in their new position before the channel opens for the passage of Na^+ ions. On this basis, the total amount of mobile charge corresponds to the presence of about 500 channels, each with three gating particles, in 1 μm^2 of membrane. This figure agrees well with estimates based on TTX binding in the squid giant axon (7,8).

A point on which there has been some controversy concerns the shutting-off time constant for the sodium conductance, $\tau_{Na,off}$, which has been stated (5,9) to be too long to fit with one-third of $\tau_{Ig,off}$. There is also a problem in explaining why $\tau_{Ig,off}$ should increase appreciably with pulse size (5,10). Later measurements of $\tau_{Na,off}$ and $\tau_{Ig,off}$ made in the same axons with particular attention to the avoidance of errors in the conductance time constant arising from incomplete compensation for the Schwann cell series resistance have confirmed that $\tau_{Ig,off}$ is dependent on pulse size while $\tau_{Na,off}$ is not, and have shown that the one-third relationship does in fact hold good for large pulses (6). This is what would be expected to occur if the opening of the channel involved two steps as in the scheme depicted in Fig. 3: first the voltage-dependent transition of the gating particles from the resting state A to an active configuration B, and second a rapid interaction between them at any channel where all three are in the appropriate position to the conducting state B_3^*. On the reasonable supposition that this interaction might modi-

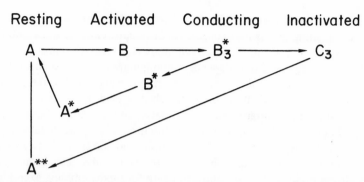

Fig. 3. A possible sequence of events during the application of a depolarizing voltage-clamp pulse.

fy the conformation of the particles in such a way as to increase their relaxation time, making $\tau_{B^*\to A^*}$ greater than $\tau_{B\to A}$, one can account rather neatly for the observed behaviour of the time constants.

Another feature of Fig. 3 to which attention should be drawn is that inactivation is shown as involving a further configurational change in the triad of gating particles from state B_3^* to state C_3. Possibly the simplest way of accounting for the observed time course of inactivation would be to suppose that in addition to the three activating m particles there is a single inactivating h particle, which moves independently but rather more slowly to a channel-blocking position when the membrane is depolarized. However, it can be calculated from the curves relating τ_h and h_∞ to potential that the movement of any such h particles would contribute a readily detectable slow component to the gating current that has not in fact been seen in any of our records. Although I would not claim that our evidence finally excludes the existence of independent h particles, I feel that it is preferable to explain the absence of a component of the gating current displaying h kinetics by supposing that the inactivation step consists in a mainly time-dependent transformation of the m particles from B_3^* to C_3. As has been shown by Goldman (11), there is no mathematical difficulty in reconciling this type of suggestion with Hodgkin-Huxley kinetics.

In constructing possible models of the sodium channel one is still limited more by one's imagination than by hard facts. One feature that seems to be more or less firmly established by the recent evidence is that the filter mechanism that determines the selectivity of the channel towards Na^+ ions is located at the outside of the membrane, because TTX applied internally has no effect, and that it operates independently of the voltage-sensitive gating mechanism, because although the gating currents are normally recorded after first blocking the sodium channels with TTX, they have been shown to be exactly the same in the absence of TTX (5). Another reasonably secure conclusion is that the gating mechanism consists of three apparently identical subunits, and that both activation and inactivation are explicable in terms of a sequence of voltage- and time-dependent configurational changes in these subunits. As to the exact nature of the structural changes within the gating particles resulting from alteration of the electric field, we are still completely in the dark. An important new fact is that since a potential change of 19 mV brings about an e-fold change in the distribution of the individual particles, their effective valency is 1.3. By this I mean that the sum of the actual charge on the particles multiplied by the fraction of the field through which each charge moves is 1.3, and unfortunately for model-building exercises there are many different ways in which this figure might be achieved. At one extreme, a globular dipole with a charge of 2 spanning the whole dielectric and rotating through 180° at 45° to the plane of the membrane would have an effective valency of 2 sin 45 = 1.3. At the other, one could envisage a helical structure carrying a larger number of charges each moving through an appropriate fraction of the field, which would open and close an aqueous channel between itself and its two neighbours either by stretching and contracting like a concertina, or by twisting along its own track so as to retract and restore a plug at one end of the channel. I look forward to a time when a combination of biophysical and biochemical approaches will reveal the innermost secrets of the molecular structure of the sodium channel, but I fear that we are still a long way from achieving this goal.

REFERENCES

1. Hodgkin, A.L., Huxley, A.F. and Katz, B. (1952) *J. Physiol. 116*, 424-448.
2. Hodgkin, A.L. and Huxley, A.F. (1952) *J. Physiol. 117*, 500-544.
3. Armstrong, C.M. and Bezanilla, F. (1973) *Nature 242*, 459-461.
4. Keynes, R.D. and Rojas, E. (1973) *J. Physiol. 233*, 28-30P.
5. Keynes, R.D. and Rojas, E. (1974) *J. Physiol. 239*, 393-434.
6. Keynes, R.D. and Rojas, E. (in press) *J. Physiol.*
7. Levinson, S.R. and Meves, H. (in press) *Phil. Trans. R. Soc. Lond. B.*
8. Keynes, R.D., Bezanilla, F., Rojas, E. and Taylor, R.E. (in press) *Phil. Trans. R. Soc. Lond. B.*
9. Armstrong, C.M. and Bezanilla, F. (1974) *J. Gen. Physiol. 63*, 533-552.
10. Meves, H. (1974) *J. Physiol. 243*, 847-867.
11. Goldman, L. (1975) *Biophys. J. 15*, 119-136.

SYNAPTIC RECEPTOR PROTEINS
AND CHEMICAL EXCITABLE MEMBRANES

E. De Robertis

Instituto de Biología Celular, Facultad de Medicina
Universidad de Buenos Aires
Buenos Aires, Argentina

INTRODUCTION

The concept of synaptic receptors originated at the end of last century from the finding that, at minute concentrations, certain neuroactive drugs produced a physiological response. The studies of Langley (1) on the effects of atropine and pilocarpine on the submaxillary gland and particularly those (2) with nicotine and curare, on the neuromuscular junction, further strengthened the concept of receptors. Langley postulated that the neural region of the muscle should contain a "receptive" substance which is activated by nerve impulses and blocked by curare. A further development was the recognition that the receptive substance, which confers the chemosensitive property to a particular cell or tissue, should reside on the cell surface. In 1926 Cook (3) observed that methylene blue, applied to the frog heart surface, could antagonize the effect of acetylcholine. More direct evidence of the localization of synaptic receptors at the cell membrane came from the work of Del Castillo and Katz (4), who applied microionophoretically acetylcholine at the motor end plate. They found that the neurotransmitter was active when applied on the surface of the neuromuscular region, but was ineffective if the injection was done into the post-synaptic region of the muscle cell. This experiment supported the idea that receptors are localized in specialized regions of the cell membrane and that the receptor sites are pointing toward the outer surface. After that time, with the development of electron microscopy, autoradiography and cell fractionation techniques, the localization of receptors at the cell membrane became definitely established and it was accepted that they are genetically determined macromolecules, localized within the structure of the postsynaptic membrane, that have specific binding sites to recognize the neurotransmitters (see ref. 5).

The recent advances on the chemical and physicochemical organization of the cell membrane (see the corresponding chapters of this Symposium) lead to the study of the primary events taking place at the cellular level, within the realm of the cell membrane. As shown in Fig. 1, we can assume that the ligand-receptor interaction triggers a conformational change in the receptor protein which initiates a series of events in the membrane that result in the translocation of ions, Ca^{2+} displacement, changes in membrane potentials, activation of adenylate cyclase and other metabolic changes that eventually will result in the physiological response. The most decisive advance made recently in this field has been the demonstration that synaptic receptors can be isolated as biochemical entities and that some of their molecular properties, particularly the primary ligand-receptor interaction and the induced conformational change, can be studied by means of biochemical and biophysical techniques (see refs. 6,7).

EXTRACTION OF RECEPTOR PROTEINS

The techniques for the isolation of synaptic receptor proteins bear some similarities with those used for enzymes; however, while in the case of the enzymes it is possible to assay *in vitro* the activity at different stages of isolation and purification, in this case, as soon as the integrity of the cell is lost, the physiological response is no longer available. The study that is generally carried out is to make the binding with specific ligands to the subcellular fractions that contain the receptor and then try to separate the speci-

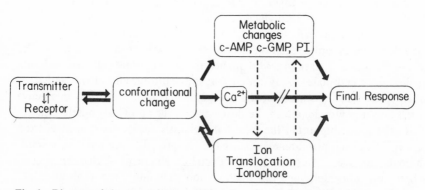

Fig. 1. Diagram of the transmitter-receptor interaction at the cell membrane with the different steps that may lead to the final response, i.e., the contraction or relaxation of muscle, the secretion of a gland, and so forth.

fic proteins from those membrane fractions. For example, in the case of the brain we separated the nerve-ending membranes, carrying the postsynaptic membrane, and also the synaptic complex formed by the two synaptic membranes (8). In other tissues, such as the electroplax, skeletal and cardiac muscle, the cell membranes, where the chemosensitive sites are localized, were separated.

Other difficulties involved in the isolation of receptor proteins come from the fact that they are present in very minute concentrations and that they are intrinsic or integral proteins according to the classification of Singer and Nicholson (9). Such proteins need the action of drastic procedures such as organic solvents, strong detergents or chaotropic agents, to be removed from the membrane structure. Integral proteins are rather hydrophobic and are surrounded by lipid molecules within the cell membrane. Once separated these proteins cannot be dissolved in water unless they are surrounded by a large number of detergent molecules which replace the lipids.

These technical problems explain the two main approaches followed to separate the receptor proteins, which involve the use of organic solvents (10) or detergents (11). There is a large literature regarding the isolation of the cholinergic receptor with detergents. Table I shows some of the proteins that have been isolated by the use of organic solvents; in these cases the proteins are solubilized in organic media, such as chloroform-methanol, butanol, N-N-dimethylformamide, etc. Folch-Pi and Lees (12) first isolated from white matter hydrophobic proteins, which they designated proteolipids. These proteins are the major constituents of myelin and are found as part of the integral proteins of all biomembranes (13). The proteolipids from myelin do not show high affinity binding for neuroactive drugs (10,14), but those proteolipids extracted from the different tissues mentioned in Table I show specific and high affinity binding for cholinergic drugs, adrenergic agents, indolamines and certain amino acids.

SEPARATION OF RECEPTOR PROTEINS AND BINDING STUDIES

In most of the works mentioned in Table I the separation of receptor proteins was done by column chromatography with the organophilic gel, Sephadex LH20. The standard procedure developed by Soto *et al.* (33) involves elution with chloroform, followed by a series of solvents of increasing polarity. This method is also used for binding studies, since the free ligand is retained by the column or is eluted only with solvents of high polarity (14). The general procedure consists of mixing the total lipid extract (TLE) with the ligand at low concentrations (10^{-8}-10^{-6}M) and, after a time, to load it on the column and to elute the various protein peaks (Fig. 2). Several of

TABLE I

Hydrophobic Receptor Proteins Isolated with Organic Solvents

Tissue	Drug Bound	Ref.
	Cholinergic Proteins	
1.* Cerebral cortex	dimethyl (^{14}C)-d-tubocurarine	(10)
2.* Cerebral cortex	(^{3}H)-atropine and atropine sulphate	(15)
3.* Cerebral cortex	(^{14}C)acetylcholine, dimethyl (^{14}C)-d-tubocurarine	(16)
4.* Insect brain	(^{14}C)-acetylcholine, (^{14}C)-decamethonium	(17)
	Cholinergic Nicotinic Proteins	
5.* Electric tissue (*Electrophorus*)	(^{14}C)-acetylcholine, (^{131}I)-bungarotoxin	(18) (19)
6. Electric tissue (*Torpedo*)	(^{14}C)-hexamethonium, (^{3}H)-Tdf**	(18)
7.* Skeletal muscle	(^{14}C)-acetylcholine, (^{14}C)-decamethonium, (^{3}H)-α-bungarotoxin	(20)
	Cholinergic Muscarinic Protein	
8.* Intestinal muscle	(^{3}H)-atropine	(21)
	Adrenergic Proteins	
9.* Basal ganglia	(^{14}C)-Sy28, (^{14}C)-dibenamine (^{14}C)-propanolol	(22,23)
10. Vas deferens	(^{14}C)-Sy28	(24)
11. Spleen capsule	(^{3}H)-norepinephrine	(25)
12.* Heart	(^{3}H)-isoproterenol	(26)
	Serotoninergic Proteins	
13.* Basal ganglia	(^{3}H)-5-hydroxytryptamine	(27)
14. Mid brain	(^{3}H)-5-hydroxytryptamine	(28)
	Amino Acid Receptor Proteins	
15. Shrimp muscle (glutamate)	(^{14}C)-glutamate	(29)
16. Shrimp muscle (GABA)	(^{14}C)-GABA	(30)

(continued)

* Isolation of receptor protein was done in whole tissue and from subcellular fractions.
** (^{3}H)-p-(trimethylammonium)-benzene diazonium fluoroborate.

TABLE I

Tissue	Drug Bound	Ref.
17. Insect muscle (glutamate)	(^{14}C)-glutamate	(31)
18.* Cerebral cortex (glutamate)	(^{14}C)-glutamate	(unpublished results)
19.* Cerebral cortex (GABA)	(^{14}C)-GABA	(unpublished results)
20.* Spinal cord (glycine)	(^{14}C)-glycine	(unpublished results)
Opiate Receptor Protein		
21. Mouse brain	$(-)(^{14}C)$-levorphanol	(32)

* Isolation of receptor protein was done in whole tissue and from subcellular fractions.

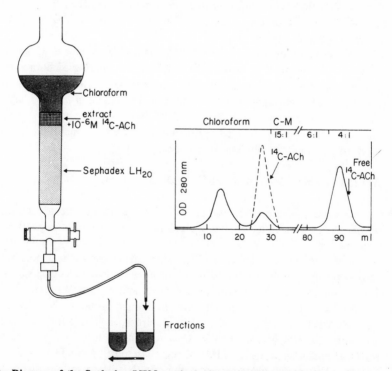

Fig. 2. Diagram of the Sephadex LH20 method of column chromatography used for the separation of hydrophobic proteins and for binding studies. The enclosure shows the binding of (^{14}C)acetylcholine to one of the protein fractions from rat diaphragm. Observe the separation between the bound and free (^{14}C)acetylcholine.

the proteins mentioned in Table I, such as the α-adrenergic protein from the spleen capsule (25), the glutamate receptor from the shrimp muscle (29), the glutamate, GABA and glycine binding proteins from the CNS, are eluted with the void volume in the chloroform. The GABA binding protein from shrimp (30), the nicotinic proteins from the electroplax (18) and skeletal muscle (20), and the muscarinic protein from smooth muscle (21) are eluted beyond the void volume with larger elution volumes. On the other hand, the cholinergic protein from the CNS and those binding indolamines and adrenergic agents appear in the chloroform-methanol at the end of the chromatogram.

The method of Sephadex LH20 chromatography for the study of ligand binding was criticized by Levinson and Keynes (34), who claimed that the coincidence between the peak of protein and the elution of radioactivity did not represent true binding but resulted from the artifactual coelution of the ligand with the protein. In their criticism, based on work on the electroplax, the authors have disregarded all the evidences from other tissues (Table I) and the fact that the ligand receptor interaction was studied in our laboratory with other methods, i.e. partition, light scattering, polarization of fluorescence, electron microscopy, etc.

Of these methods, that of partition introduced by Weber *et al.* (35) is particularly valuable since it permits the study of the binding of fluorescence probes or radiolabeled ligands in a two phase system, without the use of Sephadex LH20 chromatography. Recently, Donnellan *et al.* (36) have analyzed in detail the technique of Sephadex LH20 chromatography, answering the criticism of Levinson and Keynes and reaching the conclusion that it is valid for the study of binding of specific ligands.

AFFINITY CHROMATOGRAPHY IN ORGANIC SOLVENTS

The validity of the Sephadex LH20 technique has also been demonstrated by the use of an affinity cholinergic column that consists of Sephadex LH20, a spacer arm, −3,3 ′-iminobispropylamine−, and a quaternary ammonium of recognized cholinergic character (37). With this column it was possible to further purify the cholinergic nicotinic protein from the electroplax and skeletal muscle and the muscarinic one from smooth intestinal muscle (38).

This column retains specifically the cholinergic proteins, then the desorption is carried out with several cholinergic drugs, including the natural neurotransmitter, acetylcholine. In Fig. 3 is shown, in the upper portion, the result of applying Sephadex LH20 chromatography to the TLE of skeletal muscle. The second protein peak is the one that binds various nicotinic drugs, including (^{3}H)-α-bungarotoxin (20). This fraction contains an average of 200 μg protein/g fresh tissue. In the lower portion of Fig. 3, the previous

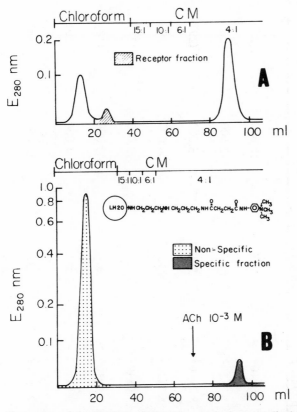

Fig. 3. Separation of the nicotinic cholinergic protein from rat diaphragm. Upper: conventional chromatography in Sephadex LH20 (20). Lower: affinity chromatography step applied to the second fraction of the preceeding step. The specific cholinergic binding fraction is eluted after application of a 10^{-3} M acetylcholine pulse. The upper chromatogram corresponds to an extract of 200 mg tissue, the lower one to 1 g dry tissue.

fraction has been passed through the affinity column, resulting in a large, non-specific protein that does not bind cholinergic ligands, and a small peak of protein desorbed by the acetylcholine pulse, which contains only 13 μg protein/g fresh tissue. This specific protein, upon rechromatography on Sephadex LH20, is able to bind (^{14}C)-acetylcholine and (^{3}H)-α-bungarotoxin. The purification has been improved by a factor of 15.4 by the affinity column, bringing the total to 15,400 times (Table II).

The results so far obtained with this column show that it can separate both nicotinic and muscarinic proteins from peripheral tissues, as well as from the CNS. In a work in progress, Saraceno and De Robertis have observed

TABLE II

Purification of Receptor Proteins by Conventional and Affinity Chromatography

Receptor protein	Source	μg Total protein/g	μg Receptor protein/g	Purification	Ref.
α-Adrenergic	Splenic capsule	1.0×10^5	6.0	16,000	(25)
β-Adrenergic	Heart	1.7×10^5	67.5	2,500	(26)
Glutamate	Shrimp muscle	8.7×10^4	27.0	3,200	(29)
GABA	Shrimp muscle	8.7×10^4	21.5	4,000	(30)
Muscarinic	Intestinal muscle	1.0×10^5	32.0	3,000	(21)
Nicotinic	*Electrophorus* electroplax	7.9×10^4	26.0	3,000	(18)
Nicotinic	Skeletal muscle	2.0×10^5	200.0	1,000	(20)
Nicotinic (affinity chromatography)	Skeletal muscle	2.0×10^5	13.0	15,400	(38)
Cholinergic (affinity chromatography)	Cerebral cortex	1.2×10^5	14.0	8,700	

that the cholinergic fractions of the cerebral cortex and nucleus caudatus are completely removed from the TLE by the affinity column to be released by the acetylcholine pulse.

The methodological approach used in the work of Barantes *et al.* (38) on skeletal and smooth muscle, in which the two chromatographic methods (Sephadex LH20 and affinity) produce additive results, regarding the purification of the binding protein, permits one to discard completely the criticisms raised by Levinson and Keynes (34) and to confirm fully the value of the Sephadex LH20 chromatography for the study of the binding of hydrophobic receptor proteins (14).

Table II shows the degree of purification achieved for the various proteins separated by conventional chromatography in non-polar systems. By determining the total content of protein of the tissue and the amount that is separated in the specific binding fraction, it is possible to obtain an enrichment which varies between 1000 and 16,000-fold for the various proteins.

A CHOLINERGIC FLUORESCENT PROBE

The study of the cholinergic binding protein from the *Electrophorus* (39, 18) was not only carried out with the use of radiolabeled drugs, but with a fluorescence probe having a cholinergic end (35). As shown in Fig. 4, the compound dansyl choline (1-dimethyl aminonaphthalene 5-sulfonamide ethyltrimethyl ammonium perchlorate) was synthesized. This drug carries a strong positive charge and resembles acetylcholine in the ethyltrimethyl-ammonium end; however, the sterophilic region of the transmitter is replaced by a bulky fluorescent end. The drug has its main excitation band at 340 nm and the emission maximum is at 520 nm (35). This cholinergic probe permits us to carry out binding studies in a two phase system, in which the protein is in

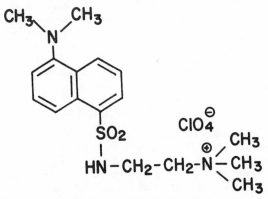

Fig. 4. Chemical formula of dansyl choline (MW 434.5). (From 35.)

the chloroform and the ligand is in a water phase. Weber *et al.* (35) have established the theory by which, knowing the partition coefficient of the ligand, it is possible to calculate the amount of ligand that is bound to the protein. The same theory has been applied to the binding of radioactive ligands having a partition coefficient favorable to water (19). With this fluorescent probe it has been possible to make competition experiments and to show that the binding of the dansyl choline to the protein from *Electrophorus* is displaced by acetylcholine, d-tubocurarine and decamethonium (35). Fig. 5 shows the binding of dansyl choline (DNETMA) to the hydrophobic receptor protein from *Electrophorus* and how this is displaced by the presence of α-bungarotoxin. It also shows that a non-receptor protein fraction from the same tissue only binds small amounts of ligands in comparison to the receptor fraction (40). Furthermore, Fiszer and De Robertis (19) have demonstrated that (^{131}I)α-bungarotoxin binds by partition to the protein from *Electrophorus*. Cohen and Changeux (41) have studied the interaction of dansyl choline with membrane fragments from *Torpedo* electroplax and the effect of nicotinic agonists and antagonists, as well as of local anesthetics,

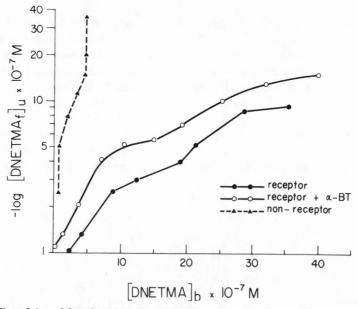

Fig. 5. Plot of -log of free dansyl choline against concentration of bound dansyl choline. Observe that α-bungarotoxin (15 µg in the upper phase) has displaced the binding from the receptor protein from the Electrophorus electroplax present in the lower phase. A non-receptor protein from the same tissue binds little dansyl choline. In this experiment the partition method of Weber et al. (35) was used. (From 40.)

on the binding. Another fluorescent dye, bis-(3-aminopyridinium)-1,10-decane has been used by Martinez-Carrion and Raftery (42) for binding with the cholinergic receptor from *Torpedo* extracted with detergents.

LIGAND-RECEPTOR INTERACTION. PHYSICAL STUDIES

The primary ligand receptor interaction was also studied with several physical methods such as light scattering (15), polarization of fluorescence (43), electron microscopy (44,45), and X-ray diffraction, which gave information about changes taking place at the macromolecular level when the protein interacted with the specific ligand. For example, upon addition of atropine sulfate, the cholinergic protein from cerebral cortex showed a sharp increase in Rayleigh scatter which followed a sigmoid curve with a Hill number of 3. The amplitude of the response was reduced by acetylcholine (non-competitive inhibition) and was competitively inhibited by dimethyl-d-tubocurarine (15); similar effects were observed studying this interaction with polarization of fluorescence (43). Both effects were explained by an increase in particle size due to the association of the protein macromolecules. In fact, under the electron microscope, with the addition of 10^{-6} M atropine sulfate or more, it was possible to observe the formation of macromolecular aggregates (44). At low ligand concentration the cholinergic proteins extracted from brain, *Electrophorus* and *Torpedo* (45) and skeletal muscle showed under the electron microscope a tendency to organize into paracrystalline arrays (Fig. 6). The dramatic changes observed in the case of the skeletal muscle protein, in the presence of hexamethonium, could be confirmed by X-ray diffraction which showed the appearance of a series of reflections, suggesting a high degree of crystallinity (see 46). While we do not know if related phenomena occur within the chemical excitable membrane, these phase transitions suggest that receptor proteins are dynamic entities that are able to undergo conformational changes upon interaction with the specific ligand.

RECEPTOR PROTEINS AND IONOPHORES

In the mechanism of action of most synaptic receptors two different but coordinated steps take place at the chemosensitive sites of the cell membrane (Fig. 1). One is the interaction of the neurotransmitter with the specific receptor macromolecule; the other is a change in ionic conductance that may either depolarize (i.e., excite) or hyperpolarize (i.e., inhibit) the postsynaptic membrane. This second event is the consequence of the ligand-receptor interaction at the cell membrane. In excitatory synapses the ionic species carrying the current are in general Na^+ and K^+, while in inhibitory synapses K^+ and

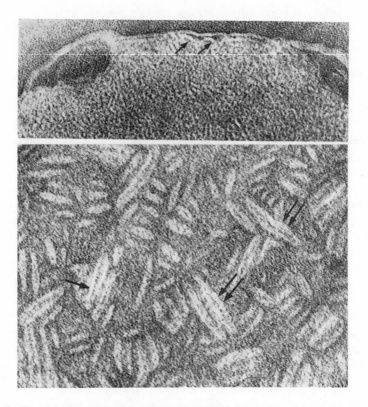

Fig. 6. Electronmicrograph with negative staining of the cholinergic receptor protein fraction from Torpedo Marmorata electroplax. Upper: part of a large micelle observed in the control showing a filamentous structure, X360,000. Lower: the same protein after interaction with hexamethonium. Spindle shaped structures are formed by the assembly of several single filaments, X400,000. (From 45.)

Cl^- may be involved. Calcium ions may also be translocated; in fact, in invertebrate muscle and in mammalian smooth muscle, Ca^{2+} may be one of the ionic carrying species.

The translocation of ions implies a rather selective mechanism of ion permeation that must be tightly coordinated with the ligand-receptor interaction, and both processes should be coupled in space and time within the realm of the cell membrane. One essential fact to be uncovered is to determine if the receptor site and the ionophore lie in the same or in different protein molecules. In their study of the permeability of vesicles (microsacs) obtained from electroplax membranes, Kasai and Changeux (47) reached the conclusion that the distinction between receptor and ionophore is justified, although "it is not clear whether the ionophore and the receptor are to be considered different polypeptide chains or as different parts of the same protein." As it will be discussed in this section, we favor a model in which the binding site and the ionophore are in the same macromolecule (see 7). At the molecular level, for the cholinergic receptor we could envision a two step mechanism:

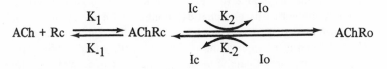

In step I, the transmitter (ACh) interacts with the receptor, which is in a closed configuration (Rc) and produces a reversible complex (AChRc). In step II, the complex changes from the closed to an open configuration (AChRo) which is also reversible. We may consider that the first step corresponds to the binding of the ligand to the receptor site and the second to the opening (I_o) and closing (I_c) of the ionophore. The kinetics of these interactions are based on the assumption that there is a simple reaction between one acetylcholine molecule and one receptor site. It may be questioned whether the ionophore is opened by a single molecule or by the cooperative action of several ACh molecules. Both in the case of the neuromuscular junction (48,49) and in the electroplax (50,51), the sigmoid shape of the dose-effect curve, with a Hill number approaching 2, suggests that at least two ACh molecules are involved in the opening of the ionophore. The neurophysiological work of Katz and Miledi (52) has brought new insight on the possible function of the cholinergic receptor at the molecular level. When a minimal but steady dose of acetylcholine is applied to the post-synaptic membrane of the myoneural junction there is a minute random fluctuation of the noise with discrete variations of only 0.22 μV. (This compares with a 700μV change in the miniature end plate potential.) These elementary steps are considered as a manifestation of the opening and closing of single ionophores by the acetylcholine interaction. The time constant of this phenomenon is about 1 ms at 22°C. This time comprises both steps in the above-mentioned reaction. According to Katz and Miledi (52) the elementary current pulse of 10 pico-amperes corresponds to a conductance of 10^{-10} mho and is associated with the translocation of approximately 5 x 10^4 univalent ions. The fact that two acetylcholine molecules could trigger the transfer of such a large number of ions across the subsynaptic membrane clearly illustrates the high degree of amplification that is produced at a chemical synapse.

RECONSTITUTION OF THE CHOLINERGIC RECEPTOR IN BLM

In our laboratory we have approached the study of the ion conducting mechanism with the use of bilayer lipid membranes (BLM)(53). Because the receptor proteins we isolate are in organic solvents, they can easily be incorporated into the membrane forming solution which, in general, contains cholesterol and synthetic phosphatidylcholine. With the apparatus shown in Fig. 7, current voltage curves can be made and the conductance can be measured before and after application of the drug with a fine capillary tube (54). As shown in the same figure, a technique was developed which permits the fixation of the membrane and the study of its planar structure under the electron microscope (55). The incorporation of small amounts of receptor protein from the *Electrophorus* electroplax (5-80 µg/ml) into the membrane forming solution, which contains lipids in mg amounts, produces a reduction in resistance which is accompanied by a smoother texture of the membrane, as can be observed with the electron microscope. These findings suggest that the presence of the protein has produced a molecular reorganization of the membrane (55). These changes are, however, unspecific in the sense that they can be observed also with other non-receptor hydrophobic proteins. Of considerable interest was the finding that the increase in conductance was proportional to the fourth power of the concentration of the protein in the mem-

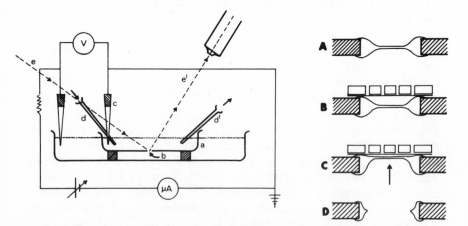

Fig. 7. Diagram of the apparatus used to study the electrical properties of the BLM. A teflon cup (a) is immersed in a Petri dish. The membranes are formed in a 1 mm hole at the bottom of the cup (b), (d) and (d′) pipettes used for fixation. The electrical measurements are made via calomel electrodes (c). V, voltmeter, µA, microamperimeter. To the right the different steps to remove the BLM for electron microscopy are indicated.

brane. This high degree of cooperativity supports the oligomeric model of the cholinergic receptor I proposed in 1971 (7). Recently, Adragna *et al.* (56) have found that, while a control membrane made of lipids has a very slight anionic permeability that is not changed by curare, the membrane containing all the proteolipids from the electroplax of *Electrophorus* shows a marked cationic conductance for K^+ and Na^+ which is blocked by curare.

Parisi *et al.* (54) found that, in certain conditions, there were discrete spontaneous fluctuations in the BLM containing the cholinergic receptor from *Electrophorus* (Fig. 8). These steps carried a current of about 30 picoamperes and the conductance was of the order of 6×10^{-10} mho. These discrete current jumps were reminiscent of those produced by the Excitability Inducing Material from *Aerobacter cloacae* (57) and alamethicin (58); both these polypeptides are also rather hydrophobic.

The most important finding in BLM was that the injection of acetylcholine and other cholinergic drugs produced a transient increase in conductance (53,54). This change can be blocked by d-tubocurarine and by other agents that block the transmission at the electroplax. Furthermore, the reaction to acetylcholine is modified with gallamine, hexamethonium and α-bungarotoxin. As shown in Fig. 9, the amplitude of the conductance change by ACh was related to the concentration of the ligand in the micro-pipette and showed a tendency to saturate at 0.1 M ACh. The amplitude of the response was also proportional to the concentration of the protein within the BLM. None of the numerous controls used, such as choline, 100 mM sodium acetate, distilled water, 300 mM KCl or NaCl, produced a significant conductance change. It was found that diffusion played a role in the kinetics of the response to ACh. In the presence of stirring, the rising phase of the conductance was the same, but the lowering phase was produced much faster.

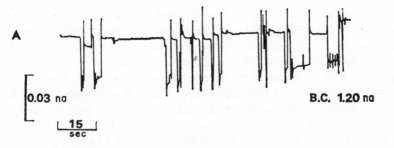

Fig. 8. Discrete spontaneous fluctuations observed in a BLM containing the receptor protein from electrophorus. BC, basal current in picoamperes. (From 54.)

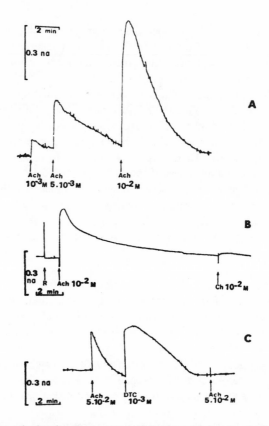

Fig. 9. Original records showing the response of BLM containing the cholinergic protein from Electrophorus. (A) Response to different doses of acetylcholine in the micro-
pipette; (B) Responses to saline (R), acetylcholine (ACh) and choline (Ch); (C) Responses to acetylcholine, dimethyl-d-tubocurarine (DTC) and blocking of the response to
ACh. (From 54.)

The electron microscope study of the membranes fixed at the height of the
conductance showed a striking change in fine structure, which indicated the
production of a conformational change of the protein within the membrane
(55). These activated membranes showed a more uneven or "corrugated" appearance with the presence of dense spots having a maximal diameter of 2nm
into which the osmium tetroxide, used in vapors, was deposited (Fig. 10).
These ultrastructural changes are transient and disappear once the membrane
regains the basal conductance. We interpreted these findings as suggesting
that the translocation of ions is probably accompanied by changes in the
protein which allow the deposition of discrete precipitates of osmium. It was
of interest to find that both the conductance and the structural changes were

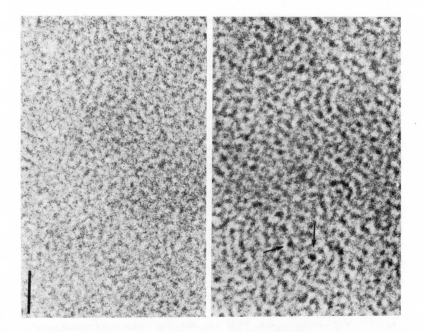

Fig. 10. Electron micrographs of BLM containing the cholinergic protein from Electrophorus electroplax. **Left:** control membrane. **Right:** membrane fixed at the height of the conductance change. Observe the change in planar structure that was produced by the acetylcholine and the presence of dense spots of about 2 nm (arrows). Bar indicates 20 nm. (From 55.)

prevented if the membrane was fixed with glutaraldehyde prior to the injection of ACh. On the other hand, if the fixation is done after the application of ACh, the conductance changes that have been initiated are not stopped and the lowering phase of the transient takes place. Discussing these findings in light of the existing literature about carrier and channel mechanism of ion translocation, Vásquez *et al.* (55) were inclined to think that the receptor ionophore is more similar to the channel produced by alamethicin than to a carrier ionophore such as valinomycin.

Other findings in BLM containing the cholinergic protein were that in the presence of uranyl ions the conductivity of the membrane increased 10-20 times, sometimes in a staircase fashion, with steps of 12×10^{-10} mho. Furthermore, uranyl ions strongly potentiated the effect of acetylcholine on the BLM containing the cholinergic receptor (59).

More recently it was found that the cholinergic response of the BLM was greatly changed by reagents that act on S-S and SH groups (60). In Fig. 11,

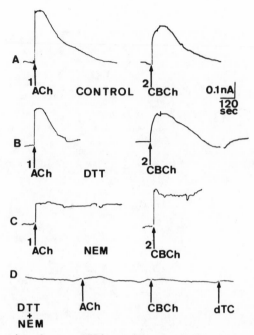

Fig. 11. Effect of various drugs on BLM containing the cholinergic receptor protein from Electrophorus electroplax. (A), effect of acetylcholine (ACh) and carbamylcholine (CBCh); (B), the same as in (A), but in the presence of 2 mM 1,4-dithiothreitol (DTT); (C), the same as in (A), but in the presence of 2 mM N-ethylmaleimide (NEM); (D), the addition of DTT and NEM blocks the response to ACh, CBCh and d-tubocurarine (dTC). (From 60.)

it may be observed that while dithiothreitol (DTT), a disulfide reducing agent, produced only a small reduction of the response to ACh and carbamyl-cho-line (CBCh), N-ethylmaleimide (NEM), an alkylating reagent of -SH groups, produced a sharp change in the kinetics of the conductance change. In this case the typical increase in conductance is produced by ACh or CBCh, but this remains at a high level without returning to the basal level. Reader and De Robertis (60) have interpreted the effect of NEM within the framework of the model presented at the beginning of this section. We think that once the channel is in the open configuration (I_o) it remains as such, without going in-to the closed configuration (I_c). Furthermore, the combined treatment of the BLM with DTT and NEM produces complete inhibition of the cholinergic response (Fig. 11). These findings are in agreement with previous observa-tions made in the living electroplax (61) and demonstrate the importance of -SH and S-S bonds in the receptor-conducting mechanism. These findings also suggest that it is the protein moiety of the proteolipid molecules, and not the lipid, which is responsible for the binding and the ionophoric properties of the receptor when this is reconstituted in the BLM.

A MODEL OF CHOLINERGIC RECEPTOR

The hydrophobic nature of the synaptic receptor proteins should be considered in relation to their possible integration within the lipoprotein framework of the cell membrane. In recent years the structural role of hydrophobic proteins in biological membranes has been stressed and several models in which segments of protein penetrate into the lipid bilayer have been produced (see 9).

In 1971 (7) I proposed the model shown in Fig. 12 for the possible macromolecular organization of the cholinergic receptor in the postsynaptic membrane. The most striking features of this model are: the oligomeric arrangement of the protein, its disposition traversing the membrane, and the fact that the binding site and the ionophore, involved in the translocation of ions, are localized and coordinated within the same macromolecule. Also, this model clearly differentiates the cholinergic receptor from acetylcholinesterase. The latter is a peripheral protein that can easily be removed from the electroplax membrane (62). The oligomeric organization — probably tetrameric — is supported by the above-mentioned finding of Parisi *et al.* (56), that in BLM the conductance increases as the fourth power of the protein concentration. Furthermore, on pharmacological evidences, a tetrameric model has been proposed in the literature (see 63).

In our model the gating mechanism for the opening of the channel would be based on a change in quaternary structure, more than in the tertiary structure of a single protein molecule, although the beginning of the conformational change could originate at this level. A small change in the degree of interaction between the monomeric units could produce the opening and closing of the channel. The conformational change would be favored by the fact that the receptor proteins are in a lipid microenvironment and held in place mainly by hydrophobic interactions (Fig. 12).

In this model an asymmetry of the individual macromolecules with a more polar region in the inner or contacting surface of the monomers is postulated. The ligand-receptor interaction may thus lead to the exposure of hydrophilic sites at this inner surface which would be available for the trapping and translocation of ions through the channel. The electron microscope observations on artificial membranes mentioned above (55) could be interpreted along these lines; the presence of dense spots of 2 nm or less in diameter in the activated membrane (Fig. 10) can be interpreted as indicating the exposure of such hydrophilic regions that can bind the osmium tetroxide vapors. The

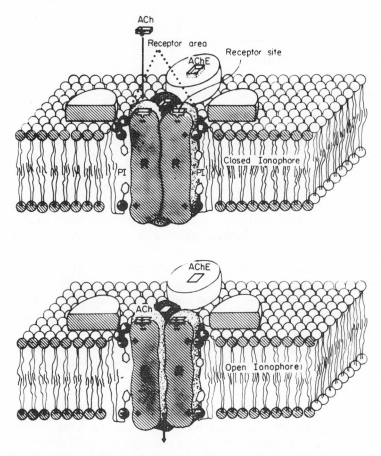

Fig. 12. Oligomeric model of a cholinergic receptor area showing the receptor protein subunits traversing the lipid bilayer. Each subunit shows the site of binding for acetyl-choline on the outer surface of the membrane. Four subunits in parallel constitute the ionophore. The presence of phosphatidylinositol (PI) attached to the receptor protein is indicated. The diagram also shows the presence of acetylcholinesterase (AChE) molecules that are peripheral to the membrane. In the upper part, the receptor is in the closed configuration. In the lower part, the receptor site, now occupied by acetylcholine, is in the open configuration. (From 7, modified.)

model of Fig. 12 also emphasizes the close relationship that phosphatidylino-sitol has with the receptor protein, which may explain the metabolic activa-tion caused by the action of the neurotransmitter (64).

REFERENCES

1. Langley, J.N. (1878) *J. Physiol. 1*, 339-358.
2. Langley, J.N. (1906) *Proc. Royal Soc. Lond., ser. B 78*, 170-194.
3. Cook, R.P. (1926) *J. Physiol. 62*, 160-165.
4. Del Castillo, J. and Katz, B. (1955) *J. Physiol. 128*, 157-181.
5. Ehrenpreis, S., Fleisch, J.H. and Mittag, T.N. (1969) *Pharmacol. Rev. 21*, 131-181.
6. De Robertis, E. (1971) *Science 171*, 963-971.
7. De Robertis, E. (in press) *Synaptic Receptors: Isolation and Molecular Biology.* Marcel Dekker, New York.
8. De Robertis, E., Azcurra, J.M. and Fiszer, S. (1967) *Brain Res. 5*, 45-56.
9. Singer, S.J. and Nicholson, G.L. (1972) *Science 175*, 720-731.
10. De Robertis, E., Fiszer, S. and Soto, E.F. (1967) *Science 158*, 928-929.
11. Changeux, J.P., Kasai, M., Huchet, M. and Meunier, J.C. (1970) *C.R. Acad. Sci. (Paris) 170*, 2864-2867.
12. Folch-Pi, J. and Lees, M. (1951) *J. Biol. Chem. 191*, 807-817.
13. Barrantes, F.J., La Torre, J.L., Llorente de Carlin, M.C. and De Robertis, E. (1972) *Biochim. Biophys. Acta 263*, 368-381.
14. De Robertis, E., Fiszer, S., Pasquini, J.M. and Soto, E.F. (1969) *J. Neurobiol. 1*, 41-52.
15. De Robertis, E., González-Rodriguez, J. and Teller, D.N. (1969) *FEBS Letters 4*, 4-8.
16. Izumi, F. and Freed, S. (1974) *FEBS Letters 41*, 151-155.
17. Cattel, K.L. and Donellan, J.F. (1972) *Biochem. J. 128*, 187-189.
18. La Torre, J.L., Lunt, G.S. and De Robertis, E. (1970) *Proc. Nat. Acad. Sci. (U.S.A.) 65*, 716-720.
19. Fiszer de Plazas, S. and De Robertis, E. (1972) *Biochim. Biophys. Acta 274*, 258-265.
20. De Robertis, E., Mosquera, M.T. and Fiszer de Plazas, S. (1972) *Life Sci. 11*, 1155-1165.
21. Ochoa, E. and De Robertis, E. (1973) *Biochim. Biophys. Acta 295*, 528-535.
22. Fiszer, S. and De Robertis, E. (1968) *Life Sci. 7*, 1093-1103.
23. De Robertis, E. and Fiszer de Plazas, S. (1969) *Life Sci. 8*, 1247-1262.
24. Mottram, D.R. and Graham, J.D.P. (1971) *Biochem. Pharmacol. 20* 1917-1919.
25. Fiszer de Plazas, S. and De Robertis, E. (1972) *Biochim. Biophys. Acta 266*, 246-254.
26. Ochoa, E., Llorente de Carlin, M.C. and De Robertis, E. (1972) *Eur. J. Pharmacol. 18*, 367-374.

27. Fiszer, S. and De Robertis, E. (1969) *J. Neurochem. 16*, 1201-1209.
28. Godwin, S. and Sneddon, J.M. (1974) *Brit. J. Pharmacol. 50*, 464-465.
29. Fiszer de Plazas, S. and De Robertis, E. (1974) *J. Neurochem. 23*, 1115-1120.
30. De Robertis, E. and Fiszer de Plazas, S. (1974) *J. Neurochem. 23*, 1121-1125.
31. Lunt, G.G. (1973) *Comp. Gen. Pharmacol. 4*, 75-79.
32. Lowney, L.I., Schultz, K., Lowery, P.J. and Goldstein, A. (1974) *Science 183*, 749-753.
33. Soto, E.F., Pasquini, J.M., Plácido, R. and La Torre, J.L. (1969) *J. Chromatog. 41*, 400-409.
34. Levinson, S.R. and Keynes, R.D. (1972) *Biochim, Biophys. Acta 288*, 241-247.
35. Weber, G., Borris, D.P., De Robertis, E., Barrantes, F.J., La Torre, J.L. and Llorente de Carlin, M.C. (1971) *Mol. Pharmacol. 7*, 530-537.
36. Donnellan, J.F., Clarke, B.S., Jewes, P.J. and Cattell, J. (in press) *J. Neurochem.*
37. Barrantes, F.J. (1973) *Proc. 9th Int. Cong. Biochem. (Stockholm)*, 443.
38. Barrantes, F.J., Arbilla, S., Llorente de Carlin, M.C. and De Robertis, E. (in press) *Biochem. Biophys. Res. Commun.*
39. De Robertis, E., Fiszer de Plazas, S., La Torre, J.L. and Lunt, G.S. (1970) In *Drugs and Cholinergic Mechanisms in the CNS* (Heilbronn, E. and Winter, A., eds.) pp. 505-521, Res. Inst. Nat. Defense, Stockholm.
40. De Robertis, E. and Barrantes, F.J. (1972) *Eur. J. Pharmacol. 17*, 303-305.
41. Cohen, J.A. and Changeux, J.P. (1973) *Biochemistry 12*, 4855-4867.
42. Martinez-Carrion, M. and Raftery, M.A. (1973) *Biochem. Biophys. Res. Commun. 55*, 1156-1159.
43. Gonzalez-Rodriguez, J., La Torre, J.L. and De Robertis, E. (1970) *Mol. Pharmacol. 6*, 122-127.
44. Vasquez, C., Barrantes, F.J., La Torre, J.L. and De Robertis, E. (1970) *J. Mol. Biol. 52*, 221-226.
45. Barrantes, F.J., Vasquez, C., Lunt, G.S., La Torre, J.L. and De Robertis, E. (1972) *J. Microscopie 13*, 391-400.
46. De Robertis, E. (1975) In *Proc. Int. Symp. Macromol.* (Mano, E.B., ed.) pp. 349-371, Elsevier, Amsterdam.
47. Kasai, M. and Changeux, J.P. (1971) *J. Membrane Biol 6*, 24-32.
48. Katz, B. and Thesleff, S. (1957) *J. Physiol. (London) 138*, 63-74.
49. Jenkinson, D.H. (1960) *J. Physiol (London) 152*, 309-312.
50. Karlin, A. (1967) *J. Theoret. Biol. 16*, 306-318.

51. Changeux, J.P. and Podlewski, T.R. (1968) *Proc. Nat. Acad. Sci. (U.S.A.) 59*, 944-949.
52. Katz, B. and Miledi, R. (1972) *J. Physiol. 224*, 665-699.
53. Parisi, M., Rivas, E. and De Robertis, E. (1971) *Science 172* , 56-57.
54. Parisi, M., Reader, T. and De Robertis, E. (1972) *J. Gen. Physiol. 60*, 454-470.
55. Vásquez, C., Parisi, M. and De Robertis, E. (1971) *J. Membrane Biol. 6*, 353-367.
56. Adragna, N.C., Salas, P., Parisi, M. and De Robertis, E. (1975) *Biochem. Biophys. Res. Commun. 62*, 110-116.
57. Muller, P. and Rudin, D.O. (1967) *Nature 213*, 603-604.
58. Muller, P. and Rudin, D.O. (1968) *Nature 217*, 713-719.
59. Reader, T.A., Parisi, M. and De Robertis, E. (1973) *Biochem. Biophys. Res. Commun. 53*, 10-17.
60. Reader, T.A. and De Robertis, E. (1974) *Biochim, Biophys. Acta 352*, 192-201.
61. Karlin, A. and Bartels, E. (1966) *Biochim. Biophys. Acta 126*, 525-535.
62. De Robertis, E. and Fiszer de Plazas, S. (1970) *Biochim. Biophys. Acta 219*, 388-397.
63. Khromov-Borisov, N.V. and Michelson, M.J. (1966) *Pharmacol. Rev. 18*, 1051-1090.
64. Lunt, G.G., Canessa, O.M. and De Robertis, E. (1971) *Nature New Biol. 230*, 187-190.

AGONIST–ANTAGONIST DISCRIMINATION BY
SOLUBILIZED ACETYLCHOLINE RECEPTOR

Alfred Maelicke and E. Reich

The Rockefeller University
New York, New York 10021

Binding of acetylcholine or cholinergic ligands to corresponding receptors situated in excitable tissue transiently increases the membrane permeability for specific cations (1). This response is blocked by cholinergic antagonists (2). Although the typical features of dose-response curves of cholinergic ligands such as sigmoidal shape and half maximal response are well established, the functional relationship between conductance or membrane potential changes and ligand concentration, is still unknown. As a first step in obtaining insight into the molecular mechanism of chemical excitation, we have solubilized and purified the nicotinic acetylcholine receptor from *Electrophorus electricus* (3) and have begun to analyze quantitatively its interaction with cholinergic ligands (3-6).

This paper summarizes experimental methods and results of competition binding studies of ten nicotinic antagonists and eight nicotinic agonists with radioactively labeled cobra toxin to the solubilized receptor published in detail elsewhere (5). Besides demonstrating agonist-antagonist discrimination on the level of the elementary binding reaction, these data contain considerable implications for the structural organization of toxin and small ligand binding sites on the receptor, for models of receptor mediated conductivity changes presently under discussion or under development, and for the mechanism of action of acetylcholinesterase in controlling synaptic transmitter levels.

METHODS

The preparation and characterization of the acetylcholine receptor from *Electrophorus electricus* and the α-neurotoxin from *Naja naja siamensis* have been described previously (2,4,7). The α-cobratoxin was radioactively labeled by reaction with pyridoxal phosphate followed by reduction of the Schiff's bases with sodium [^{3}H] -borohydride. Mono-, di- and tri-substituted toxins were separated from unlabeled toxin by chromatography on phosphocellulose. Repeated chromatography of the heterogeneous monolabeled toxin using shallow gradients of potassium phosphate resulted in a complete separation of the different products (4). The main monolabeled toxin used throughout these binding studies was a single homogeneous species with respect to chromatography on phosphocellulose, Biorex 70 and Sephadex G-75, and polyacrylamide gel electrophoresis. In addition, its homogeneity is demonstrated by the single first-order complex dissociation rate measured under conditions of large receptor excess (5,6). Two preparations with specific radioactivities of 3.9 and 2.75 Ci/mmole were used in the present studies. The dissociation constant of the main monolabeled toxin was 1.5 times the K_D of unlabeled toxin when measured by direct competition (4). All compounds used for ligand binding studies were analyzed chemically and by mass spectrometry (8) prior to use.

Binding assay

The principal assay method has been described previously (3). The reaction mixtures of receptor, toxin and small ligands in 0.13 M NaCl, 0.02 M Tris-HCl, 1% w/v Tween 80, pH 7.4, were filtered through two apposed DEAE-cellulose filter discs and washed with buffer consisting of 0.02 M NaCl in 0.02 M Tris-HCl, pH 7.4. The receptor and all bound toxin were retained on the filter discs, even after excessive washing (e.g., 200 ml of buffer) whereas free toxin and small ligands were washed through. Since the assay measures only the concentration of stable receptor-toxin complexes, any dissociation of small ligands during filtration does not influence the actual experimental result. This is because the receptor-toxin complex formation rate is so slow that a change in the concentration of free receptor due to dissociation of small ligands does not result in a detectable increase in complex concentration during the time of filtration. To ensure such conditions, the reaction mixture was diluted approximately 5 to 10 times prior to filtration. A series of experiments with filter stacks of variable size and small DEAE-cellulose columns in which the filtrates were analyzed with respect to their concentration of receptor and toxin showed unambiguously complete retention of receptor in free and bound form by the cationic resin, and highly reproducible recovery of toxin-receptor complexes independent of the period of filtration varied between 1 and 20 min.

Data treatment

Receptor concentrations are expressed in moles of toxin binding sites. Data were analyzed by an apparatus of mathematical equations developed for mutually exclusive binding (5). This assumes exclusion of ternary complexes of the type (RTI) in which binding of the ligand I does not coincide with displacement of the toxin T from its receptor binding site. The experimental data showed that this assumption was fulfilled in all cases. The functional relationship between bound toxin C_B and free toxin C_F is then similar to the equation of a classical double reciprocal plot, except that the apparent dissociation constant for each concentration of competing ligand is now a function of this concentration:

$$\frac{1}{C_B} = \frac{1}{R_0} + \frac{K_D \text{(app)}}{R_0} \cdot \frac{1}{C_F} \tag{1}$$

with $$K_D \text{(app)} = K_D (1 + I^n/K_I) \tag{2}$$

R_0 denotes the total concentration of receptor, I is the concentration of competing ligand, $K_{D \text{(app)}}$ and K_D are the dissociation constants of toxin in presence and absence of competing ligand, K_I is the dissociation constant of the competing ligand and n is the ratio of ligand to toxin binding sites in the case of linear double reciprocal plots.

Experimentally, K_I and n can be deduced from double reciprocal plots and secondary plots of binding data representing toxin binding (variable ligand) for several fixed concentrations of small ligand. Data analyzed in this manner are shown in Figs. 1 and 3.

To determine the ratio of binding sites more accurately, experiments were also performed under conditions of large excess of toxin over receptor sites so that in the absence as well as in the presence of competing ligand the concentration of free receptor sites was below 2% of the total sites R_0. Under these conditions the concentration of total sites R_0 is equal to the sum of sites occupied by toxin (RT) and small ligand (RI). Since RT is measured experimentally and R_0 can be determined in the absence of small ligand, the concentration of sites occupied by small ligand (RI) can be calculated and independent binding curves for the small ligand can be obtained. The mathematical expression for such binding plots is

$$\log \left[\frac{(RI)}{(RT)} \quad (1 + {}^T/K_D) \right] = n \log I - \log K_I \tag{3}$$

In the case of a single class of binding sites for both radioactively labeled toxin and competing ligand, a plot of the left side of equation (3) as a function of log I produces a straight line, the slope of which gives the ratio of binding sites n, the intercept yielding K_I. K_I can also be obtained more conveniently from the log I-axis intercept.

Validity of binding data

The main consideration in this context is whether the receptor-toxin interaction is a valid probe for binding of small ligands. The basic requirements for such a probe, namely reversibility and mutually exclusive binding with respect to the ligands under study, was unambiguously fulfilled (3-7). Both purified receptor and radioactively labeled toxin were homogeneous species by a number of criteria including their interaction with each other (4-6). The final equilibrium of receptor-toxin interaction is, however, reached only after incubation times inconveniently long for competition binding studies. We have therefore used constant periods of incubation (200 min) for each experimental point which, on the basis of reaction rates (5,6), guaranteed the formation of at least 95% of the equilibrium concentration of receptor toxin complexes even for the smallest toxin concentrations applied (5). As is outlined in detail elsewhere (5,6) two receptor-toxin complexes with different affinities were formed under these conditions and the K_D values for toxin reflected, therefore, the contributions of both species. Due to the different degrees of equilibrium reached for different toxin concentrations and consistent with a detailed mathematical analysis of the slopes of double reciprocal plots under the conditions (9), these plots appeared linear over wide ranges of concentrations and yielded a single, though too large, K_D-value for toxin. This is a convenient analytical artifact which did not perturb the binding data of competing ligands since K_I- and n-determinations from double reciprocal plots depend only on ratios of $K_{D (app)}$ and K_D but not on their absolute values (equation 2). The experimental conditions for double logarithmic plots (equation 3) resembled more closely equilibrium conditions since much larger constant concentrations of toxin were applied. The validity of these theoretical considerations was also verified by assessing binding curves for several agonists and antagonists after 51 hrs of incubation instead of the usual 200 min. No difference in the binding pattern was observed.

RESULTS

Binding of antagonists

Fig. 1a shows a double reciprocal plot for binding of $[^3H]$-cobratoxin to the acetylcholine receptor in the presence of three different concentrations of dimethyl-d-tubocurarine, the prototype of a nicotinic antagonist. Intersection of all four curves at the same point on the $1/C_B$-axis shows that receptor bound dimethyl-d-tubocurarine could be completely replaced by toxin at sufficiently high concentrations. Hence, binding of these two ligands was mutually exclusive. The apparent dissociation constants of toxin in the presence of curare were linear functions of the square root of dimethyl-d-tubocurarine concentra-

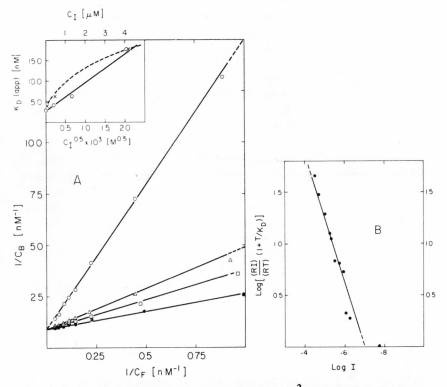

Fig. 1. Competitive binding of dimethyl-d-tubocurarine and ^{3}H-cobratoxin to the receptor. (A) Double-reciprocal plot: dimethyl-d-tubocurarine concentrations: ($\bullet$—$\bullet$) I = 0, ($\square$—$\square$) I = 4.10 x 10^{-8} M, ($\triangle$—$\triangle$) I = 4.55 x 10^{-7} M, ($\circ$—$\circ$) I = 4.17 x 10^{-6}M. Insert: A replot of the apparent dissociation constants as a function of the concentration (x--x), and of the square root of the concentration ($\circ$—$\circ$) of dimethyl-d-tubocurarine. Intrinsic dissociation constant per site of dimethyl-d-tubocurarine calculated from the linear replot is K_I = 1.2 x 10^{-7} M. Note that the experimental point C_I = C_I = 0 is the same for both functions and it is represented only by a single symbol. (B) Logarithmic plot of the binding of dimethyl-d-tubocurarine to the receptor. Receptor concentration: 4.9 x 10^{-9} M. Toxin concentration (fixed): 3.55 x 10^{-8} M: Slope m = n = 0.58, K_I per mole of inhibitor K_I = 1.0 x 10^{-7} M.

tions implying that there were only half as many receptor sites for the low molecular weight antagonist as for toxin. A similar result is obtained from the double logarithmic plot of Fig. 1b, the slope of which yields a ratio of binding sites near 0.5.

Similar complete sets of binding data have been obtained for benzoquinonium, alloferin, hexamethonium, and other bismethonium antagonists. The patterns of these binding curves are all identical in that binding is mutually exclusive, only one single class of binding sites for low molecular weight antagonists is observed and the ratios of their binding sites to toxin sites is approximately 1:2. As an example, the double logarithmic plots obtained for benzoquinonium and hexamethonium are shown in Fig. 2. Although their K_I values differed by approximately two orders of magnitude (see log I-axis intercepts), the slopes of their binding curves were very similar, indicating the same number, if not topographically identical binding sites, for these antagonists.

Binding of acetylcholine and agonists

Fig. 3a shows a double reciprocal plot for binding of toxin to the receptor in the presence and absence of the neurotransmitter, acetylcholine. The binding curves again intersected in one point on the $1/C_B$-axis, showing that binding of these two ligands to the receptor was mutually exclusive. However, the affinity of acetylcholine for the receptor was concentration dependent, being highest at low concentrations of bound acetylcholine and decreasing with increasing concentrations. The same was reflected in the strong curvature of the double logarithmic plot (Fig. 3b) which suggested the existence of either several distinct sites with different affinities or anti-cooperative interaction between two or more identical sites for acetylcholine.

The initial slopes of these curves which determine the ratio of acetylcholine to toxin binding sites were difficult to assess, not only because of their pronounced curvature, but also because of the lower than usual reproducibility of

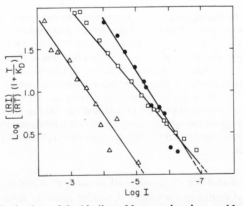

Fig. 2. Double logarithmic plots of the binding of benzoquinonium and hexamethonium to the receptor. (●——●) benzoquinonium $R_0 = 1.0 \times 10^{-9}$ M, $T_0 = 4.0 \times 10^{-8}$ M, slope n = 0.49, $K_I = 6.3 \times 10^{-8}$ M. (△——△) hexamethonium; $R_0 = 2.43 \times 10^{-9}$ M, $T_0 = 4.85 \times 10^{-8}$ M, n = 0.53, $K_I = 4.0 \times 10^{-6}$ M. (○——○) dimethyl-d-tubocurarine (see Fig. 1).

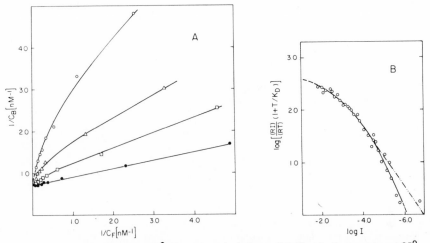

Fig. 3. Competitive binding of [^{3}H]-neurotoxin and acetylcholine to the receptor at 25°. (A) Double-reciprocal plot. Ligand concentrations: (●—●) I = 0, (□—□) I = 4.1 x 10^{-5} M, (△—△) I = 2.0 x 10^{-4} M, (○—○) I = 7.5 x 10^{-4} M. (B) Logarithmic plot: R_0 = 1.88 x 10^{-9} M, T_0 = 2.69 x 10^{-8} M, sample volume = 1040 μliters. Reaction was performed in presence of 1.1 x 10^{-3} M DFP. Initial slope m = n = 0.95. Initial dissociation constant per mole of acetylcholine K_I = 7.9 x 10^{-7} M. (−··−··) second extrapolation (see text) n = 0.60, K_I = 1 x 10^{-7} M.

binding data. We attribute this to the relatively low stability of acetylcholine in solution. This is even the case after preincubation of the receptor with 1.1 x 10^{-3} M diisopropylfluorophosphate, which completely abolishes the residual traces of esterases (less than 0.02% in our preparation (3)) without impairing the binding and kinetic properties of the receptor with respect to toxin, antagonists and agonists. Fig. 3b shows, therefore, two extrapolated initial slopes yielding 0.95 and 0.62, respectively, for the ratio of acetylcholine to toxin binding sites. We have repeated this experiment several times with different receptors, toxin and acetylcholine preparations but were not able to settle the ratio of binding sites unambiguously for this specific ligand. Though all investigated agonists gave ratios of binding sites near 0.5, our experiments still are more indicative of an identical number of acetylcholine and toxin sites than a ratio of 1:2.

Similar curved binding patterns were obtained for all pharmacological agonists of acetylcholine investigated in the present study. Fig. 4 shows as an example the double logarithmic plots obtained for carbamylcholine, nicotine, succinylcholine and decamethonium. All these agonists bound mutually exclusively with respect to α-cobratoxin to the receptor; their binding was suggestive of

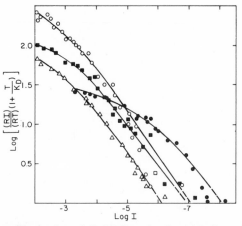

Fig. 4. Double logarithmic plots of the binding of carbamylcholine, nicotine, succinyl-
choline, and decamethonium to the receptor. ($\triangle$—$\triangle$) carbamylcholine; R_0 = 2.3 x 10^{-9}
M, T_0 = 2.9 x 10^{-8} M, n = 0.61, K_I = 7.1 x 10^{-7} M. ($\blacksquare$—$\blacksquare$) nicotine; R_0 = 1.4 x 10^{-9} M,
T_0 = 4.3 x 10^{-8} M, n = 0.52, K_I = 1.2 x 10^{-7} M. ($\bullet$—$\bullet$) decamethonium; R_0 = 1.23 x
10^{-9} M, T_0 = 4.06 x 10^{-8} M. ($\circ$—$\circ$) acetylcholine; extrapolation for n = 0.6, K_I = 1 x
10^{-7} M (see Fig. 3).

anticooperative interaction between two or more identical binding sites, and
they had approximately half as many binding sites as toxin at the receptor. In
addition, all binding reactions are accompanied by substantial reaction entropies (5) which make it more attractive to interpret the concentration dependent
affinities as the result of anticooperative interactions of agonist binding sites.
All binding data are summarized in Table I.

Binding of homologous bis-methonium compounds

The binding curves obtained for homologous bis-methonium compounds
with methylene bridges ranging from two to ten can serve as a demonstration
of how small structural differences may decisively influence binding to the
receptor. One representative curve from double reciprocal plots for each of
these compounds is presented in Fig. 5. Bis-methonium compounds with
methylene chains from two to six gave linear binding patterns typical for antagonists whereas those with seven or more methylene groups bound with concentration dependent affinity typical for agonists.

DISCUSSION

The data presented here show distinct phenomenological differences between

TABLE I

Number of Binding Sites n, Dissociation Constant K_I and Binding Mode
for Agonists and Antagonists
of Acetylcholine with Solubilized Acetylcholine Receptor

Agonists	n	K_I (M)*
Acetylcholine	0.95	7.9×10^{-7}
Carbamylcholine	0.61	7.1×10^{-7}
Succinylcholine	0.74	1.8×10^{-7}
Nicotine	0.52	1.2×10^{-7}
Decamethonium	0.52	6.0×10^{-9}
Antagonists		
[^{3}H]-α-neurotoxin	1.0	4.0×10^{-11}
Dimethyl-d-tubocurarine	0.58	1.0×10^{-7}
Benzoquinonium	0.50	6.3×10^{-8}
Alloferine	0.51	9.0×10^{-10}
Hexamethonium	0.56	4.0×10^{-6}

* per mole of ligand, all K_I values except for toxin and alloferine from logarithmic plots.
The K_I for agonists refers to the initial dissociation constant at low ligand concentrations.
All agonists bound with concentration dependent affinity to the receptor; all antagonists
bound to a single class of non-interacting sites.

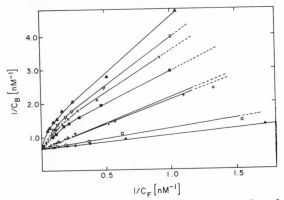

Fig. 5. Competitive binding of toxin and bis-methonium compounds to the receptor.
(●—●) toxin, (□—□) tetra 1 x 10^{-4} M, (+—+) penta 9 x 10^{-5} M, (○—○) hexa 9 x 10^{-6}
M, (■—■) hepta 8 x 10^{-5} M, (x—x) octa 8 x 10^{-5} M, ($\triangledown$—$\triangledown$) ennea 4.9 x 10^{-5} M, (▲—▲)
deca 2.16 x 10^{-5} M.

agonist and antagonist binding to the solubilized acetylcholine receptor. Although additional experiments in different time ranges and by different methods are required to understand the structural basis of the observed phenomena, they may have important implications for receptor function in solution and in membranes.

1. Detailed equilibrium binding and kinetic studies of the interaction of α-cobratoxin with the solubilized receptor (5,6) lead to the conclusion that toxin sites are organized as pairs of adjacent and/or strongly interacting identical sites with only transient asymmetry. We have presented evidence here for directly competitive and mutually exclusive binding of α-cobratoxin and small cholinergic ligands to the receptor and for half as many small ligand sites (with the possible exception of acetylcholine) as for toxin at the receptor. Both lines of evidence together seem to support a mechanism in which each small ligand molecule displaces one pair of toxin molecules upon binding to the receptor. This can be visualized by a direct overlapping of one small ligand site with two toxin sites oriented as antiparallel pair or by small ligand induced conformational changes of the receptor leading to "closing" of a second toxin binding site. These results are fundamentally different from those of Fu *et al.* (10) who found in competition binding studies of small ligands and α-bungarotoxin to membrane bound receptor that there exist as many small ligand sites as toxin sites at the receptor but that only half of these sites are directly competitive.

2. We did not find any indication of an equilibrium between two or more preformed states of the receptor (5) as was postulated for allosteric models of receptor-mediated conductivity changes (11,12). With the reservation that our assay methods would not detect minority populations of 5% or less of the total concentration of receptor molecules in solution, we assume that the solubilized receptor initially forms only one single population of unoccupied ligand binding sites. Since the binding pattern of agonists and antagonists differed only at higher degrees of receptor saturation (the range in which the binding affinity for agonists decreases, see Figs. 3 and 4) and since all binding processes to the solubilized receptor are accompanied by substantial reaction entropies, the following mechanism for ligand binding seems plausible: Binding of cholinergic ligands leads, at low degrees of saturation, to an activated form of the receptor which is similar or identical for agonists, antagonists, or toxin. This is the final state for antagonist binding at any degree of receptor saturation. Only agonists, at higher degrees of saturation corresponding to their anticooperative binding mode, can move the receptor into a third conformation which corresponds to the high conductivity or open channel form. Such a mechanism would allow for a common recognition and binding step for all cholinergic ligands followed by a conformational change to the open channel form specific only for agonists

and induced by agonists only due to additional features of their chemical structure. This could be visualized by a multipoint attachment between cholinergic ligands and the receptor, some of which are shared by all ligands, whereas some of which are specific for agonists, antagonists or toxin only. Such a mechanism could also offer functional relationships between electrical response and ligand binding in which the sigmoidal shape of dose-response curves is a result of agonist binding in only the anticooperative range of their interaction with the receptor (13).

3. The anticooperative binding of agonists also suggests a plausible mechanism for the role of acetylcholine esterase in regulating transmitter concentrations: Although it is generally accepted that transmitter molecules are rapidly removed by enzymatic hydrolysis, thereby freeing the receptor for subsequent response, this mechanism is unlikely to apply at low concentrations of receptor bound acetylcholine since all reported K_m values for acetylcholine esterase are in excess of 10^{-4} M. Hence, acetylcholine esterase is unlikely to control transmitter levels or influence receptor response in the range of tight binding. In the anticooperative binding region, however, the dissociation constant K_I for acetylcholine and the K_m value of the esterase are of comparable size and transmitter levels could be regulated efficiently by the hydrolyzing enzyme. Steep increases in transmitter concentration would be required for binding in the anticooperative region, but these need only be short-lived. At high local concentrations some of the transmitter would become susceptible to rapid enzymatic hydrolysis; this combined with the effects of diffusion could ensure that transmitter molecules associated with the initial high affinity sites remain bound for much longer periods without impairing the ability of the receptor to respond rapidly to further stimuli.

ACKNOWLEDGEMENTS

We thank Emmanuel Dumaguing for excellent technical assistance. This work was supported in part by the Deutsche Forschungsgemeinschaft and Grant CA 08290-07 from the National Institutes of Health, United States Public Health Service.

REFERENCES

1. Katz, B. (1966) *Nerve, Muscle and Synapse*, McGraw-Hill, New York.
2. Nachmansohn, D. (1959) *Chemical and Molecular Basis of Nerve Activity*, Academic Press, New York.
3. Klett, R.P., Fulpius, B.W., Cooper, D., Smith, M., Reich, E. and Possani, L.D. (1973) *J. Biol. Chem. 248*, 6841-6853.

4. Fulpius, B.W., Maelicke, A., Klett, R.P. and Reich, E. (1975) In *Cholinergic Mechanism* (Waser, P.G., ed.) p. 375, Raven Press, New York.
5. Maelicke, A., Fulpius, B.W., Klett, R.P. and Reich, E. (in press) *J. Biol. Chem.*
6. Maelicke, A. and Reich, E. (in press) *Symp. Quant. Biol.*, Vol. XL, Cold Spring Harbor.
7. Cooper, D. and Reich, E. (1972) *J. Biol. Chem. 247*, 3008-3013.
8. Shabanowitz, J., Brynes, P., Maelicke, A., Bowen, D.W. and Field, F.H. (in press) *Biomed. Masspectr.*
9. Cha, S. Manuscript in preparation.
10. Fu, J.L., Donner, D.B. and Hess, G.P. (1974) *Biochem. Biophys. Res. Commun. 60*, 1072-1080.
11. Karlin, A. (1967) *J. Theoret. Biol. 16*, 306-320.
12. Changeux, J.-P. and Podleski, T.R. (1968) *Proc. Nat. Acad. Sci. (U.S.A.) 59*, 944-950.
13. Maelicke, A. Manuscript in preparation.

VII
Microsonal and Related Membranes

THERMODYNAMIC CONTROL FEATURES
OF THE *PSEUDOMONAS PUTIDA* MONOXYGENASE SYSTEM

I.C. Gunsalus

School of Chemical Sciences
University of Illinois
Urbana, Illinois 61801

This, the final session of a symposium on membranes and related phenomena so beautifully organized and managed, requires of me two tasks; both simple and pleasant, but essential and brief. The first, to introduce two casts of speakers; the second, to reserve time for a compact, clear and concise summary of the proceedings by Frank Huennekens.

The speakers, if unknown to you, are from groups who have contributed with prime significance to understanding monoxygenase systems, components, and organization. Lars Ernster will speak for the Stockholm group, principally, on the biological consequences of the reactive intermediates of aromatic oxidation. Minor Coon, leader of the Michigan effort, will discuss their recent demonstration of multiple hepatic microsomal P-450 cytochromes, their preparation in pure forms, and the reductase stoichiometry. Ronald Estabrook, for the Dallas group, will primarily refocus new data on the electron transport problems of hepatic microsome systems. After a refreshment break, we shall discuss these three presentations. Then Tokuji Kimura, from Wayne State, Detroit, will report on the state of knowledge of the synthetic adrenal cortex mitochondrial P-450 cytochrome-adrenodoxin system.

Those of you who have for the past two days punched holes in membranes, excited, energy coupled, and oxidative phosphorylated them, can now relax while I talk for a few minutes to the young biochemists, of whatever age, about the marvelous problems which lie ahead: The biochemistry of clean proteins; enzymes and enzyme aggregates, the reactions they catalyze and their regulation; a trace of genetics; a few words about evolution, permeability, and conformation.

Changes are fast in biochemistry; the tools are generated and picked up more slowly as pictures give way to numbers, models to testable hypotheses and, less frequently, to laws. Time can yield both insight and dogma; the former we must stress. The central biochemical characters of this play are one class of cytochromes, termed P-450, their redox changes with two electrons that cleave a molecule of dioxygen, reducing one atom to water and incorporating, by a mechanism still undisclosed, the other atom into a carbon-hydrogen bond to form an alcohol. Although variations on this primary theme exist, the form of monoxygenase activity thus illustrated is preserved irrespective of the systems, whether synthetic, as formation of steroid hormones in the adrenal mitochondria; dissimilatory, as in the microbes, with supply of carbon and energy for growth; or detoxifying, as in the hepatic microsomal systems.

All but the very young know that P-450 stands for the protein with a prominent 450 nm absorption following reduction and carbon monoxide addition. This discovery came in 1958 from a young biochemist named Klingenberg, and almost simultaneously by Garfinkel (1,2). Martin Klingenberg talked here yesterday on another theme. This tale, first told in 1958, bears a history barely half that of the oxidative phosphorylation Paul Boyer recalled for us from his graduate days in Wisconsin; the symposium on respiratory enzymes in the summer of 1940 which truly provided, as we hope will occur here, stimulation and encouragement to many eager, young scientists and students through a first personal and intellectual access to Europe's recently arrived leaders: Meyerhof, Ochoa, Lipmann, Stern, the Coris, Kalckar and so many others.

The four monoxygenase papers that follow turn quickly to cytochrome P-450. Two carry P-450 in the title, one modestly awaits the first line of the text, and the fourth, the second line. One reconstitutes, two pursue electron transport, and the fourth is concerned with a potentially deleterious verse of the P-450 story, that highly reactive intermediates of polyaromatic molecules alter nucleic acid, proteins, etc., to damage tissues and alter genetic patterns, including regulation of cell proliferation.

All tissues and organs so far studied contain P-450 cytochromes. Some are inducible, as in the hepatic microsomes and in bacteria; some appear to be constitutive, as in the adrenal cortex mitochondria. All the P-450 cytochrome systems investigated have fallen into a single class in spectral properties, stable intermediate states, and reaction cycle. Although the electron transport may or may not include an iron-sulfide protein, as shown in Fig. 1 to be discussed shortly, two reducing equivalents are required. The iron-sulfide component connecting flavoprotein to cytochrome in two separate one-electron steps occurs in the microbial and the adrenal mitochondrial systems. Ligation of substrate and protein components controls the equilibrium in the first electron transfer. Oxygen addition follows to form a heme protein with properties

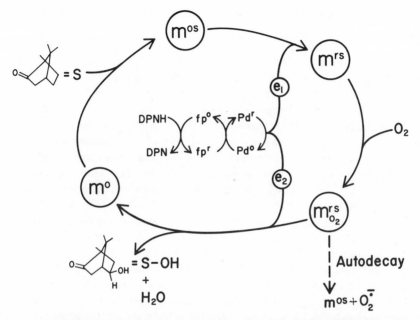

Fig. 1. Abbreviated P-450 monoxygenase reaction cycle. Cytochrome P-450 is abbreviated m, for monoxygenase. Superscripts o, r and s refer to oxidized (ferric), reduced (ferrous) and substrate (camphor) bound forms, respectively. $m_{O_2}^{rs}$ is the oxygenated ferrous intermediate.

similar to the oxygen-carrying ferro-heme proteins, oxyhemoglobin and oxymyoglobin. A more complex series of reactions in the oxygenated cytochrome is then initiated by a second electron transfer. The microsomal cytochromes appear to accept both electrons 1 and 2 directly from flavoprotein, with a phospholipid regulation.

Boyd (3) summarized the intermediate states of dioxygen reduction at a symposium in Edinburgh in 1970 in the following way. Dioxygen, a stable paramagnetic compound with two unpaired electrons, has an oxygen-oxygen bond distance of about 1.2 Å. The introduction of a single electron requires a considerable quantity of energy; even after modulation by substrates in the biological system, the potential is near -179 mV. The product, at the reduction level of superoxide, O_2^-, exhibits an oxygen-oxygen bond distance of about 1.3 Å. Although the subsequent steps are less clear, the input of a second reducing equivalent brings the molecule to the reduction level of peroxy anion, O_2^{2-}, with a bond distance extended to approximately 1.5 Å. This two-electron reduction of dioxygen, perhaps followed by [O] transfer to substrate, comprises the molecular events of monoxygenase action.

We turn now to the molecular properties, reactions and regulation of the cytochrome-bearing monooxygenases. For illustration, we choose the system that has yielded most readily to purification, provided pure components in quantity, and thus the chemical and physical properties required for quantitation and classification of the reaction sequence, namely, the camphor methylene hydroxylase of the soil bacterium, *Pseudomonas putida* (4). This system is isomorphic to the adrenal steroid hydroxylase in all respects of components, spectra, dynamics, and the resonance properties of the active sites. Each exhibits a high degree of substrate specificity and an essential association of components into multienzyme aggregates. They differ by a thousand fold in turnover number in favor of the microbial system, and lack any reciprocal complementation of protein components for catalysis. The components of the *Pseudomonas putida* system include the P-450 cytochrome, designated $P-450_{cam}$ or cytochrome m, a specific $Fe_2S_2^*Cys_4$ iron-sulfide protein termed putidaredoxin, abbreviated Pd, and a FAD flavoprotein dehydrogenase-reductase that connects NADH to the iron-sulfur protein. The cytochrome was isolated in homogeneous form and crystallized in our laboratory in the late 1960's. The amino acid composition of these three proteins is known (5) as is the primary structure of the redoxin, Pd (6); some structural features of the cytochrome have been determined (4). The analogous iron-sulfur protein of the adrenal system, adrenodoxin, has a similar composition but a different sequence (7).

A mechanistic understanding of a multi-intermediate reaction cycle must encompass both dynamic and equilibrium aspects. Great strides have been made during the past year toward a complete description of the hydroxylation process and varied equilibrium states of the oxygenase components, including the oxidized, reduced, liganded and unbound parameters in the form of the thermodynamic principles laid down by Willard Gibbs a century ago. A thermodynamic model for the control of equilibrium processes in this monooxygenase via ligation of substrate and protein components has been proposed by Sligar et al. (8). The data and the formulations presented here are mainly from Dr. Sligar's thesis (9) and the publications derived therefrom. These studies were made possible by an earlier blending of skills by Drs. Lipscomb and Sligar, an abundant supply of pure components, and support by our chemical group. The superb physical definition of the states of heme and sulfide-bound iron in the cytochrome and redoxin arose from the physics and biomolecular interests of Professors Debrunner and Frauenfelder and their students in the Physics Department at Illinois, see reference 4.

We shall consider separately the two redox processes. The first, a ferric-ferrous reduction of a heme iron, occurs by the Pd-mediated transfer of an electron to cytochrome m. The accumulated evidence supports the assignment of an essential catalytic role to a complex between the two components. An examination of the details of the equilibrium oxidation/reduction potentials of redoxin and cytochrome reveals two liganding processes; that is, cytochrome combined with substrate, S_1, and/or redoxin, Pd. Sligar (9) has achieved the separate examination of these equilibrium associations by careful titration. The cytochrome m ligand states are graphically shown in Fig. 2. Their dissociation constants are convertible to free energies — the camphor binding via the relation $\Delta G = RT\ln(K_D)$ and the redox equilibrium with potential $\Delta E_0'$ as given by $\Delta G = -\Delta E_0'$. Presentation of the data in this form dramatizes the conditions for free energy conservation. The energy levels of the processes, drawn to scale in Fig. 3, demonstrate an asymmetry in the dissociation constants for

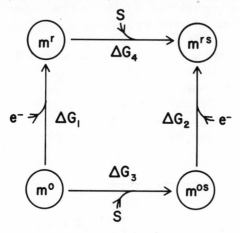

$$\Delta G_1 - \Delta G_2 - \Delta G_3 + \Delta G_4 = 0$$

Fig. 2. Cytochrome m: substrate and redox states.

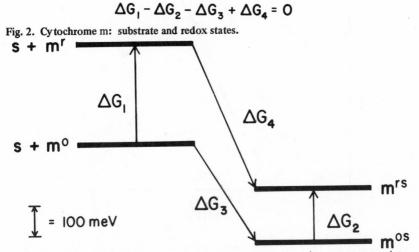

Fig. 3. Substrate modulation of cytochrome m redox potential. $\Delta G_1 = 340$ meV, $\Delta G_2 = 170$ meV, $\Delta G_3 = -325$ meV, $\Delta G_4 = -495$ meV (9).

oxidized and reduced cytochrome on substrate binding, as related to the observed shift in redox potential.

The association of Pd to cytochrome m presents a more complex case resulting from the two redox centers. Fig. 4 shows, in analogy to the binding of camphor, an asymmetry for association of oxidized and reduced Pd and the corresponding energy diagram is in Fig. 5. The equilibrium of this coupling occurs with a shift in the redox potential of Pd on ligation to cytochrome m, as verified experimentally in Fig. 6 (9). The reduction of cytochrome m by Pd is then viewed as shown in Fig. 7. This presentation of the equilibrium data in terms of free energy diagram illuminates in clear and precise form the coupling between ligation and redox flow for the first electron transfer to cytochrome m in the process of dioxygen cleavage.

We now turn to the second input of a reducing equivalent, that is, the reactions of the oxygenated intermediate, abbreviated $m\overset{rs}{O_2}$. This electron transfer initiates the catalytic events that are coupled to the release of the reaction products. The kinetics of product formation in the complete or hydroxylase system, with either pyridine nucleotide or chemical reductant, indicate clearly a $Pd \cdot m\overset{rs}{O_2}$ dienzyme complex as intermediate in the hydroxylation event (12).

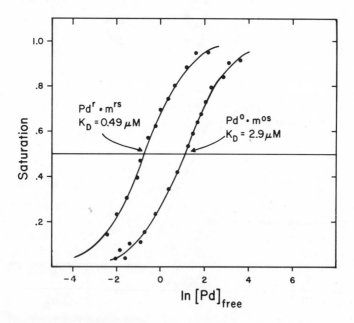

Fig. 4. Putidaredoxin-cytochrome m binding, monitored by the quenching of a fluorescein-isothiocyanate label covalently attached to a sulfhydryl on the cytochrome m surface (11).

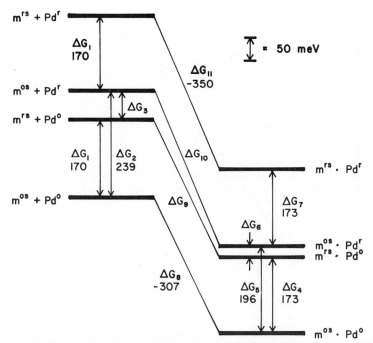

Fig. 5. Cytochrome m-Pd: free energy states. The redox energies of free states are shown on the left, those of the bound forms at right. $\Delta G_{8\text{-}11}$ are the binding energies for the indicated processes in meV/molecule (1 meV/molecule = 0.023 Kcal/mole).

This $Pd \cdot m^{rs}$ binding site is distinct from that of the first reduction and is monitored by an induced optical difference spectrum, Fig. 8, which parallels the spin state changes observed in EPR and Mössbauer. The equilibrium association parameters agree remarkably well with the Michaelis constant derived from a kinetic analysis of product formation (10). The rate limiting step in the entire hydroxylation cycle is the breakdown of the $Pd \cdot m^{rs}_{O_2}$ complex to generate hydroxylated camphor. Fig. 9 shows a schematic of the decay of the $m^{rs}_{O_2}$ intermediate in the autoxidation and the product forming reactions. Clearly, a complete description of the monoxygenase cycle in equilibrium and dynamic terms will require additional insights and measurement.

Two crucial points remain and need to be stressed: a necessity for consistently pure components in reasonable quantity, and a system adequate to the criteria required to reveal mechanistic details. In multicomponent protein systems assembled from soluble enzymes, a clearly critical feature is their association into higher order aggregates. The extension to membrane-bound proteins, where content and structure are less well understood, requires also a quantitative characterization of the catalysts. Finally, although a great desire

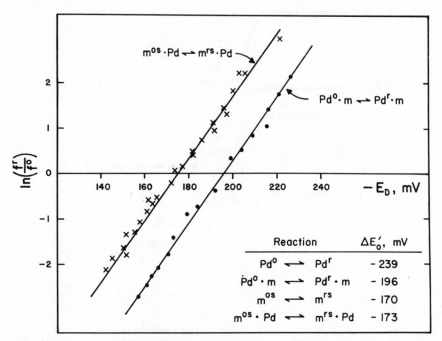

Fig. 6. Redox potentials in the Pd-cytochrome m complex. The log of fraction oxidized/ reduced is plotted vs. total system potential monitored by Brilliant Alizarin Blue ($E_0' = -173$ mV).

$$Pd^r + m^{os} \overset{k_1}{\underset{k_{-1}}{\rightleftharpoons}} Pd^r \cdot m^{os} \overset{k_2}{\underset{k_{-2}}{\rightleftharpoons}} Pd^o \cdot m^{rs} \overset{k_3}{\underset{k_{-3}}{\rightleftharpoons}} Pd^o + m^{rs}$$

ΔG, meV	-350	23	-307
k_+/k_-	$2.0\ \mu M^{-1}$	2.6	$.34\ \mu M^{-1}$

Fig. 7. Summary of m^{os} reduction by Pd^r.

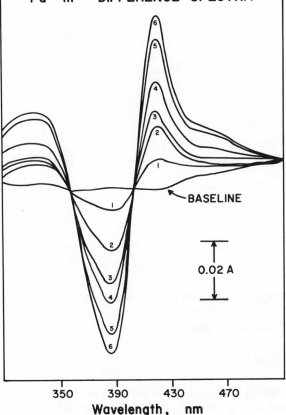

Fig. 8. Induced optical difference spectra of cytochrome m on Pd binding. Cytochrome m at 10 μM is dissolved in 50 mM potassium phosphate buffer, pH 7.0, containing 200 μM D-camphor. Redoxin in the same buffer is added at concentrations (1) 3.4 μM, (2) 6.8 μM, (3) 10.2 μM, (4) 20.4 μM, (5) 37.5 μM, and (6) 54.4 μM. A dissociation constant of 26 μM is derived.

exists for a "moving picture" of the catalytic event, i.e., "cysteine no. 1348 moves 1.235 Å to the left," the most powerful techniques for describing a system are fundamental thermodynamic and kinetic considerations too often overlooked in the quest for cartoons. We have, thus, as in the Virginia Slim's advertisement, "come a long way," and the next few years will hold much excitement as the fundamental aspects of catalysis and electron transfer in the oxygenase systems yield to the current and developing techniques of physics,

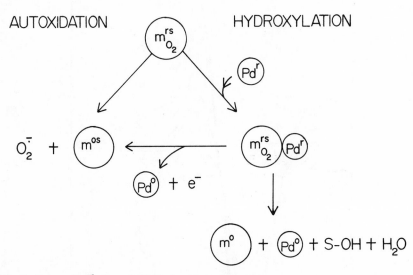

Fig. 9. Decay of $m_{O_2}^{rs}$ in autoxidation and in hydroxylation.

chemistry and molecular biology.

ACKNOWLEDGMENTS

The author wishes to thank Dr. S.G. Sligar for his generous and effective help and discussions, including his contribution to the art and writing of this manuscript. Supported in part by grants HEW PHS GM 21161 and AM 562.

REFERENCES

1.   Klingenberg, M. (1958) *Arch. Biochem. Biophys. 75*, 376-386.
2.   Garfinkel, D. (1958) *Arch. Biochem. Biophys. 71*, 493-509.
3.   Boyd, G.S. (1972) In *Biological Hydroxylation Mechanisms* (Boyd, G.S., ed.) pp. 1-21, Academic Press, New York.
4.   Gunsalus, I.C., Meeks, J.R., Lipscomb, J.D., Debrunner, P.G. and Münck, E. (1974) In *Molecular Mechanisms of Oxygen Activation* (Hayaishi, O., ed.) pp. 559-613, Academic Press, New York.
5.   Dus, K., Katagiri, M., Yu, C.-A., Erbes, D.L. and Gunsalus, I.C. (1970) *Biochem. Biophys. Res. Commun. 40*, 1423-1430.

6. Tanaka, M., Haniu, M., Yasunobu, K.T., Dus, K. and Gunsalus, I.C. (1974) *J. Biol. Chem. 249*, 3689-3701.

7. Tanaka, M., Haniu, M. and Yasunobu, K.T. (1973) *J. Biol. Chem. 248*, 1141-1157.

8. Sligar, S.G. and Gunsalus, I.C. In preparation.

9. Sligar, S.G. (1975) *A Kinetic and Equilibrium Description of Camphor Hydroxylation by the P-450$_{cam}$ Monoxygenase System*, Ph.D. Thesis, University of Illinois, Urbana, 147 pp.

10. Lipscomb, J.D. (1975) *Energy Transfer and Segregation During Mixed Function Oxidation by P-450$_{cam}$ and Putidaredoxin*, Ph.D. Thesis, University of Illinois, Urbana, 326 pp.

11. Sligar, S.G., Debrunner, P.G., Lipscomb, J.D., Namtvedt, M.J. and Gunsalus, I.C. (1974) *Proc. Nat. Acad. Sci. (U.S.A.) 71*, 3906-3910.

12. Sligar, S.G., Debrunner, P.G., Namtvedt, M.J. and Gunsalus, I.C. (1975) *Fed. Proc. 34*, 662.

THE METABOLISM OF POLYCYCLIC HYDROCARBONS
AND ITS RELATIONSHIP TO CHEMICAL CARCINOGENESIS

Lars Ernster, Jorge Capdevila, Gustav Dallner, Joseph W. DePierre
Sten Jakobsson and Sten Orrenius

Department of Biochemistry
Arrhenius Laboratory
University of Stockholm
and
Department of Forensic Medicine
Karolinska Institutet
Stockholm, Sweden

The purpose of this paper is to illustrate how studies on the structural basis of membrane function — the subject matter of this Symposium — can be useful in approaching some theoretical and practical aspects of a problem with important bearing on human health and welfare — namely, the problem of chemical carcinogenesis.

A large number of different environmental factors have been implicated in converting normal cells into tumor cells (Table I)(for review, see ref. 1). There is good evidence that ultraviolet radiation, radioactivity, physical irritation, many inorganic substances, an even greater number of organic chemicals, and viruses are involved in the development of various forms of cancer, both in experimental animals and in human beings. Viruses, or at least a part of the genetic information they contain, must be present when the cancer develops; but exposure to other initiating factors may take place some time, even decades, before the disease actually appears.

Of these factors, the chemical carcinogens seem to be quantitatively most important. Some 1000 compounds have already been shown to be oncogenic; and a considerable and growing number of these are used in large quantities in our factories and pollute the biosphere. Estimates are that 70% or more of all tumors in man arise as a result of exposure to chemicals (1).

At first glance, a list of the organic chemicals thought to cause cancer seems

TABLE I

Environmental Factors Implicated in Causing Various Forms of Cancer

Factor	Example of tissue in which cancer may arise
Ultraviolet radiation	Skin
Radioactivity	Blood, Lung
Physical irritation	Connective tissue
Inorganic substances	
asbestos	Lung
chromate	Lung
nickel	Nasal sinus
Organic chemicals	
aniline dyes	Bladder
alcohol	Larynx, esophagus
vinyl chloride	Liver
polycyclic hydrocarbons	Lung
Viruses	Blood, mammary gland

From A. Braun (1974) The Biology of Cancer (ref. 1).

to reveal a wide spectrum of structurally unrelated compounds. However, it has become more and more apparent that the active forms of many of these substances have a common characteristic, namely, they are all strong electrophiles. In some cases the parent carcinogen itself has this property. In many other cases it does not, but a strong electrophile is produced during metabolism of the original compound (1,2). As we will see, it is a membrane-bound enzyme system, namely, the microsomal aryl hydrocarbon monooxygenase, that is responsible for transforming many xenobiotics into active carcinogens.

Many of these strong electrophiles have been shown to react readily with the strongly nucleophilic groups found in nucleic acids and proteins (2). Presumably as a result of such reactions, numerous chemical carcinogens (or their metabolites) effectively cause mutations in bacteria and in certain animal cells in culture (2). This interesting observation has been developed, chiefly by Ames and his collaborators (3), into a simple and promising screening system for potential carcinogens. The chemical in question is incubated with a mixture of rat liver microsomes (in order to achieve mammalian metabolism) and histidine-requiring mutants of *Salmonella*. Most carcinogens produce a large number of revertants to the non-histidine-requiring phenotype in this model system.

Thus, many chemical carcinogens (or their metabolites) bind covalently to DNA and are powerful mutagens. In addition, chemically transformed cells — just as tumor cells in general — transmit their characteristic properties to their progeny throughout many generations (1). These observations constitute the major support for the theory which maintains that chemicals (or their metabolites) transform normal cells into tumor cells by direct reaction with the genetic material to cause somatic mutations. This theory is far from proven, but it is used as a working hypothesis by many investigators, including ourselves.

These general conclusions about chemical carcinogens are based on extensive studies with substances such as the polycyclic hydrocarbons, some of whose structures are illustrated in Fig. 1. These many-ringed aromatic compounds are found in cigarette smoke, in coal tar, among the products of the incomplete combustion of fossil fuels, and in other waste materials of technology (4-6). They are implicated as a major cause of lung cancer (7-17).

From their structures it is clear that the polycyclic hydrocarbons are chemically rather inert. Indeed, this is one of the cases where it is not the compounds themselves, but rather a much more reactive metabolic intermediate, that is the true carcinogenic agent. For this reason it is very important that we understand the metabolism of polycyclic hydrocarbons in mammalian cells.

Our present knowledge of this metabolism is pictured in Fig. 2. The first step is the formation of an arene oxide, and it is catalyzed by an enzyme sys-

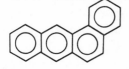

benz(α)anthracene

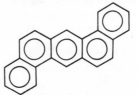

dibenz(a,h)anthracene

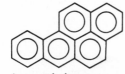

benzo(α)pyrene

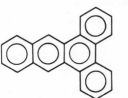

dibenz(a,c)anthracene

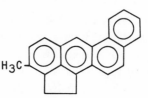

3-methylcholanthrene

Fig. 1. Structures of some polycyclic hydrocarbons.

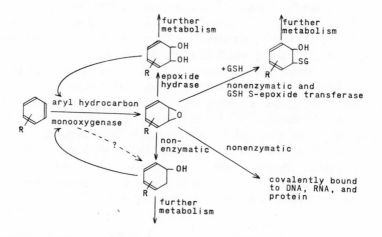

Fig. 2. The metabolism of polycyclic hydrocarbons in mammalian cells.

tem called aryl hydrocarbon monooxygenase (18,19). This enzyme system
seems to be almost ubiquitous. Nebert and Gelboin (20) detected aryl hydrocarbon monooxygenase activity in 90% of the tissues they examined, including
liver, lung, gastrointestinal tract and kidney in rats, mice, hamsters and mon-
keys. This activity has also been demonstrated in human lung (21), placenta
(22), lymphocytes (23-27), and alveolar macrophages (23).

Upon closer examination it became clear that the formation of arene oxides
is catalyzed by the cytochrome P-450-linked monooxygenase system, the same
system that catalyzes the first step in the metabolism of many other xenobiotics, as well as various steps in the metabolism of steroids and fatty acids.
Our present conception of the redox cycle catalyzed by cytochrome P-450 is
represented in Fig. 3. This reaction scheme has been arrived at by research in
a number of different laboratories, in particular in Dr. Estabrook's laboratory
(30-33). Recent studies in our laboratory by Eugene Hrycay in collaboration
with Jan-Åke Gustafsson and Magnus Ingelman-Sundberg at the Karolinska
Institute (28,29) have employed organic hydroperoxides, sodium periodate
and sodium chlorite instead of NADPH to support the cytochrome P-450-
catalyzed hydroxylation of various steroid substrates in rat liver microsomes.
These studies suggest that a ferryl ion form of cytochrome P-450 may be the
common intermediate hydroxylating species in the monooxygenase reactions
catalyzed by this protein.

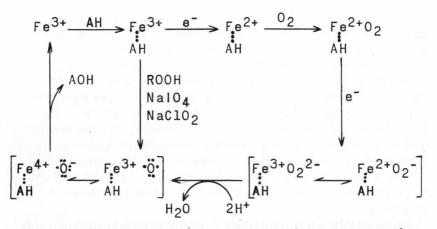

Fig. 3. A proposed mechanism for cytochrome P-450-catalyzed monooxygenase reactions.
Fe = the heme iron of cytochrome P-450; AH = substrate undergoing monooxygenation;
ROOH = organic hydroperoxides. (From E.G. Hrycay, J.-Å. Gustafsson, M. Ingelman-
Sundberg and L. Ernster (28,29).)

A simplified scheme of the microsomal cytochrome P-450-linked monooxygenase system is depicted in Fig. 4. Various properties of this system may
be summarized as follows. First, aryl hydrocarbon monooxygenase activity is
localized on the endoplasmic reticulum of mammalian cells and is generally
studied *in vitro* by using microsomes. The system involves at least two protein
components, cytochrome P-450 and a flavoprotein called NADPH-cytochrome
P-450 reductase. It has long been postulated that there is a third component
involved, perhaps an iron-sulfur protein similar to that found in the cytochrome
P-450 system of *Pseudomonas putida* (e.g., 34) and of adrenal mitochondria
(e.g., 35); but the existence of such a component in microsomes has not yet
been demonstrated. Reconstitution experiments by Coon, Lu and their coworkers (36-39) have shown that phospholipids of the endoplasmic reticulum

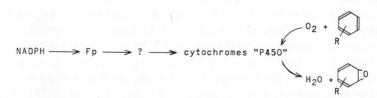

Fig. 4. A simplified scheme of the microsomal aryl hydrocarbon monooxygenase system.

membrane are also required for the functioning of aryl hydrocarbon monooxygenase.

Electrons are accepted from NADPH by the reductase, and eventually passed on to cytochrome P-450, which incorporates one atom of an oxygen molecule into the polycyclic hydrocarbon, and uses the other atom, two protons, and two electrons to make a molecule of water. Since the substrate binds to cytochrome P-450, it is this cytochrome that determines the substrate specificity of the system. Cytochrome c can short-circuit the system by accepting electrons from NADPH-cytochrome P-450 reductase and is thus an inhibitor. In fact, the NADPH-cytochrome c reductase activity catalyzed by the flavoprotein is used as a convenient means of quantitating this component. Carbon monooxide inhibits by binding directly to the reduced form of cytochrome P-450. 7,8-Benzoflavone and related compounds are also powerful inhibitors of aryl hydrocarbon monooxygenase (40-43), but their mode of action is not yet understood.

One of the most important features of aryl hydrocarbon monooxygenase is that it can be induced by its substrates. Such induction has been shown in a large number of mammalian tissues, including the liver, lung, kidney and gastrointestinal tracts of rats; certain mammalian cell types grown in culture; and human placenta and lymphocytes (20,22-27,43-48). The induction requires continued RNA and protein synthesis (48-52). Large quantitative differences in the inducibility of aryl hydrocarbon monooxygenase have been found: this activity can be increased 2- to 10-fold in rat liver, 3- to 20-fold in rat lung, and 200- to 300 fold in certain other tissues and strains (18,43). Furthermore, the inducibility of this enzyme system appears to be regulated genetically both in mice and in men (24,25,43,53-56).

As is well known, induction of the cytochrome P-450-linked monooxygenase system can also be achieved with other xenobiotics. For instance, phenobarbital injected intraperitoneally into rats once a day for four or five days can increase rat liver levels of cytochrome P-450, NADPH-cytochrome c reductase, and aminopyrine demethylase 5-fold in specific activity and 10- to 12-fold in total activity (e.g., 57). Treatment with the polycyclic hydrocarbon 3-methylcholanthrene also increases cytochrome P-450 levels several-fold; but a striking feature of this induction is that microsomal levels of NADPH-cytochrome c reductase are unaffected or even slightly decreased (e.g., 18,58).

Christina Lind in our laboratory is carrying out experiments designed to explain this observation. She is studying DT diaphorase, a flavoprotein that catalyzes the oxidation of NADH and NADPH by various redox dyes and quinones (59). This enzyme is found in the cytoplasm, in mitochondria and in microsomes (60). She observed that DT diaphorase is co-induced with aryl hydrocarbon monooxygenase by methylcholanthrene (see Fig. 5)(59), an observation also made earlier by Huggins and coworkers (61,62). In addition, she

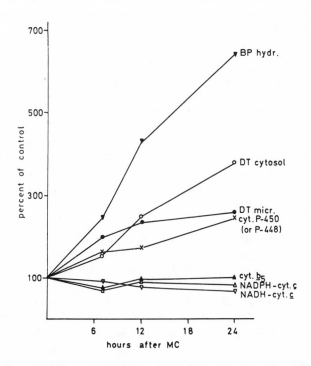

Fig. 5. Changes in certain hepatic enzyme parameters following injection of rats with a single dose of 3-methylcholanthrene. (From C. Lind and L. Ernster (58).)

discovered that 7,8-benzoflavone is a powerful inhibitor of DT diaphorase, with a K_i (3.5 x 10^{-5} M) similar to that found for the inhibition of the methylcholanthrene-induced aryl hydrocarbon monooxygenase activity by this compound. Both of these findings suggest that DT diaphorase may be cytochrome P-450 reductase in the case of induced polycyclic hydrocarbon metabolism, and Christina Lind is actively investigating this possibility.

It should come as no surprise to workers in the field if there turns out to be a family of cytochrome P-450 reductases, since it is already well established that there is a family of cytochrome P-450's. Evidence for the existence of a number of different cytochrome P-450 species may be summarized as follows. In the first place, the cytochrome P-450-linked systems in various organs demonstrate different substrate specificities. The system in the adrenal cortex, in ovaries, and in testicular tissue is specialized for steroid metabolism (18); the

kidney system preferentially metabolizes fatty acids (63); while liver microsomes metabolize drugs and other xenobiotics, steroids, and fatty acids (18,
64). Secondly, different substrates induce cytochrome P-450's with different
substrate specificities and spectral characteristics (18,65). Phenobarbital induces a cytochrome which metabolizes many different xenobiotics but not polycyclic hydrocarbons and which, when reduced and complexed with carbon
monooxide, exhibits an absorption maximum at 450 nm. Methylcholanthrene,
on the other hand, induces a cytochrome species with a corresponding absorption maximum at 448 nm and specialized for polycyclic hydrocarbon metabolism. In addition, these cytochromes P-450 and P-448 can be isolated and used
to reconstitute functional drug metabolism which demonstrates the same substrate specificity as that observed *in situ* (36-39,66). Fourth, spectral studies
reveal that there are at least 3 different cytochrome P-450's with different affinities for ethylisocyanide and *n*-octylamine (67). And finally, investigation
with SDS disc gel electrophoresis shows that microsomes contain at least 3 or
4 cytochrome P-450 species with molecular weights between 49 and 55 thousand daltons (68,69). It seems quite clear that there is a family of cytochromes
"P-450" but it is not yet clear how large this family is. Dr. Coon will discuss
this question in greater detail later.

We may conclude our characterization of the aryl hydrocarbon monooxygenase system by noting that it contains a specialized cytochrome P-450 almost
certainly structurally related to the cytochromes "P-450" involved in the metabolism of other xenobiotics. Indeed, aryl hydrocarbon monooxygenase intro-
duces epoxide groups at a number of different points on the rings of a number
of different hydrocarbons and even metabolizes the dihydrodiols and, as we
shall see, phenols. It is thus possible that this system itself contains more than
one type of cytochrome P-450.

Let us now return to our reaction scheme of polycyclic hydrocarbon metabolism (cf. Fig. 2). The arene oxides formed in the first step of polycyclic
hydrocarbon metabolism are short-lived, highly reactive intermediates that are
transformed into more stable substances both enzymatically and non-enzymatically. On one branch the arene oxides are hydrated by epoxide hydrase to
yield dihydrodiols (70-74). Epoxide hydrase has been measured in various
organs of the rat and guinea pig and has been shown to be high in liver, low in
kidney, very low in lung and intestine, and not detectable in brain, heart, spleen
or muscle (74). This enzyme has been found exclusively in the microsomal
fraction from the liver and kidney of guinea pigs and has even been suggested
as an excellent marker for the endoplasmic reticulum (71). Recent studies in
our laboratory have shown that the specific activity of epoxide hydrase (measured using styrene oxide as substrate) in rat liver microsomes increases 160%
after induction with phenobarbital but only 33% after treatment with methylcholanthrene. These results are in agreement with an earlier report by Oesch

and his coworkers (71). However, there are several indications for the existence of more than one species of epoxide hydrase (75); and the induction pattern might have been different if this enzyme had been measured using an arene oxide as substrate.

Arene oxides may also be conjugated with glutathione by glutathione S-epoxide transferase, an activity found in the cytoplasm of various organs of the rat and other vertebrates (72,74,76-81). When measured with benzanthracene 5,6-oxide as substrate this enzyme was found to be present at higher levels in the lung than in the liver of methylcholanthrene-treated rats (82). Glutathione S-epoxide transferase also appears to be a family of different enzymes and is also inducible by polycyclic hydrocarbons (83).

In addition, as diagrammed in Fig. 2, arene oxides may react non-enzymatically with glutathione (19,84), nucleic acids, and proteins (85-93) and may isomerize in water to form phenols (19,94). Dihydrodiols, phenols and glutathione conjugates can all be further metabolized in various ways (19,74,95-101). The dihydrodiols and phenols may themselves serve as substrates for the aryl hydrocarbon monooxygenase system, with the result that a new epoxide group is introduced at another position on the same molecule.

There is substantial evidence that it is the arene oxides that are directly responsible for causing cancer that results from exposure to polycyclic hydrocarbons (102-107). In the first place, it is the epoxides rather than the parent hydrocarbons or other metabolites that bind to nucleic acids (92). Brookes and Lawley (89) found a positive correlation between the carcinogenicity of a series of polycyclic hydrocarbons and the extent of their binding to DNA in mouse skin; there was no correlation between carcinogenicity and binding to protein. Secondly, K-region epoxides of benzanthracene, dibenzanthracene and methylbenzanthracene were shown by Ames and coworkers (105) to cause frame-shift mutations in *Salmonella typhimurium*, whereas the parent hydrocarbons, K-region dihydrodiols and phenols were inactive (105). Third, Grover and his coworkers (103) have examined the effects of epoxide, dihydrodiol and phenol derivatives of benzanthracene, dibenzanthracene and methylcholanthrene on hamster embryo cells. The epoxides were most potent by far in causing malignant transformations. Similar experiments using cell cultures derived from mouse prostate gave identical results (107).

Finally, it has been found that in certain cases inhibition of the first step of polycyclic hydrocarbon metabolism can protect against the carcinogenic effects of these compounds. Thus, 7,8-benzoflavone inhibits the formation of tumors produced by application of 7,12-dimethylbenzanthracene to mouse skin (43, 108). However, a number of contradictory observations have been reported in this area (109-112). For instance, 7,8-benzoflavone does not have any effect on the promotion of mouse skin tumors by benzpyrene (43). These apparent contradictions are probably due to the fact that it is the steady state level of arene oxide which is the important factor in carcinogenesis; and this steady state level is affected not only by the synthesis of arene oxides via aryl hydrocarbon monooxygenase, but also by their breakdown via epoxide hydrase and glutathione S-epoxide transferase as well as non-enzymatically.

In order to further understand the biochemical events underlying chemical initiation of lung cancer we are presently investigating the synthesis and breakdown of arene oxides in rat liver and lung. It seems to us that the key observation is that polycyclic hydrocarbons, whether injected intraperitoneally or breathed in as cigarette smoke, cause cancer in the lung but not in the liver of experimental animals. If we can find a difference in the metabolism of polycyclic hydrocarbons in these two tissues, we might have a clue as to why this is so.

The first problem we encountered was that aryl hydrocarbon monooxygenase is a rather difficult activity to assay reliably. Incubation of benzpyrene (which is the compound most often used as substrate in these assays) with rat liver microsomes yields at least 8 major metabolites extractable with ethyl acetate (113), not to mention the water-soluble conjugates. All of these products must be determined in order to obtain a true measure of monooxygenase activity. However, the most widely used assay procedures are based on the fluorescence exhibited by a few of the products, especially 3-hydroxybenzpyrene (e.g., 114); and these procedures detect only fluorescent metabolites that can be extracted into hexane.

Using rat liver microsomes we have developed an assay for benzpyrene monooxygenase based on the use of tritium-labeled substrate (115). This procedure is illustrated in Fig. 6. The reaction is stopped with 0.5 N NaOH in 80% ethanol and the mixture is then extracted with hexane. After this extraction, approximately 99.5% of the remaining substrate is found in the upper phase and can easily be quantitated by scintillation counting.

We have used this new radioactive assay to investigate the metabolism of benzpyrene in isolated, intact rat-liver cells (116). These cells are prepared as follows (117,118): first, the portal vein of the liver in an anaesthetized animal is cannulated and the liver is then removed. After perfusion of this organ for 5 min with a salt solution containing EGTA, BSA and bovine erythrocytes, perfusion was carried out for an additional 10 min with a salt solution containing collagenase. The disintegrating liver is then incubated for 5 more min in a medium containing collagenase, and the cells are collected and washed. Cells prepared in this manner look intact under the electron microscope, and 90-95% of them exclude trypan blue for up to 2 hr after preparation.

Benzpyrene monooxygenase activity in these cells was linear for up to 60 min. Compared to cells from untreated rats, metabolism of benzpyrene by corresponding amounts of isolated microsomes was twice as rapid; but after

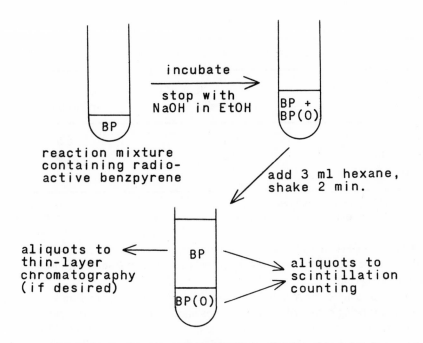

Fig. 6. Radioactive assay of benzpyrene (BP) monooxygenase. For details see J.W. DePierre, M.S. Moron, K.A.M. Johannesen and L. Ernster (115).

induction with methylcholanthrene metabolism in the two systems was about the same. The apparent K_m for benzpyrene was found to be 34 μM in induced cells and 6.7 μM in induced microsomes.

The effects of various inhibitors of aryl hydrocarbon monooxygenase – including SKF 525-A, metyrapone, α-naphthoflavone and hexobarbital – on the metabolism of benzpyrene by intact cells were about as expected from studies with isolated microsomes. Most recently, we have set out to identify the products of benzpyrene metabolism in these cells. As shown in Table II, conjugated products are most abundant, followed by dihydrodiols, in cells from both normal and methylcholanthrene-induced animals. We are now engaged in identifying the products in detail and in examining the effects of various treatments – such as the addition of an inhibitor of epoxide hydrase or a lowering of the level of glutathione in the cell – on the metabolite pattern.

We feel that with isolated, intact liver cells we can investigate polycyclic hydrocarbon metabolism under conditions much closer to those found *in vivo* than is the case with isolated microsomes, and with more control over conditions than can be obtained with organ perfusions. With an intact cell system the effect of complex metabolic variables such as the oxidation-reduction level

TABLE II

The Products of Benzpyrene Metabolism by Isolated, Intact Rat-Liver Cells

Product	Rate of formation* in	
	control cells	induced cells
Conjugated	9.2	26.5
Dihydrodiols	3.5	21.8
Phenols	0.8	8.4

* nanomoles/45 min/3 x 10^6 cells.
From H. Vadi, P. Moldeus, J. Capdevila and S. Orrenius (116).

of cellular NADPH on aryl hydrocarbon monooxygenase activity can be examined. We hope eventually to be able to carry out analogous studies with isolated intact rat lung cells.

For the present, our research on polycyclic hydrocarbon metabolism in lung tissue is being done with isolated microsomes. In order to obtain suitable lung microsomes we have had to utilize a number of special techniques (119). In the first place lung microsomes prepared in the same manner as liver microsomes are very heavily contaminated with hemoglobin; and this contamination renders spectral studies of lung cytochromes P-450 and b_5, which are present in rather low levels, difficult if not impossible. We have gotten around this problem by separating the particles in the post-mitochondrial supernatant from soluble components by gel filtration of a Sepharose 2B column (119,120). This procedure results in a microsomal preparation that contains only 1-2% as much hemoglobin as microsomes prepared in the usual manner by ultracentrifugation.

In addition we wanted to subfractionate lung microsomes into rough and smooth vesicles in order to make possible certain types of studies on the subcellular localization of aryl hydrocarbon monooxygenase. This subfractionation has been achieved by the use of a Cs^+-containing discontinuous sucrose gradient similar to that designed by Dallner to prepare rough and smooth microsomes from liver (121). This procedure is illustrated in Fig. 7. CsCl is added to the post-mitochondrial supernatant to give a final concentration of 20 mM, and this supernatant is layered over 1.2 M sucrose also containing 20 mM CsCl. After centrifugation at 102,000 g for 90 min the rough microsomes — which are selectively aggregated by the Cs^+ ions — are recovered in the pellet while smooth microsomes remain behind at the interface. Not only does this separation now allow further studies on the subcellular localization of polycyclic

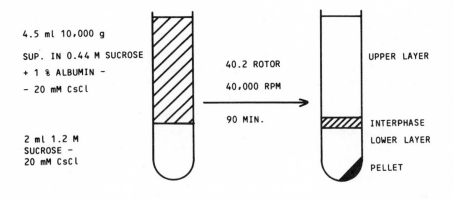

Fig. 7. Preparation of rough and smooth microsomes from rat lung using a Cs^+-containing discontinuous sucrose gradient.

hydrocarbon metabolism, it also provides a lung microsomal fraction, the rough microsomes, which is composed chiefly of fragments of the endoplasmic reticulum.

We have carried out various studies on the aryl hydrocarbon monooxygenase system in lung microsomes (119). Similar studies are going on in Dr. Estabrook's laboratory in Dallas, and the results of our two groups are in good agreement. The most interesting of these results may be summarized as follows. First, treatment of rats with methylcholanthrene has been found to induce lung cytochrome P-450 2- to 4-fold and lung benzpyrene monooxygenase activity about 10-fold. Secondly, the methylcholanthrene-induced form of cytochrome P-450 in lung microsomes gives, when reduced and complexed with carbon monooxide, an absorption maximum that shows a 2 nm blue shift relative to the control cytochrome. In lung this shift is from 453 to 451 nm. This finding suggests that a special form of cytochrome P-450 is also involved in the aryl hydrocarbon monooxygenase system of lung tissue. And finally, cytochrome P-450 from lung microsomes can be solubilized, partially purified, and reconstituted with partially purified NADPH-cytochrome *c* reductase to give an active benzpyrene monooxygenase system using methods very similar to those developed for the liver. All three of these observations suggest that aryl hydrocarbon monooxygenase in lung tissue closely resembles the corresponding activity in liver.

Recently, we have obtained a rather exciting result (122) that we would like to take up as the last point in this presentation. Some of our studies on lung aryl hydrocarbon monooxygenase were carried out with the fluorescent assay, and this assay proved to be linear for only a very few minutes. We were curious as to why this was so, so we incubated 3-hydroxybenzpyrene, which is the major fluorescent metabolite, with lung microsomes and found that this compound was rapidly converted into a non-fluorescent form. This conversion was inhibited by carbon monooxide and by α-naphthoflavone, and the rate of conversion could be induced 10- to 15-fold by treatment of rats with methylcholanthrene. Thus, 3-hydroxybenzpyrene is apparently metabolized further by aryl hydrocarbon monooxygenase in lung.

We then asked ourselves if the product of this further metabolism could bind to DNA; and, as shown in Table III, it could Lung microsomes were incubated with tritium-labeled 3-hydroxybenzpyrene in the presence of calf thymus DNA, and the DNA was subsequently reisolated. In the presence of boiled microsomes only 0.12 pmoles of radioactive metabolite were bound per mg DNA. With control microsomes after 60 min the corresponding value was 1.55 pmoles per mg; and this binding could be increased about 14-fold by induction. Thus, lung aryl hydrocarbon monooxygenase apparently metabolizes 3-hydroxybenzpyrene to another arene oxide. We have indications that liver microsomes also metabolize 3-hydroxybenzpyrene further via a cytochrome P-450-linked system, but as far as we know the new metabolite does not bind to DNA. This may be the first example of a major difference between polycyclic hydrocarbon metabolism in liver and in lung.

TABLE III

DNA-Bound Radioactivity after Incubation of Rat-Lung Microsomes with ^{3}H-3-Hydroxybenzo(α)pyrene in the Presence of Calf Thymus DNA

Material		Counts/min/mg DNA	pMoles bound radioactive metabolite /mg DNA
Boiled microsomes		821	0.12
Control microsomes,	60 min	10,569	1.55
MC microsomes*,	30 min	96,236	14.05
MC microsomes*,	60 min	148,436	21.70
MC microsomes*, (-NADPH)	60 min	1,800	0.26

* Microsomes from rats treated with 3-methylcholanthrene (20 mg/kg body weight administered intraperitoneally during 3 days). (From J. Capdevila, B. Jernström and S. Orrenius (122)).

In conclusion, we hope that the compactness and generality of this presentation have not made it difficult to get a survey of this multifaceted field of research. We thought it appropriate in a meeting with broad perspectives such as this to give an overview of polycyclic hydrocarbon metabolism and its relationship to chemical carcinogenesis as we now understand it, as well as to give you some idea of how workers in the field are trying to improve this understanding. The situation is, of course, much more complicated than we have described it. Many factors such as the nutritional and hormonal state, age, sex and genetic makeup of an organism influence its susceptibility to chemical carcinogenesis. But we hope that you are convinced, as we are, that enough is now known about the microsomal cytochrome P-450-linked monooxygenase system at the basic level to make a goal-oriented attempt to understand the biochemical background of polycyclic hydrocarbon-induced lung cancer protifable.

ACKNOWLEDGEMENTS

We would like to acknowledge that our work is supported by National Cancer Institute Contract No. NO1 CP 33363, National Institutes of Health, Bethesda, Maryland.

REFERENCES

1. Braun, A.C. (1974) *The Biblogy of Cancer*, Addison-Wesley, London.
2. Miller, J.A. and Miller, E.C. (1971) *J. Nat. Cancer Inst. 47*, VXIV.
3. Ames, B.N., Durston, W.E., Yamasaki, E. and Lee, F.D. (1973) *Proc. Nat. Acad. Sci. (U.S.A.) 70*, 2281-2285.
4. Kennaway, E.L. and Lindsey, A.J. (1958) *Brit. Med. Bull. 14*, 124-131.
5. Wynder, E.L. and Hoffman, D. (1959) *Cancer 12*, 1079-1086.
6. Anonymous (1972) *Particulate Polycyclic Organic Matter*, National Academy of Sciences, Washington.
7. Blacklock, J.W.S. (1957) *Brit. J. Cancer 11*, 181-191.
8. Clayson, D.B. (1962)*Chemical Carcinogenesis*, pp. 135-140, Churchill, London.
9. Auerbach, O., Hammond, E.C., Kirman, D. and Garfinkel, L. (1970) *Arch. Environ. Health 21*, 754-768.
10. Hanna, M.G., Nettesheim, P. and Gilbert, J.R., eds. (1970) *Inhalation Carcinogenesis*, U.S. Atomic Energy Commission, Division of Technical Information.
11. Nettesheim, P., Hanna, M.G. and Deatherage, J.W., eds. (1970) *Morphology of Experimental Respiratory Carcinogenesis*, U.S. Atomic Energy Commission, Division of Technical Information.
12. Inui, N. and Takayama, S. (1971) *Brit. J. Cancer 25*, 574-583.

13.   Hirao, F., Fujisawa, T., Tsubara, E. and Yamamura, Y. (1972) *Cancer Res. 32*, 1209-1217.

14.   Saffiotti, U., Montesano, R., Sellakumar, A.R., Cefis, F. and Kaufman, D.G. (1972) *Cancer Res. 32*, 1073-1081.

15.   Shubik, P. (1972) *Proc. Nat. Acad. Sci. (U.S.A.) 69*, 1052-1055.

16.   Sterling, T.D. and Pollack, S.V. (1972) *Am. J. Public Health 62*, 152-158.

17.   Henry, M.C., Port, C.D., Bates, R.R. and Kaufman, D.G. (1973) *Cancer Res. 33*, 1585-1592.

18.   Conney, A.H. (1967) *Pharmacol. Rev. 19*, 317-366.

19.   Daly, J.W., Jerina, D.M. and Witkop, B. (1972) *Experientia 28*, 1129-1149.

20.   Nebert, D.W. and Gelboin, H.V. (1969) *Arch. Biochem. Biophys. 134*, 76-89.

21.   Grover, P.L., Hewer, A. and Sims, P. (1973) *Biochem. Pharmacol. 23*, 323-332.

22.   Nebert, D.W., Winker, J. and Gelboin, H.V. (1969) *Cancer Res. 29*, 1763-1769.

23.   Cantrell, E.T., Warr, G.E., Busbee, D.L. and Martin, R.R. (1973) *J. Clin. Invest. 52*, 1881-1884.

24.   Kellermann, G., Luyten-Kellermann, M. and Shaw, C.R. (1973) *Am. J. Human Genet. 25*, 327-331.

25.   Kellermann, G., Shaw, C.R. and Luyten-Kellermann, M. (1973) *New Eng. J. Med. 289*, 934-937.

26.   Kellermann, G., Cantrell, E. and Shaw, C.R. (1973) *Cancer Res. 33*, 1654-1656.

27.   Booth, J., Keysell, G.R., Pal, K. and Sims, P. (1974) *FEBS Letters 43*, 341-344.

28.   Hrycay, E.G., Gustafsson, J.-Å., Ingelman-Sundberg, M. and Ernster, L. (in press) *FEBS Letters.*

29.   Hrycay, E.G., Gustafsson, J.-Å., Ingelman-Sundberg, M. and Ernster, L. (in press) *Biochem. Biophys. Res. Commun.*

30.   Estabrook, R.W., Hildebrandt, A.G., Baron, J., Netter, K.J. and Leibman, K. (1971) *Biochem. Biophys. Res. Commun. 42*, 132-139.

31.   Estabrook, R.W., Matsubara, T., Mason, J.I., Werringloer, J. and Baron, J. (1973) *Drug Metab. Dispos. 1*, 98-110.

32.   Estabrook, R.W., Baron, J., Peterson, J. and Ishimura, Y. (1972) In *Biological Hydroxylation Mechanisms* (Boyd, G.S. and Smellie, R.M.S., eds.) pp. 159-168, Academic Press, New York.

33.· Rahimtula, A.D., O'Brien, P.J., Hrycay, E.G., Peterson, J.A. and Estabrook, R.W. (1974) *Biochem. Biophys. Res. Commun. 60*, 695-702.

34.   Gunsalus, I.C., Meeks, J.R., Lipscomb, J.D., Debrunner, P. and Münck, E. (1974) In *Molecular Mechanisms of Oxygen Activation* (Hayaishi, O., ed.)

pp. 559-563, Academic Press, New York.

35. Baron, J., Taylor, W.E. and Masters, B.S.S. (1972) *Arch. Biochem. Biophys. 150*, 105-115.

36. Lu, A.Y.H., Junk, K.W. and Coon, M.J. (1969) *J. Biol. Chem. 244*, 3714-3721.

37. Lu, A.Y.H., Kuntzmann, R., West, S. and Conney, A.H. (1971) *Biochem. Biophys. Res. Commun. 42*, 1200-1206.

38. Lu, A.Y.H. and Levin, W. (1972) *Biochem. Biophys. Res. Commun. 46*, 1334-1339.

39. Levin, W., Ryan, D., West, S. and Lu, A.Y.H. (1974) *J. Biol. Chem. 249*, 1747-1754.

40. Diamond, L. and Gelboin, H.V. (1969) *Science 166*, 1023-1025.

41. Selkirk, J.K., Huberman, E. and Heidelberger, C. (1971) *Biochem. Biophys. Res. Commun. 43*, 1010-1016.

42. Wiebel, F.L., Leutz, J.C., Diamond, L. and Gelboin, H.V. (1971) *Arch. Biochem. Biophys. 144*, 78-86.

43. Gelboin, H.V., Kinoshita, N. and Wiebel, F.J. (1972) *Fed. Proc. 31*, 1298-1309.

44. Conney, A.H., Miller, E.C. and Miller, J.A. (1957) *J. Biol. Chem. 228*, 753-766.

45. Wattenberg, L.W. and Leong, J.L. (1962) *J. Histochem. Cytochem. 10*, 412-421.

46. Gelboin, H.V. and Blackburn, N.R. (1964) *Cancer Res. 24*, 356-360.

47. Juchau, M.R., Cram, R.L., Plaa, G.L. and Fouts, J.R. (1965) *Biochem. Pharmacol. 14*, 473-478.

48. Grundin, R., Jakobsson, S. and Cinti, D.L. (1973) *Arch. Biochem. Biophys. 158*, 544-555.

49. Conney, A.H. and Gilman, A.G. (1963) *J. Biol. Chem. 238*, 3682-3685.

50. Nebert, D.W. and Gelboin, H.V. (1968) *J. Biol. Chem. 243*, 6242-6249.

51. Nebert, D.W. and Gelboin, H.V. (1968) *J. Biol. Chem. 243*, 6250-6261.

52. Nebert, D.W. and Gelboin, H.V. (1970) *J. Biol. Chem. 245*, 160-168.

53. Nebert, D.W., Benedict, W.F., Gielen, J.E., Oesch, F. and Daly, J.W. (1972) *Mol. Pharmacol. 8*, 374-379.

54. Thomas, P.E., Kouri, R.E. and Hutton, J.J. (1972) *Biochem. Genet. 6* 157-168.

55. Kouri, R.E., Salerno, R.A. and Whitmire, C.E. (1973) *J. Nat. Cancer Inst. 50*, 363-368.

56. Nebert, D.W., Considine, N. and Owens, I.S. (1973) *Arch. Biochem. Biophys. 157*, 148-159.

57. Ernster, L. and Orrenius, S. (1965) *Fed. Proc. 24*, 1190-1199.

58. Lind, C. and Ernster, L. (1974) *Biochem. Biophys. Res. Commun. 56*, 392-400.

59. Ernster, L., Danielson, L. and Ljunggren, M. (1962) *Biochim. Biophys. Acta.58*, 171-188.
60. Ernster, L. (1967) *Methods Enzymol. 10*, 309-317.
61. Williams-Ashman, H.G. and Huggins, C. (1961) *Med. Experimentalis 4*, 223-236.
62. Huggins, C. and Fukunishi, R. (1964) *J. Exper. Med. 119*, 923-942.
63. Orrenius, S., Ellin, Å., Jakobsson, S.V., Thor, H., Cinti, D., Schenkman, J.B. and Estabrook, R.W. (1973) *Drug Metab. Disp. 1*, 350-357.
64. Kupfer, D. and Orrenius, S. (1970) *Eur. J. Biochem. 14*, 317-322.
65. Gnosspelius, Y., Thor, H. and Orrenius, S. (1969/1970) *Chem.-Biol. Interactions 1*, 125-137.
66. Lu, A.Y.H., Kuntzmann, R., West, S., Jacobson, M. and Conney, A.H. (1972) *J. Biol. Chem. 247*, 1727-1734.
67. Comai, K. and Gaylor, J.L. (1973) *J. Biol. Chem. 248*, 4947-4955.
68. Alvares, A.P. and Siekevitz, P. (1973) *Biochem. Biophys. Res. Commun. 54*, 923-929.
69. Welton, A.F. and Aust, S.D. (1974) *Biochem. Biophys. Res. Commun. 56*, 898-906.
70. Pandov, H. and Sims, P. (1970) *Biochem. Pharmacol. 19*, 299-303.
71. Oesch, F., Jerina, D.M. and Daly, J.W. (1971) *Biochim. Biophys. Acta 227*, 685-691.
72. Mandel, H.G. (1971) In *Fundamentals of Drug Metabolism and Drug Disposition* (La Du, B.N., Mandel, H.G. and Way, E.L., eds.) pp. 149-158, Williams and Wilkins, Baltimore.
73. Sims, P. (1972) *Biochem. J. 130*, 27-35.
74. Oesch, F. (1972) *Xenobiotica 3*, 305-333.
75. Oesch, F., Jerina, D.M. and Daly, J.W. (1971) *Arch. Biochem. Biophys. 144*, 253-261.
76. Boyland, E., Ramsay, G.S. and Sims, P. (1961) *Biochem. J. 78*, 376-384.
77. Boyland, E. and Williams, K. (1965) *Biochem. J. 94*, 190-197.
78. Boyland, E. and Sims, P. (1965) *Biochem. J. 95*, 788-792.
79. Boyland, E. and Sims, P. (1965) *Biochem. J. 97*, 8-16.
80. Boyland, E. and Chasseaud, L.F. (1969) *Adv. Enzymol. 32*, 173-219.
81. Jakoby, W.B. and Fjellstedt, T.A. (1973) *The Enzymes VII*, 199-212.
82. Grover, P.L. (1974) *Biochem. Pharmacol. 23*, 333-343.
83. Kaplowitz, N., Kuhlenkamp, J. and Clifton, G. (1975) *Biochem. J. 146*, 351-356.
84. Booth, J., Keysell, G.R. and Sims, P. (1973) *Biochem. Pharmacol. 22*, 1781-1791.
85. Miller, E.C. (1951) *Cancer Res. 11*, 100-108.
86. Wiest, W.G. and Heidelberger, C. (1953) *Cancer Res. 13*, 246-249.

87. Heidelberger, C. and Moldenhauer, M.G. (1956) *Cancer Res. 16*, 442-449.
88. Heidelberger, C. and Davenport, G.R. (1961) *Acta Unio Intern. Contra Cancrum 17*, 55-63.
89. Brookes, P. and Lawley, P.D. (1964) *Nature 202*, 781-784.
90. Grover, P.L. and Sims, P. (1968) *Biochem. J. 110*, 159-160.
91. Gelboin, H.V. (1969) *Cancer Res. 29*, 1271-1276.
92. Grover, P.L. and Sims, P. (1973) *Biochem. Pharmacol. 22*, 661-666.
93. Rogan, E.G. and Cavalieri, E. (1974) *Biochem. Biophys. Res. Commun. 58*, 1119-1126.
94. Swaisland, A.J., Grover, P.L. and Sims, P. (1973) *Biochem. Pharmacol. 22*, 1547-1556.
95. Ayengar, P.K., Hayaishi, O., Nakojima, M. and Tomida, J. (1959) *Biochim. Biophys. Acta 33*, 111-119.
96. Inscoe, J.K., Daly, J. and Axelrod, J. (1965) *Biochem. Pharmacol. 14*, 1257-1263.
97. Jerina, D.M., Daly, J.W. and Witkop, B. (1967) *J. Am. Chem. Soc. 89*, 5488-5489.
98. Jerina, D.M., Daly, J.W., Witkop, B., Zaltzman-Nirenberg, P. and Udenfriend, S. (1970) *Biochemistry 9*, 147-155.
99. Creveling, C.R., Dalgard, N., Shimizu, H. and Daly, J.W. (1970) *Mol. Pharmacol. 6*, 691-696.
100. McCormick, J.I., Flanagan, R. and Lloyd, A.G. (1972) *Biochem. J. 130*, 83P-84P.
101. Meyer, T. and Scheline, R.R. (1972) *Xenobiotica 2*, 391-398.
102. Cookson, M.J., Sims, P. and Grover, P.L. (1971) *Nature New Biol. 234*, 186-187.
103. Grover, P.L., Sims, P., Huberman, E., Marquardt, H., Kuroki, T. and Heidelberger, C. (1971) *Proc. Nat. Acad. Sci. (U.S.A.) 68*, 1098-1101.
104. Huberman, E., Aspiras, L., Heidelberger, C., Grover, P.L. and Sims, P. (1971) *Proc. Nat. Acad. Sci. (U.S.A.) 68*, 3195-3199.
105. Ames, B.N., Sims, P. and Grover, P.L. (1972) *Science 176*, 47-49.
106. Huberman, E., Kuroki, T., Marquardt, H., Selkirk, J.K., Heidelberger, C., Grover, P.L. and Sims, P. (1972) *Cancer Res. 32*, 1391-1396.
107. Marquardt, H., Kuroki, T., Huberman, E., Selkirk, J.K., Heidelberger, C., Grover, P.L. and Sims, P. (1972) *Cancer Res. 32*, 716-720.
108. Kinoshita, N. and Gelboin, H.V. (1972) *Proc. Nat. Acad. Sci. (U.S.A.) 69*, 824-828.
109. Huggins, G., Grand, L. and Fukunishi, R. (1964) *Proc. Nat. Acad. Sci. (U.S.A.) 51*, 737-742.
110. Wattenberg, L.W. and Leong, J.L. (1968) *Proc. Soc. Exp. Biol. Med. 128*, 940-943.
111. Wheatley, D.N. (1968) *Brit. J. Cancer 22*, 787-797.

112. Wattenberg, L.W. and Leong, J.L. (1970) *Cancer Res. 30*, 1922-1925.
113. Selkirk, J.K., Croy, R.G. and Gelboin, H.V. (1974) *Science 184*, 169-170.
114. Wattenberg, L.W., Page, M.A. and Leong, J.L. (1968) *Cancer Res. 28*, 934-937.
115. DePierre, J.W., Moron, M.S., Johannesen, K.A.M. and Ernster, L. (1975) *Analyt. Biochem. 63*, 470-484.
116. Vadi, H., Moldeus, P., Capdevila, J. and Orrenius, S. (in press) *Cancer Res.*
117. Questorff, B., Bondesen, S. and Grunnet, N. (1973) *Biochim. Biophys. Acta 320*, 503-516.
118. Moldeus, P., Grundin, R., von Bahr, C. and Orrenius, S. (1973) *Biochem. Biophys. Res. Commun. 55*, 937-944.
119. Capdevila, J., Jakobsson, S., Jernström, B. and Orrenius, S. (in press) *Cancer Res.*
120. Tangen, O., Jonsson, J. and Orrenius, S. (1973) *Analyt. Biochem. 54*, 597-603.
121. Dallner, G. (1963) *Acta Pathol. Microbiol. Scand., Suppl. 166*.
122. Capdevila, J., Jernström, B., Vadi, H. and Orrenius, S. (in press) *Biochem. Biophys. Res. Commun.*

LIVER MICROSOMAL MEMBRANES:
RECONSTITUTION OF THE HYDROXYLATION SYSTEM CONTAINING CYTOCHROME P-450*

M.J. Coon, D.A. Haugen[†], F.P. Guengerich[†], J.L. Vermilion and W.L. Dean

Department of Biological Chemistry
Medical School
The University of Michigan
Ann Arbor, Michigan

Cytochrome P-450 is apparently the chief membranous enzyme of liver cells, for it occurs at levels as high as ten percent of the microsomal protein, and the endoplasmic reticulum from which the microsomes are derived represents as much as ninety percent of the total membranes. This pigment is also unusual in its metabolic versatility, for in liver microsomes in the presence of NADPH and molecular oxygen it catalyzes the hydroxylation or other chemical modifications of steroids and fatty acids, as well as of drugs, insecticides, anesthetics, petroleum products, carcinogens, and numerous other foreign compounds. Although much was learned about the inducibility and other properties of this enzyme system through work with microsomal suspensions, it was clear that many of the more difficult questions concerning the system could not be answered without isolation and characterization of the individual components.

The present paper is concerned with the purification and properties of liver microsomal cytochrome P-450 (P-450$_{LM}$) and the other components of this mixed function oxidase system, with their interactions, and with the question of whether one or more forms account for the broad substrate specificity of this cytochrome in liver microsomal membranes. Our laboratory accomplished the solubilization, resolution and reconstitution of this enzyme system some years ago (1,2) and has subsequently been concerned with the

* This research was supported by Grant BMS71-01195 from the National Science Foundation and Grant AM-10339 from the United States Public Health Service.
† Postdoctoral Fellow of the United States Public Health Service.

characterization of the components. The solubilization of liver microsomes with deoxycholate in the presence of glycerol and other protective agents, followed by column chromatography on DEAE-cellulose, yielded two frac- tions (A and B) which were found to be necessary for the hydroxylation of fatty acids. The first fraction contained both cytochrome P-450 (A_1) and NADPH-cytochrome P-450 reductase (A_2), which could be partially separated because the latter was eluted at higher ionic strength; the second fraction contained a heat-stable component which was called Factor B (3). The factor was subsequently shown to be a phospholipid, and phosphatidylcholine was identified as the active component and shown to be required along with A_1 and A_2, as well as NADPH and molecular oxygen, for the hydroxylation of a variety of substrates, including fatty acids, hydrocarbons and drugs (4-6). More recently, we have obtained evidence that cytochrome P-450 purified to a state where it is free of significant amounts of other known electron accep- tors such as cytochrome b_5, flavins, nonheme iron, or other metals, accepts two electrons from dithionite per molecule of heme (7,8). One electron re- duces the heme iron atom, and the other is taken up by an unidentified ac- ceptor. This finding was unexpected since certain other well-characterized heme proteins which have been examined in this manner, such as cytochrome *c*, take up only one electron. The unidentified acceptor, which is referred to as Factor C, is closely associated with the cytochrome P-450, but, as describ- ed below, its role in the hydroxylation reactions catalyzed by this enzyme system remains uncertain.

CHARACTERIZATION OF COMPONENTS OF THE MICROSOMAL HYDROXYLATION SYSTEM

Cytochrome P-450

Cytochrome P-450 has recently been purified to apparent homogeneity from phenobarbital-induced rabbit liver microsomes (8,9). As shown in Table I, the procedures used include pyrophosphate extraction to remove hemoglobin, solubilization with cholate and fractionation with polyethylene glycol, and column chromatographic steps carried out in the presence of Renex 690, a nonionic detergent. The purified protein, containing up to 18 nmol of cytochrome P-450 per mg of protein, exhibited a single polypep- tide band having a molecular weight of 50,000 daltons when exposed to SDS and mercaptoethanol and submitted to polyacrylamide gel electrophoresis. Such results indicate apparent homogeneity but do not rule out the presence of apocytochrome P-450 or other polypeptides of identical molecular weight. As isolated, the protein is an aggregate with an apparent molecular weight of about 300,000, in agreement with estimates made earlier by several techniques

TABLE I

Purification of Cytochrome P-450
from Phenobarbital-Induced Rabbit Liver Microsomes

Preparation	Cytochrome P-450 Content
	nmol per mg protein
Pyrophosphate-treated microsomes	3.1 (2.6- 3.6)[a]
Polyethylene glycol 6000 precipitate	
(8-10%) of sodium cholate-	
solubilized preparation	5.9 (5.4- 7.0)
DEAE-cellulose column eluate	
(0.5% Renex 690 present)	13.7 (9.0-15.2)
Hydroxylapatite column eluate	
(0.1% Renex present)	17.4 (13.0-18.0)

[a] The values in parentheses indicate the range of cytochrome P-450 concentrations found in a series of such purifications.

with less purified preparations (8,10). No cytochrome P-420, cytochrome b_5, NADPH-cytochrome P-450 reductase (measured by its activity toward cytochrome c), or NADH-cytochrome b_5 reductase (measured by its activity toward ferricyanide) could be detected in the purified protein. The absolute spectra are shown in Fig. 1. In the oxidized spectrum maxima occur at 568, 534, and 417 nm, and upon reduction the Soret band shifts to 414 nm with a decrease in absorbance and a single band at 542 nm replaces the α and β bands seen with the oxidized form. The spectrum of the CO complex of the reduced form has maxima at 552 and 451 nm. Although the purified P-450$_{LM}$ and that from *Pseudomonas putida* (P-450$_{cam}$)(11) differ markedly in substrate specificity, solubility and the requirement of the former for a phospholipid and the latter for an iron-sulfur protein for hydroxylation activity, they show immunological cross reaction by competitive binding and by inhibition of catalytic activity and are of similar subunit molecular weight and amino acid composition (12). Furthermore, upon treatment with cyanogen bromide, P-450$_{LM}$ and P-450$_{cam}$ yield heme-containing peptides of similar amino acid composition (12). It should be noted that the purified P-450$_{LM}$ retains the ability to catalyze the hydroxylation of a variety of substrates in a system containing the reductase and phospholipid. Cytochrome b_5 obviously plays no obligatory role in substrate hydroxylation, since none of this cytochrome was detected in the purified P-450$_{LM}$ and reductase preparations used.

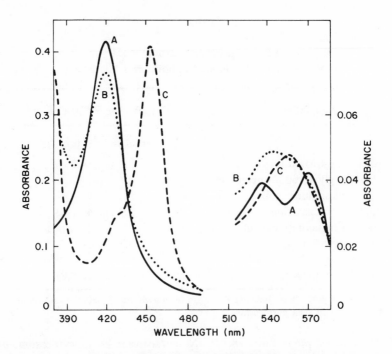

Fig. 1. Absolute spectra of purified cytochrome P-450 diluted to a concentration of 3.5 nmoles per ml in 0.05 M phosphate buffer, pH 7.4, containing 20% glycerol and 1.0 mM EDTA: A, oxidized; B, dithionite-reduced; C, dithionite-reduced CO complex. The spectra at higher wavelengths are shown with a 5-fold expanded scale.

On the other hand, a facilitating role of cytochrome b_5 in electron transfer for hydroxylation in microsomal membranes is not ruled out.

NADPH-Cytochrome P-450 Reductase

As shown earlier (3), the steapsin-solubilized NADPH-cytochrome c reductase described by Williams and Kamin (13) is incapable of replacing our detergent-solubilized reductase as a component of the reconstituted hydroxylation system. This enzyme solubilized with bile salts has recently been purified from phenobarbital-induced rabbit (8) and rat liver microsomes (14); the methods used to obtain the reductase in a highly purified state from the latter source are shown in Table II. The procedures are similar to those used in purifying the cytochrome P-450, including the use of Renex 690. Since the ratio of activities toward cytochrome c and cytochrome P-450 did not vary significantly throughout the fractionation procedures, it appears that a single enzyme is responsible for both activities. The purified enzyme gives a single major band with a molecular weight of about 79,000 daltons on SDS-poly-

TABLE II

Purification of NADPH-Cytochrome P-450 Reductase
from Phenobarbital-Induced Rat Liver Microsomes

Preparation	Cyt. c reduction[a]	Cyt. P-450 reduction[a]	Ratio of activities
Microsomes	0.30		
DEAE-cellulose column eluate (0.05% deoxycholate present)	1.88	0.11	17.1
DEAE-cellulose column eluate (0.4% Renex present)	15.1	0.89	17.0
DEAE-Sephadex A-50 column eluate (0.1% Renex present)	21.5	1.22	17.6
Hydroxylapatite column eluate (0.1% Renex present)	33.2	2.06	16.1

[a] The activities were determined at 30° and are expressed as μmol per min per mg of protein. NADPH-cytochrome P-450 reductase activity was estimated from the rate of NADPH disappearance in a reconstituted system containing benzphetamine, purified cytochrome P-450, and phosphatidylcholine, with reductase as the rate-limiting component.

acrylamide gel electrophoresis, whereas in the absence of detergents it has a higher apparent molecular weight. No cytochrome P-450, cytochrome b_5, or NADH-cytochrome b_5 reductase (measured by its activity toward ferricyanide) could be detected in the purified preparation. FMN and FAD are present in about equimolar amounts in the purified rat liver reductase; both flavin nucleotides are also present in purified reductase preparations from rabbit liver microsomes, as shown by Iyanagi and Mason (15) and confirmed in this laboratory (8). Selective removal of the FMN by treatment with KBr results in an extensive loss in the ability of the enzyme to transfer electrons to cytochrome P-450, cytochrome c, or dichlorophenolindophenol, whereas ferricyanide reduction is unaffected. Subsequent reconstitution with FMN largely restores the initial activities (16,17).

Phospholipid

After phosphatidycholine had been identified as the active component in Fraction B, several synthetically prepared compounds were tested for activity in the reconstituted hydroxylation system (4). Dilauroylglyceryl-3-phosphorylcholine (dilauroyl-GPC) and dioleoyl-GPC were at least as effective as the isolated microsomal phosphatidylcholine fraction, whereas dipalmitoyl-GPC and lysolauroyl-GPC were relatively poor. Stopped flow measurements showed that a substrate (18) and the phospholipid (4,19) must both be present for the rapid enzymatic reduction of cytochrome P-450 by NADPH. The dilauroyl compound is used routinely in hydroxylation assays because of its high activity and because, unlike the microsomal phosphatidylcholine fraction containing unsaturated fatty acids, it does not undergo chemical or enzymatic peroxidation.

Unidentified Electron Acceptor: Factor C

The observation that P-450$_{LM}$ accepts two electrons from dithionite (7,20) has been confirmed and extended with more highly purified enzyme preparations (21). The spectral changes observed upon the stepwise reduction of an apparently homogeneous sample of the rabbit liver cytochrome under highly anaerobic conditions in an atmosphere of CO are presented in Fig. 2. Upon the addition of dithionite the expected peak at 451 nm appeared with the concomitant loss of the Soret band at 417 nm, and a single band at 552 nm appeared with the loss of the distinct α and β bands characteristic of the oxidized heme protein. Several clear isosbestic points are evident in the titration curves. As shown in the plot of the data (inset), following a short lag, 1.05 molecules of dithionite, or 2.1 electrons, were consumed per molecule of P-450$_{LM}$ converted to the reduced CO complex. Similar results were obtained in a variety of other experiments: with NADPH as the electron donor in the presence of NADPH-cytochrome P-450 reductase as a catalyst, with other ligands as well as with substrate or phosphatidylcholine present, after denaturation to form cytochrome P-420, or with cytochrome P-450 partially purified from mouse or rat liver microsomes. Since the reduced P-450$_{LM}$ donates two electrons to oxidizing agents, including molecular oxygen, and the resulting oxidized protein is then again capable of accepting two electrons, the process is clearly reversible.

Although these unexpected findings provide evidence for the presence of an electron acceptor in addition to the heme iron atom, significant amounts of nonheme iron, other metals or cofactors, or disulfide bonds were not found, nor were free radicals detected by electron paramagnetic resonance spectrometry. The important question then arises as to whether the unidentified electron acceptor plays a role in substrate hydroxylation. The scheme in Fig. 3 shows a possible reaction sequence. The binding of the substrate to the oxidized cytochrome is shown in Step 1. In Step 2, the iron atom and

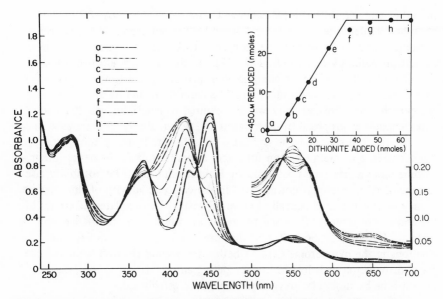

Fig. 2. Anaerobic titration of highly purified P-450LM$_2$ with dithionite. The reaction mixture, in a final volume of 3.0 ml under an anaerobic atmosphere of CO, contained P-450LM$_2$ (9.5 μM; 18.0 nmol per mg of protein), dilauroyl-GPC (0.15 mM), phosphate buffer, pH 7.7 (0.1 M), glycerol (20%), EDTA (0.67 mM), benzphetamine (0.67 mM), glucose oxidase (0.3 unit per ml), catalase (0.3 unit per ml), methylviologen (1 μM), and glucose (67 mM). Small aliquots of dithionite were added at intervals, and the spectra were recorded after equilibrium was reached. The inset is a plot of the amount of P-450LM$_2$ reduced, determined at 450 nm, vs. the amount of dithionite added.

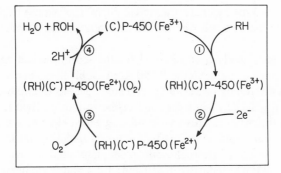

Fig. 3. Proposed scheme for hydroxylation reactions catalyzed by P-450LM where RH is the substrate, ROH the product, and C the unidentified electron acceptor.

the other factor (arbitrarily indicated as C) are reduced by NADPH in the phospholipid-dependent reaction catalyzed by the reductase. Next, oxygen is pictured as combining with the reduced cytochrome in a manner similar to that proposed by Estabrook *et al.* (22). In the last step, electron transfer is proposed within the active complex with proton uptake to yield one molecule of water and one of product with regeneration of the oxidized form of the cytochrome. Further work will be required to determine whether Factor C participates as proposed in these reactions. Alternatively, the presence of C in the reduced state may be necessary for substrate hydroxylation without the occurrence of rapid electron turnover; a similar role for heme a_3 in cytochrome oxidase has been postulated by Malmström (23). The possibility also remains that the factor is not involved in the catalytic activity of this enzyme. It may be noted that chemically reduced $P\text{-}450_{LM}$ hydroxylates substrates in the absence of both NADPH and the reductase when air is admitted to the system. The yield is about 50% of that predicted on the basis of the two electrons in the cytochrome (21). Whether the second electron is unavailable for hydroxylation or is lost in the competing autooxidation of the reduced cytochrome by molecular oxygen remains to be established.

SUBSTRATE SPECIFICITY: SEPARATION AND CHARACTERIZATION OF MULTIPLE FORMS OF CYTOCHROME P-450

The question of whether the numerous activities attributed to $P\text{-}450_{LM}$ reside in one or more forms of this pigment has been the subject of much investigation. Although some kinetic data appeared to support the presence of a single enzyme, spectral and genetic evidence suggested the involvement of two forms. The application of procedures for the resolution and reconstitution of the system has proven to be a valuable approach to the study of this problem. The cytochrome fractions from animals treated with phenobarbital or with 3-methylcholanthrone (or β-naphthoflavone) exhibited different absorption maxima and were found to determine the substrate specificity (24-26).

Evidence has recently been reported for the separation and characterization of multiple forms of cytochrome P-450 from rabbit liver microsomes (27). Fig. 4 shows the results of SDS-polyacrylamide gel electrophoresis of proteins in normal and induced microsomes and in the various purified fractions. Beginning with the major microsomal band of greatest electrophoretic mobility, the bands are numbered according to decreasing mobility and increasing molecular weight. In other electrophoretic experiments, staining with benzidine in the presence of H_2O_2 indicated that bands 1,2,4 and 7 are heme-peptides (27). As seen in the figure, $P\text{-}450_{LM_2}$, which has been purified to apparent

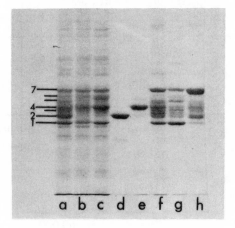

Fig. 4. Electrophoresis of P-450$_{LM}$ fractions on polyacrylamide slab gel in a discontinuous buffer system. The preparations were treated with SDS and mercaptoethanol at 100° and submitted to electrophoresis by the method of Laemmli (28) with a 7.5% separating gel. Migration was from top to bottom. The gel was fixed in 65:25:10 water–isopropanol–acetic acid, stained with 0.05% Coomassie Blue in the same solvent, and destained with an 80:10:10 mixture. The samples were analyzed at the protein levels indicated: (a) phenobarbital-induced microsomes, 6 μg; (b) normal microsomes, 6 μg; (c) β–naphthoflavone-induced microsomes, 6 μg; (d) LM$_2$, 1 μg; (e) LM$_4$, 1 μg; (f) fraction containing chiefly LM$_1$ and LM$_7$, 3 μg; (g) fraction enriched in LM$_1$, 2 μg; and (h) fraction enriched in LM$_7$, 2 μg.

homogeneity (9), clearly corresponds to the phenobarbital-inducible band in microsomes, and P-450$_{LM_4}$ corresponds to the β-naphthoflavone-inducible band. Another fraction contains bands 1 and 7 as the major cytochrome components along with lesser amounts of the intermediate bands, and fractions are also shown which are enriched in P-450$_{LM_1}$ or P-450$_{LM_7}$.

Table III summarizes the extent of purification and some of the properties of the different forms of P-450$_{LM}$. The similarities in the absorption maxima of the reduced CO complexes of P-450$_{LM_4}$ and LM$_{1,7}$ indicate that the terms "P-450" and "P-448" are no longer adequate to refer to these species. It is also unsuitable to refer to the various forms by reference to the inducing agents used, since some forms (LM$_1$ and LM$_7$) are not yet known to be induc-

TABLE III

Properties of Different Forms
of Rabbit Liver Microsomal Cytochrome P-450

Form	Inducing agent	Absorption maximum of reduced CO complex	Apparent molecular weight	Specific content
		nm		nmol/mg/protein
LM_2	Phenobarbital	451.0	50,000	18.0
LM_4	β–Naphthoflavone	447.5	54,000	11.7
$LM_{1,7}$a		448.5	47,000 60,000	3.3

[a] $P-450_{LM_{1,7}}$ was obtained from animals administered phenobarbital, but this agent is not known to induce the synthesis of these forms.

ed, and others (LM_4 and perhaps also LM_2) are present, although at lower levels, in the absence of inducing agents. Furthermore, it is not yet certain whether the endogenous and induced polypeptides are identical in the case of LM_2 and in the case of LM_4. When the functions and chemical properties of the various forms of $P-450_{LM}$ are better understood, it may be possible to name them on such more suitable grounds. Meanwhile, it seems best to use a system which identifies the subunits without reference to the complicated induction patterns observed with this particular enzyme system. A further advantage of the proposed nomenclature is that it refers to the individual polypeptide subunits rather than to the aggregates which may conceivably contain more than one type of subunit.

A comparison of the immunochemical properties of the various forms of $P-450_{LM}$ by Ouchterlony diffusion analysis is shown in Fig. 5. In Expt. a, a strong precipitin line was observed from the reaction of anti-$P-450_{LM_2}$ antibodies with $P-450_{LM_2}$ but not with $P-450_{LM_4}$ or $LM_{1,7}$. On the basis of the known immunochemical similarity of $P-450_{LM_2}$ and $P-450_{cam}$ (12) a precipitin line would have been expected with the latter cytochrome, but the solubility of the product prevented its being seen by the Ouchterlony method. In Expt. b, with deoxycholate added to solubilize the microsomal preparations, the phenobarbital-induced microsomes also gave a strong interaction, as expected from the high content of $P-450_{LM_2}$, but normal and β–naph-

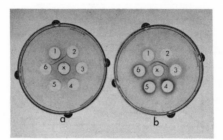

Fig. 5. Ouchterlony diffusion analysis showing reaction of rabbit antibody against
$P-450_{LM_2}$ with microsomes or various purified cytochrome P-450 preparations. The agar
contained a saturating concentration of camphor in Expt. a and 0.5% sodium deoxycholate
in Expt. b. The central wells contained anti-$P-450_{LM_2}$ γ-globulin (1.5 mg of protein). In
Expt. a, the wells contained, respectively: (1), $P-450_{LM}$ (0.3 nmol); (2), $P-450_{LM_{1.7}}$
(0.3 nmol); (3), $P-450_{LM_4}$ (0.3 nmol); (4), $P-450_{cam}$ (0.3 nmol) with a saturating level
of camphor; (5), $P-450_{cam}$ (1 nmol) with a saturating level of camphor; and (6), no added
protein. In Expt. b, the wells contained, respectively: (1), phenobarbital-induced rabbit
liver microsomes (0.3 nmol of $P-450_{LM}$); (2), $P-450_{LM_2}$ (0.3 nmol); (3), $P-450_{LM_{1.7}}$
(0.9 nmol); (4), $P-450_{LM_4}$ (0.9 nmol); (5), β-naphthoflavone-induced microsomes (0.9
nmol of $P-450_{LM}$); and (6), normal microsomes (0.9 nmol of $P-450_{LM}$). The plates were
incubated at 5° for 24 hr, placed in 1% NaCl containing 0.1% DOC for 24 hr, and then
stained with 0.05% Coomassie Blue.

thoflavone-induced microsomes did not. In other studies not presented here
the precipitate formed between either phenobarbital-induced microsomes or
$P-450_{LM_2}$ and the anti-$P-450_{LM_2}$ γ-globulin was submitted to SDS-polyacrylamide gel electrophoresis and shown to contain only the expected $P-450_{LM_2}$
band. Fig. 6 shows a radioimmune assay of competition of the unlabeled
cytochromes with ^{125}I-labeled $P-450_{LM_2}$ for binding to anti-$P-450_{LM_2}$ antibodies. Under the conditions employed, $P-450_{cam}$ exhibited extensive cross-
reactivity with the rabbit antibodies elicited against $P-450_{LM_2}$. The lack of
reactivity of $P-450_{LM_4}$ and $P-450_{LM_{1.7}}$ with the antibody both in these experiments and in the Ouchterlony diffusion experiments suggests that these forms
are chemically different from $P-450_{LM_2}$. Although $P-450_{LM_2}$ and $P-450_{LM_4}$
apparently have major structural differences, their carbohydrate composition
is similar. Each contains close to two molecules of mannose, one of glucos-
amine, a variable amount (not exceeding one molecule) of glucose, and no
sialic acid.

Table IV shows the activities of the different $P-450_{LM}$ fractions toward
various substrates. LM_2 is most effective with benzphetamine, ethylmorphine,
and *p*-nitroanisole, as well as with the 4- position of biphenyl and the 16α-
position of testosterone, and $LM_{1.7}$ is most effective with benzpyrene and

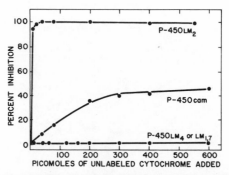

Fig. 6. Competition of unlabeled cytochrome P-450 preparations with ^{125}I-labeled P-450$_{LM2}$ for binding to anti-P-450$_{LM2}$ antibodies. To an unlabeled antigen in 0.15 M NaCl-0.02 M potassium phosphate buffer, pH 7.4, containing 10% glycerol and 0.08% bovine serum albumin, 0.025 ml of antibody was added to give a total volume of 0.4 ml. After a 1-hr incubation at 4°, 44 pmol of ^{125}I-labeled P-450$_{LM2}$ (2.2 x 10^4 cpm) were added, and after an additional 24 hrs the precipitates were collected by centrifugation, washed once with a similar buffer solution from which the albumin was omitted, and counted. The inhibition of binding is based on maximal P-450$_{LM2}$-anti-P-450$_{LM2}$ binding expressed as 100%.

the 6β- position of testosterone. The individual forms may have the ability to bind many or all of the potential substrates but differ in their relative efficiency in the hydroxylation of such compounds. Variations in the level of as many as four forms of P-450$_{LM}$ might account for the differences in catalytic activity observed by various investigators with microsomal suspensions from animals differing in sex, age, method of induction, and so forth. On the other hand, the occurrence of additional forms not distinguishable by electrophoresis cannot yet be ruled out.

ROLE OF PHOSPHOLIPID IN INTERACTIONS OF COMPONENTS OF THE MIXED FUNCTION OXIDASE SYSTEM

Although P-450$_{LM}$ and the reductase individually form small aggregates, as already described, mixing the two proteins and the phospholipid does not cause the formation of large aggregates or membrane-like structures, as shown by sedimentation studies, gel exclusion chromatography, and electron microscopy (10). Such findings of course do not rule out the possibility that a dissociable complex containing P-450$_{LM}$, reductase, phosphatidylcholine and substrate functions in catalysis.

The data obtained in experiments designed to determine the extent of substrate binding to P-450$_{LM2}$ are shown in Fig. 7. The gel filtration method devised by Hummel and Dreyer (29) was used with varying concentra-

TABLE IV

Substrate Specificity of Forms of P-450$_{LM}$

Substrate		Activity		
		LM$_2$	LM$_4$	LM$_{1,7}$
		(nmol product/min/nmol P-450$_{LM}$)		
Benzphetamine		66	3.0[b]	7.5
Ethylmorphine		6.1	3.0[b]	3.0[b]
p-Nitroanisole		6.2	0.6[b]	0.6[b]
Aniline		1.0	0.4	0.6
Biphenyl	(2- position)	0.7	0.3	0.4
Biphenyl	(4- position)	5.4	0.4	0.6
Testosterone	(6β- position)	0.02[b]	0.02[b]	0.32
Testosterone	(7α- position)	0.02[b]	0.04	0.02[b]
Testosterone	(16α- position)	0.43	0.02[b]	0.07
Benzpyrene[a]		0.04	trace	0.5

[a] Determined by Dr. Daniel W. Nebert. Experiments in collaboration with Drs. H.V. Gelboin, F. Wiebel and J. Selkirk of the National Cancer Institute show that the relative amounts of different products formed from benzpyrene differ with the various P-450$_{LM}$ preparations.

[b] These values were at the minimal level of detection.

tions of tritiated benzphetamine. The maximal value found for r, the moles of substrate bound per mole of P-450$_{LM_2}$ subunit, is 1.1, indicating that, under the conditions employed, including the presence of phosphatidylcholine, the cytochrome probably has a single binding site for this substrate. the Ks of benzphetamine for P-450$_{LM_2}$ was determined from the same data to be 1.0 x 10^{-4} M. Similar experiments were carried out using ^{14}C-labeled dilauroyl-GPC in the absence of substrate, as shown in Fig. 8, and it was found that about 22 moles of phosphatidylcholine are bound per mole of P-450$_{LM_2}$ subunit; the apparent Ks of dilauroyl-GPC is 7.0 x 10^{-6} M. It should be noted that the critical micellar concentration for dilauroyl-GPC, determined under these conditions according to the method of Bonsen *et al.* (31), is considerably higher (about 4.5 x 10^{-5} M). Preliminary studies on the binding of the reductase to P-450$_{LM_2}$ using both a gel filtration method and an immunochemical procedure, suggest that the molar ratio of reductase to cytochrome, calculated on the basis of subunit molecular weights, is in the range of 1.1 to 2.1. Optimal hydroxylation activity in the reconstituted

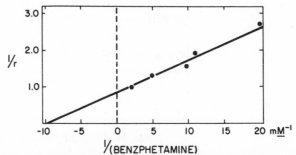

Fig. 7. Extent of benzphetamine binding to cytochrome P-450 determined by gel filtration. P-450$_{LM2}$ (56 nmol; 5.2 mg of protein) in 1.5 ml of 0.05 M Tris-acetate buffer, pH 7.5, containing 0.83 mM dilauroyl-GPC, 0.1 mM EDTA, 10% glycerol, and d-[N-methyl-^{3}H]benzphetamine (0.54 μCi per μmol; at the same concentrations as in the equilibrated columns) were applied at 23° to a Bio-Gel P-6 column, 0.6 x 60 cm, which had been equilibrated with the same buffer solution containing radioactive benzphetamine at the particular concentrations shown in the figure. Each column was eluted with the same solution as used for equilibration, and 1.2-ml fractions were collected and counted. The concentration of free substrate was determined from the base line level of radioactivity, and that of the bound substrate was determined from the radioactivity in the protein peak corrected for the base line level. The reciprocal of r (where r is defined as moles of substrate bound per mole of P-450$_{LM2}$ subunit) is plotted against the reciprocal of the free benzphetamine concentration.

system requires that the reductase be present at about 2.5 times the concentration of the P-450$_{LM2}$. Since the ratio of reductase to cytochrome P-450 in phenobarbital-induced rabbit liver microsomes (again calculated on the basis of subunit concentrations) is only about 0.05:1 for total P-450$_{LM}$ and 0.1:1 for P-450$_{LM2}$, it appears that the reductase is the rate-limiting component in such membranes.

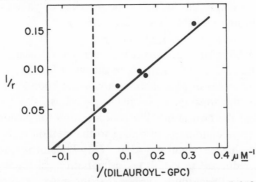

Fig. 8. Extent of phosphatidylcholine binding to cytochrome P-450 determined by gel filtration. Experiments similar to those in Fig. 7 were carried out with P-450$_{LM2}$ (11.8 nmol; 1.1 mg of protein) using radioactive dilauroyl-GPC (0.13 μCi per μmol, having 1-^{14}C-labeled lauroyl residues) and with substrate absent.

Evidence for the interaction of NADPH-cytochrome P-450 reductase with P-450$_{LM2}$ in the presence of the phospholipid and substrate is shown by the gel filtration experiment in the upper part of Fig. 9. Dilauroyl-GPC, which exceeded the critical micellar concentration, appeared first from the column. Cytochrome P-450 and the reductase were eluted together in a broad peak,

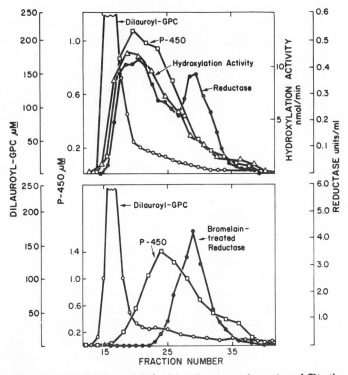

Fig. 9. Interaction of cytochrome P-450 with reductase as shown by gel filtration. In the experiment shown in the upper part of the fig., a mixture of P-450$_{LM2}$ (25 nmol; 2.7 mg of protein), NADPH-cytochrome P-450 reductase (specific activity, 16.7; 0.72 mg of protein), radioactive dilauroyl-GPC (1.55 μmol; 0.2 μCi), and benzphetamine (2.2 μmol) in 2.2 ml, final volume, of 0.05 M Tris buffer, pH 7.5, containing 15% glycerol, was allowed to stand at 23° for 10 min and then applied to a column of Bio-Gel A-1.5 m, 1.0 x 50 cm, previously equilibrated with Tris buffer, pH 7.5, containing 10 mM MgCl$_2$ and 1.0 mM benzphetamine. The column was eluted with the equilibrating buffer, and 1.3-ml fractions were collected and analyzed for cytochrome P-450 by the CO difference spectrum, the reductase by cytochrome c reduction at 30°, phospholipid by the radioactivity, and hydroxylation activity at 30° (after the addition of NADPH) by formaldehyde formation from benzphetamine. In the experiment shown in the lower part of the figure, highly purified bromelain-solubilized NADPH–cytochrome c reductase (specific activity, 57; 0.70 mg of protein) was substituted for the usual cytochrome P-450 reductase.

apparently as a complex having a molecular weight of 470,000, and additional reductase was eluted subsequently in a peak with an apparent molecular weight of 230,000. Hydroxylation activity, measured by the demethylation of benzphetamine upon the addition of NADPH, corresponded closely to the concentration of the cytochrome P-450. In a control experiment, not shown, P-450$_{LM_2}$ in the absence of the reductase had an apparent molecular weight of about 270,000. The interaction of the two proteins presumably resulted in the formation of a complex of higher molecular weight. In the experiment shown in the lower part of Fig. 9, highly purified protease-solubilized reductase was substituted for the purified detergent-solubilized cytochrome P-450 reductase and showed no interaction with the cytochrome P-450. These results are particularly interesting since the inability of the bromelain-treated reductase to form a complex with cytochrome P-450 under these conditions correlates with its inability to couple with partially purified cytochrome P-450 to effect substrate hydroxylation in a reconstituted system (32). Presumably, proteolytic solubilization by bromelain (30) or steapsin (33) removes part of the polypeptide chain (34) which is necessary for binding to cytochrome P-450.

The phospholipid appears to play a role in the binding of both the substrate and the reductase to cytochrome P-450. In experiments not presented here, the addition of dilauroyl-GPC decreased the Ks of benzphetamine for P-450$_{LM_2}$, as determined from the magnitude of the type I spectral change at varying concentrations, from 6.2×10^{-4} M to 1.0×10^{-4} M, and decreased the apparent Ks of the reductase, as measured from a frontal boundary gel filtration procedure (35,36) from 4.8×10^{-7} M to 1.2×10^{-7} M. These effects may explain in part the phosphatidylcholine requirement for rapid electron transfer from NADPH to cytochrome P-450, which is known to require the presence of substrate as well as reductase, but they do not rule out other functions of the lipid in substrate hydroxylation, such as at the stage of oxygen insertion.

SUMMARY

The enzyme system in liver microsomal membranes which hydroxylates fatty acids, steroids, and a variety of foreign compounds was previously solubilized and resolved into three components: cytochrome P-450, NADPH-cytochrome P-450 reductase, and phosphatidylcholine. The reconstituted system retains the ability to catalyze the hydroxylation of a variety of substrates in the presence of NADPH and molecular oxygen. Cytochrome P-450 has been purified to apparent homogeneity from phenobarbital-induced rabbit liver microsomes. Such preparations, which are free of cytochrome b_5

and other known electron acceptors, take up two electrons from NADPH or dithionite per molecule of heme, thereby indicating the presence of an acceptor (called Factor C) in addition to the heme iron atom. Both electrons are readily transferred from the reduced cytochrome P-450 to molecular oxygen and other acceptors. The detergent-solubilized reductase, which has also been highly purified, contains both FMN and FAD and retains the ability to couple with cytochrome P-450, unlike the protease-solubilized enzyme.

Evidence has recently been obtained for the occurrence of at least four distinct forms of liver microsomal cytochrome P-450 (P-450$_{LM}$). They are designated by their relative electrophoretic mobilities. P-450$_{LM2}$, which is induced by phenobarbital, has a subunit molecular weight of 50,000, and P-450$_{LM4}$, which is induced by β-naphthoflavone, has a subunit molecular weight of 54,000. P-450$_{LM1.7}$, which is induced by neither of these agents, is a mixture of forms with subunit molecular weights of 47,000 and 60,000. Some preparations have been obtained containing primarily P-450$_{LM1}$ or P-450$_{LM7}$. These forms differ in their catalytic properties as well; for example, P-450$_{LM2}$ is more active toward benzphetamine and P-450$_{LM1.7}$ toward benzpyrene.

P-450$_{LM2}$ appears to bind one molecule of substrate (benzphetamine), one to two molecules of reductase, and over 20 molecules of phosphatidylcholine per subunit. The presence of the phospholipid is necessary for rapid electron transfer from NADPH to cytochrome P-450 and for maximal hydroxylation activity. These effects are believed to be due to the ability of the phospholipid to facilitate the binding of substrate and reductase to the cytochrome. The results obtained argue against the idea that microsomal cytochrome P-450 is functional after solubilization only when reinserted into a membrane. Although the occurrence of additional forms of the cytochrome with highly similar electrophoretic behavior is not ruled out, it appears that the presence of these forms differing in subunit molecular weight and immunochemical properties may account for the variety of catalytic activities attributed to this pigment in the hepatic endoplasmic reticulum.

REFERENCES

1. Lu, A.Y.H. and Coon, M.J. (1968) *J. Biol. Chem. 243*, 1331-1332.
2. Coon, M.J. and Lu, A.Y.H. (1969) In *Microsomes and Drug Oxidations* (Gillette, J.R. *et al.*, eds.) pp. 151-166, Academic Press, New York.
3. Lu, A.Y.H., Junk, K.W. and Coon, M.J. (1969) *J. Biol. Chem. 244*, 3714-3721.
4. Strobel, H.W., Lu, A.Y.H., Heidema, J. and Coon, M.J. (1970) *J. Biol. Chem. 245*, 4851-4854.
5. Lu, A.Y.H., Strobel, H.W. and Coon, M.J. (1970) *Mol. Pharmacol. 6*,

213-220.

6. Coon, M.J., Strobel, H.W., Heidema, J.K., Kaschnitz, R.M., Autor, A.P. and Ballou, D.P. (1972) In *The Molecular Basis of Electron Transport* (Shultz, J. and Cameron, B.F., eds.) pp. 231-250, Academic Press, New York.

7. Ballou, D.P., Veeger, C., van der Hoeven, T.A. and Coon, M.J. (1974) *FEBS Letters 38*, 337-340.

8. van der Hoeven, T.A. and Coon, M.J. (1974) *J. Biol. Chem. 249*, 6302-6310.

9. van der Hoeven, T.A., Haugen, D.A. and Coon, M.J. (1974) *Biochem. Biophys. Res. Commun. 60*, 569-575.

10. Autor, A.P., Kaschnitz, R.M., Heidema, J.K. and Coon, M.J. (1973) *Mol. Pharmacol. 9*, 93-104.

11. Yu, C.-A., Gunsalus, I.C., Katagiri, M., Suhara, K. and Takemori, S. (1974) *J. Biol. Chem. 249*, 94-101.

12. Dus, K., Litchfield, W.J., Miguel, A.G., van der Hoeven, T.A., Haugen, D.A., Dean, W.L. and Coon, M.J. (1974) *Biochem. Biophys. Res. Commun. 60*, 15-21.

13. Williams, C.H., Jr. and Kamin, H. (1962) *J. Biol. Chem. 237*, 587-595.

14. Vermilion, J.L. and Coon, M.J. (1974) *Biochem. Biophys. Res. Commun. 60*, 1315-1322.

15. Iyanagi, T. and Mason, H.S. (1973) *Biochemistry 12*, 2297-2308.

16. Vermilion, J.L. and Coon, M.J. (1975) *Fed. Proc. 34*, 729.

17. Vermilion, J.L. and Coon, M.J. (in press) In *Fifth International Symposium on Flavins and Flavoproteins.*

18. Guengerich, F.P. and Coon, M.J. (1975) *Fed. Proc. 34*, 622.

19. Coon, M.J., Autor, A.P., Boyer, R.F. and Strobel, H.W. (1973) In *Oxidases and Related Redox Systems (Proc. 2nd Int. Symp.)* (King, T.E., Mason, H.S. and Morrison, M., eds.) pp. 529-553, University Park Press, Baltimore.

20. Coon, M.J., van der Hoeven, T.A., Haugen, D.A., Guengerich, F.P., Vermilion, J.L. and Ballou, D.P. (in press) In *Second Philadelphia Conference on Heme Protein P-450*, Plenum Press, New York.

21. Guengerich, F.P., Ballou, D.P. and Coon M.J. (in press) *J. Biol. Chem.*

22. Estabrook, R.W., Hildebrandt, A.G., Baron, J., Netter, K.J. and Leibman, K. (1971) *Biochem. Biophys. Res. Commun. 42*, 132-139.

23. Malmström, B.G. (1973) *Quart. Rev. Biophys. 6*, 389-431.

24. Lu, A.Y.H., Kuntzman, R., West, S., Jacobson, M. and Conney, A.H. (1972) *J. Biol. Chem. 247*, 1727-1734.

25. Lu, A.Y.H. and Levin, W. (1972) *Biochem. Biophys. Res. Commun. 46*, 1334-1339.

26. Nebert, D.W., Heidema, J.K., Strobel, H.W. and Coon, M.J. (1973) *J. Biol. Chem. 248*, 7631-7636.
27. Haugen, D.A., van der Hoeven, T.A. and Coon, M.J. (1975) *J. Biol. Chem. 250*, 3567-3570.
28. Laemmli, U.K. (1970) *Nature 227*, 680-685.
29. Hummel, J.P. and Dreyer, W.J. (1962) *Biochim, Biophys. Acta 63*, 530-532.
30. Pederson, T.C., Buege, J.A. and Aust, S.D. (1973) *J. Biol. Chem. 248*, 7134-7141.
31. Bonsen, P.P.M., de Haas, G.H., Pieterson, W.A. and van Deenen, L.L.M. (1972) *Biochim. Biophys. Acta 270*, 364-382.
32. Coon, M.J., Strobel, H.W. and Boyer, R.F. (1973) *Drug Metab. Disp. 1*, 92-97.
33. Masters, B.S.S., Williams, C.H., Jr. and Kamin, H. (1967) *Meth. Enzymol. 10*, 565-573.
34. Satake, H., Imai, Y. and Sato, R. (1972) *Abstr. Ann. Meeting Jap. Biochem. Soc.*
35. Nichol, L.W. and Winzor, D.J. (1965) *Biochim. Biophys. Acta 94*, 591-594.
36. Gilbert, G.A. (1966) *Nature 210*, 299-300.

THE MICROSOMAL MEMBRANE: A SERAGLIO FOR
UNIQUE ELECTRON TRANSPORT CARRIERS*

R.W. Estabrook, J. Werringloer, B.S.S. Masters, H. Jonen[+], T. Matsubara[†],
R. Ebel, D. O'Keeffe and J.A. Peterson

Department of Biochemistry
Southwestern Medical School
University of Texas Health Science Center
Dallas, Texas 75235

The endoplasmic reticulum of many types of cells contains a unique electron transport system functional in the oxidative metabolism of a wide variety of organic compounds. A fuller understanding of this enzyme system is necessary to further our knowledge of the mechanism of drug toxicity, the initiating reactions of chemical carcinogenesis, the maintenance of steroid hormone balance in the organism, and cellular and metabolic alterations resulting from exposure to many environmental pollutants.

The endoplasmic reticulum (microsomes) is recognized (1-3) as containing at least two flavoproteins and two hemoproteins as electron transfer components, i.e. an NADPH dehydrogenase (NADPH–cytochrome c reductase), an NADH dehydrogenase (NADH–cytochrome b_5 reductase), cytochrome b_5 and cytochrome P-450. Cytochrome P-450 functions (4,5) as a terminal oxidase where it participates in oxygen activation and substrate interaction in a cyclic manner, as shown in Fig. 1. Experiments have been reported (6,7) suggesting that an integrated electron transport complex exists in the microsomal membrane where NADPH dehydrogenase functions in the donation of the first

* Supported in part by research grants from the National Institutes of Health (GM 16488, HL 13619 and GM 19036) and from The Robert A. Welch Research Foundation (I-405 and I-453). J.A. Peterson is the recipient of a Research Career Development Award from the United States Public Health Service (GM 30962).
+ Present address: University of Mainz, Department of Pharmacology, D-65 Mainz, Germany.
† Present address: Shionogi Research Laboratory, Shionogi & Co., Ltd., Fukushima-ku, Osaka, 553 Japan.

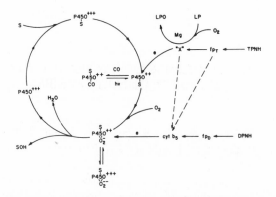

Fig. 1. Schematic representation of the cyclic function of cytochrome P-450 of microsomes and the relationship to other microsomal electron transport carriers. The valence of
iron of cytochrome P-450 is indicated by the +2 or +3 charges. Substrates for hydroxylation are termed S while electrons are designated by e. The flavoproteins TPNH (NADPH)–
cytochrome c reductase and DPNH (NADH)–cytochrome b_5 reductase are indicated by
f_{PT} and f_{PD}, respectively. A postulated electron transport carrier, X, is presumed to function between f_{PT} and cytochrome P-450.

electron for the reduction of cytochrome P-450 (with the postulated role (8,9)
for an uncharacterized electron carrier, X), while cytochrome b_5 is proposed to
participate in the reduction of oxycytochrome P-450. In addition, two other
electron transport carriers have been reported as components of the microsomal
fraction of some tissues, e.g. a cyanide sensitive factor functional in fatty acid
desaturation reactions (10,11) and a flavoprotein required for N-oxidation reactions (12). Further, the peroxidative degradation of unsaturated fatty acids
is also catalyzed by an unknown reaction mechanism when NADPH or ascorbate
are added to microsomes under appropriate conditions. A representation of the
multitude of oxidative reactions occurring in the microsomal fraction is presented in Fig. 2.

A study of electron transfer reactions associated with the microsomal membrane offers interesting opportunities for evaluating membrane structure–function relationships. In particular, interest has centered on the role of cytochrome
P-450 since it is the hemoprotein of highest concentration in the liver (13) and
it is relatively easy to alter the membrane composition of this electron transport
component by treatment of animals with various "inducing agents" such as
barbiturates, polycyclic hydrocarbons, etc. (14).

A number of challenging and perplexing questions can be asked concerning
the molecular organization of these electron transfer proteins as they reside in
or on the microsomal membrane. The influence of spatial relationships on their
functionality in carrying out physiologically important reactions remains to be
defined. Greatest emphasis has been placed on a study of the microsomal fraction prepared from liver because of the ease in obtaining large amounts of mate-

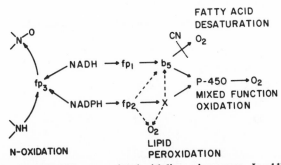

Fig. 2. Types of oxidative reactions associated with liver microsomes. In addition to cytochrome P-450 dependent mixed function oxidation reactions, fatty acid desaturation involving a cyanide sensitive factor, lipid peroxidation and flavoprotein (fp_3) catalyzed N-oxidation reactions occur in the presence of reduced pyridine nucleotides.

rial for study, the rapid alteration of hemoprotein composition upon treatment of animals with "inducers" and the broad spectrum of substrates oxidatively metabolized by cytochrome P-450. Therefore, the present discussion will be restricted to a consideration of our current knowledge of the electron transport carriers associated with the microsomal fraction of liver.

THE COMPOSITION OF LIVER MICROSOMES

The microsomal fraction of liver is composed principally of protein and phospholipid in a ratio of about 3 (15). In addition, the rough endoplasmic reticulum does contain ribonucleic acid associated with bound ribosomes. Pretreatment of animals with various inducing agents results in only a slight alteration in the ratio of phospholipid to protein. Of greater interest, however, are the large changes in the content of the hemoprotein, cytochrome P-450, associated with the microsomal fraction from liver, which occurs after treatment of an animal (16) with a drug such as phenobarbital (Fig. 3). Further, such data lead to a consideration of the stoichiometry of microsomal electron transport carriers, and indicate that a simple linear chain of interacting proteins is inadequate to explain the nature of the electron transfer reactions occurring during the function of cytochrome P-450 (Fig. 4). The data presented in Fig. 3 (17) show that 1 molecule of the NADPH flavoprotein dehydrogenase (NADPH- cytochrome *c* reductase) presumably interacts with 20 to 30 molecules of cytochrome P-450 if a direct interaction occurs between these two electron transport carriers. Such a hypothesis certainly requires a unique arrangement of proteins within the microsomal membrane.

When considering alternative schemes to describe the possible organization of these microsomal respiratory chains it is important to consider the following

RAT LIVER MICROSOMES

	Control	Phenobarbital Induced
	m μmoles/mg protein	
Cytochrome b_5	0.47	0.56
Cytochrome P-450	0.71	2.1
FMN	0.075	0.08
FAD	0.15	0.132
TPNH-c*	0.03	0.06
DPNH-b_5*	0.015	0.015

*calculated from activity measurements

Fig. 3. The content of electron transport carriers in liver microsomes prepared from untreated (control) or phenobarbital treated rats. The concentration of cytochromes b_5 and P-450 were determined spectrophotometrically. The content of FMN and FAD was established by fluorometric analysis of trichloroacetic acid extracts of microsomes. The concentration of NADPH (TPNH)–cytochrome c reductase and NADH (DPNH)–cytochrome b_5 reductase was calculated from activity measurements and related to the turnover numbers established for the purified flavoproteins (17).

points:

(A) The microsomal membrane is a single membrane of thickness approximately 70-80 angstroms (18). This suggests a limit of 2 or 3 molecules of molecular weight 50,000 spanning the thickness of the membrane;

(B) Using liver microsomes from phenobarbital treated animals, cytochrome P-450, a protein of molecular weight approximately 50,000, can account for as much as 15 to 20% of the protein of the liver microsomal membrane (13,17);

(C) Studies with proteolytic enzyme digestion of liver microsomal membranes (19-21) indicates that the NADPH flavoprotein dehydrogenase (NADPH–

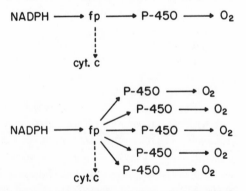

Fig. 4. Representation of a linear sequence (Scheme A) or a branched chain arrangement (Scheme B) for electron transfer from the flavoprotein, NADPH-cytochrome c reductase, (fp), to cytochrome P-450.

cytochrome *c* reductase) and cytochrome b_5 are rapidly released during short term proteolysis, suggesting that they are on the exposed surface of the membrane. Further, studies by Sato *et al.* (22,23) and Strittmatter *et al.* (24,25) have shown the presence of a lipophilic residue in these proteins (i.e. cytochrome b_5 and NADPH–cytochrome *c* reductase) which presumably serves as the anchor point for association of these electron transport carriers with the membrane;

(D) Cytochrome P-450 is relatively resistant to proteolytic digestion, in particular in the presence of a substrate to be hydroxylated (26), suggesting that it is more deeply buried in the membrane;

(E) Direct evidence to demonstrate the role of a lipophilic mobile electron carrier, comparable to ubiquinone of the mitochondrial respiratory chain, has not been obtained, although anaerobic titration studies of the microsomal cytochromes do suggest (27) the possible existence of additional electron acceptors associated with microsomes.

An example of the possible organization of electron transport carriers in the microsomal membrane has been proposed by Franklin and Estabrook (28) as shown in Fig. 5. It was suggested that the flavoprotein, NADPH–cytochrome *c* reductase, is located at the core of a complex surrounded by molecules of cytochrome P-450 and cytochrome b_5. Such a visualization also accounts for the interaction of a second flavoprotein, NADH–cytochrome b_5 reductase, with cytochrome b_5. This type of geometric arrangement suggests the presence of clusters of electron transport carriers as discrete complexes within the membrane.

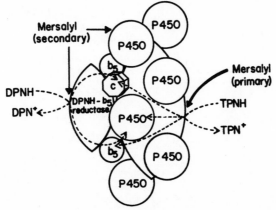

Fig. 5. Theoretical representation of the relationship of the flavoproteins, NADPH–cytochrome c reductase and NADH–cytochrome b_5 reductase to cytochromes b_5 and P-450. The figure purports to represent a side view cross section of one complex in the membrane, omitting any consideration of polarity for sidedness of the membrane.

An alternative representation of the organization of these respiratory carriers is shown in Fig. 6. The concept of clusters has been retained but the need to consider the flavoprotein, NADPH–cytochrome *c* reductase, and the hemoprotein, cytochrome b_5, as amphipathic proteins on the surface of the membrane attached by hydrophobic residues has been included. Further, this diagram includes a consideration of the need to accommodate molecules of cytochrome P-450 contiguous with the flavoprotein reductase as well as molecules of cytochrome P-450 which are more distant. The latter is a requirement dictated by the physical impossibility of surrounding a protein of 80,000 molecular weight by more than eight or ten molecules of molecular weight 50,000 when restricted by the condition that the flavoprotein is bound on the surface of the membrane.

INTERACTION OF CYTOCHROME P-450 MOLECULES IN THE MICROSOMAL MEMBRANE

In an earlier report, Franklin and Estabrook (28) described experiments designed to evaluate the possible types of interactions which might occur between the NADPH flavoprotein dehydrogenase and cytochrome P-450 when associated with the microsomal membrane. For these studies the effect of mercurial inhibition was evaluated. In the absence of any added protective pyridine nucleotide

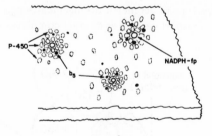

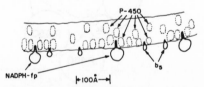

Fig. 6. An alternative schematic representation of the arrangement of oxidation-reduction pigments of the microsomal membrane. The bottom figure represents a cross-section of the membrane while the top is a surface view. Molecules of cytochrome b_5 and NADPH– cytochrome c reductase are shown on the surface of the membrane attached by a hydrophobic residue. Distribution of cytochrome P-450 molecules within the membrane are indicated by dotted circles.

(29), the flavoprotein (NADPH–cytochrome c reductase) is very sensitive to inhibition by mercurials such as mersalyl, as shown in Fig. 7. When liver microsomes are incubated with varying concentrations of mersalyl, the rate and *extent* of enzymatic reduction of cytochrome P-450, using anaerobic condition in the presence of carbon monoxide, was determined as shown in Fig. 8. The rather unexpected result was obtained which showed that increased inhibition of the flavoprotein dehydrogenase inhibits both the rate and the extent of enzymatic reduction of cytochrome P-450. It should be noted that the amount of cytochrome P-450 reduced in *ten* minutes was taken as the point of measurement for extent of reduction. The inhibition of cytochrome P-450 reduction was parallel with the inhibition of NADPH–cytochrome c reductase activity as measured by reduction of cytochrome c, as well as the overall activity of oxidative metabolism such as measured by the N-demethylation of the drug, ethylmorphine (28).

These results were interpreted to favor the existence of clusters or aggregates of cytochrome P-450 surrounding a flavoprotein dehydrogenase with a minimal extent of interaction between electron carriers on separated clusters. The resultant mosaic type of pattern of proteins within the microsomal membrane would suggest a *limited mobility* of these flavoproteins and hemoproteins.

A different way of representing this limitation of interaction observed between the complexes of flavoprotein and cytochrome P-450 is shown in Fig. 9. If a non-rigid system existed one would expect rapid interaction to occur and

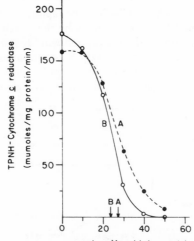

Fig. 7. The effect of varying concentrations of mersalyl on the activity of liver microsomal NADPH–cytochrome c reductase. Curve A was obtained in the presence of 6.5 mM ethylmorphine and Curve B in the absence. The mersalyl concentration required for 50% inhibition is shown by the arrows. Other conditions as described in ref. 28.

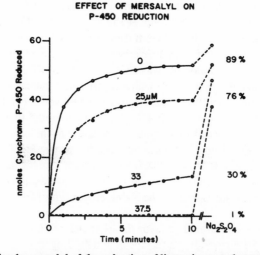

EFFECT OF MERSALYL ON
P-450 REDUCTION

Fig. 8. Inhibition by mersalyl of the reduction of liver microsomal cytochrome P-450. Rat liver microsomes (2 mg per ml) were incubated with varying concentrations of mersalyl as indicated and the contents of the reaction cuvette gassed with carbon monoxide. The reaction was initiated by the addition of NADPH at zero time and the increase in absorbance at 450 minus 490 nm measured. At ten minutes sodium dithionite was added to the reaction mixture and the extent of reduction of cytochrome P-450 determined as indicated by the dashed lines. The percent of enzymatically reducible cytochrome P-450 compared to chemical reduction is tabulated on the right side of the figure.

all the molecules of cytochrome P-450 to be rapidly reduced even though 50% of the flavoprotein molecules are inhibited by the mercurial. Since this is not what is observed experimentally, it was proposed (28) that a rather rigid arrangement might exist where little or only very slow and limited interactions occur.

If the scheme proposed above is correct, similar results should be obtained when the *extent* of reduction of cytochrome b_5 is measured in the presence of

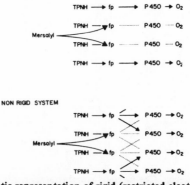

Fig. 9. Diagrammatic representation of rigid (restricted electron transfer between adjacent chains) and a non-rigid arrangement of microsomal electron transport carriers.

varying concentrations of mersalyl. As shown in Fig. 10, using anaerobic conditions, increasing the concentration of mersalyl inhibits both the rate and the extent of cytochrome b_5 reduction. This is observed whether NADH or NADPH is the reductant, i.e. whether the flavoprotein NADPH–cytochrome c reductase or the flavoprotein NADH–cytochrome b_5 reductase is the functional electron donor for cytochrome b_5. Again, it should be noted that the amount of cytochrome b_5 reduced in *ten minutes* is taken as the end point for the reaction.

It should be noted that increasing the concentration of reduced pyridine nucleotide added to the reaction mixture does result in the slow, progressive reduction of the cytochromes with time in the presence of mersalyl. This observation can be interpreted to mean that the inhibitor may dissociate from the flavoprotein reductase, allowing reduced pyridine nucleotide access to the enzyme for subsequent transfer of electrons to cytochrome P-450. Thus, using our basic diagram (Fig. 11), representing a postulated cross-section of the microsomal membrane, it is proposed that inhibition of a molecule of flavoprotein reductase associated with a cluster of cytochrome P-450 molecules prevents, or at least severly limits, the reduction of hemoprotein molecules in that cluster; in other words, *rapid movement* of protein molecules in the membrane resulting in their interaction is not readily observed.

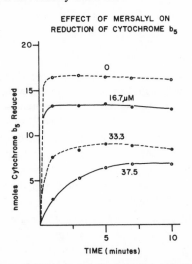

EFFECT OF MERSALYL ON
REDUCTION OF CYTOCHROME b_5

Fig. 10. The inhibition of cytochrome b_5 reduction by mersalyl. Liver microsomes were suspended in a reaction mixture containing 50 mM Tris-chloride buffer (pH 7.4), 5 mM $MgCl_2$ and 150 mM KCl. Glucose and glucose oxidase were added to the nitrogen gassed sample to ensure anaerobiosis. Varying concentrations of mersalyl were added and the extent of reduction of cytochrome b_5 determined after addition of NADH. Repetitive spectra were recorded from 500 to 600 nm every minute and the reduction of cytochrome b_5 evaluated from the increase in absorbance at 556 nm relative to 540 nm.

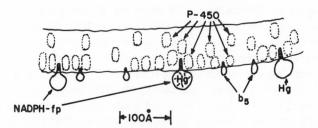

Fig. 11. Schematic representation of the side view of the microsomal membrane showing the interaction with a mercurial such as mersalyl.

Mersalyl, however, is recognized as a rather non-specific reagent and it has been shown that mercurials of this type can interact with cytochrome P-450, changing the optical as well as electron paramagnetic resonance spectral characteristics of this hemoprotein (30). Therefore, an alternative method to test the hypothesis that microsomal electron transport carriers have only limited mobility in the membrane was sought.

EFFECT OF LIMITED PROTEOLYTIC DIGESTION

Digestion of liver microsomes with crude pancreatic lipase has been the method of choice (31) for the isolation of the flavoprotein, NADPH–cytochrome c reductase. Incubation of microsomes with proteolytic enzymes, such as trypsin, subtilisin and nagarse, has been used to examine the accessibility of cytochrome b_5 and NADPH–cytochrome c reductase on the exposed surface of the microsomal membrane and studies by Ito and Sato (20) have demonstrated the time course of release of these electron carriers after limited digestion. Therefore studies were carried out to evaluate the effect of short term digestion of microsomes with trypsin on the ability of NADPH to reduce cytochrome P-450. As shown in Fig. 12, incubation of microsomes for only a brief period of time (1 min) with varying amounts of trypsin, resulted in a marked decrease in the rate and extent of reduction of cytochrome P-450 (Fig. 13). Prolonged incubations of microsomes with trypsin were not carried out because of the eventual alteration of cytochrome P-450 to P-420, even in the presence of a substrate, such as benzphetamine, known to protect the hemoprotein. The pattern of kinetic curves obtained during the measurement of cytochrome P-450 reduction suggests, however, the *very slow* time course for the eventual attainment of full enzymatic reduction of the hemoprotein. These observations suggest a *limited*, albeit slow, interaction between cytochrome molecules associated with different clusters of electron transport complexes. This undoubtedly reflects the fact that greater than 10% of the protein of the membrane is cytochrome P-450 and the

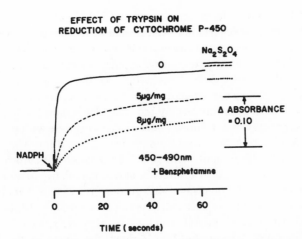

Fig. 12. The kinetics of cytochrome P-450 reduction after brief incubation of microsomes with trypsin. Liver microsomes were diluted in a buffer mixture similar to that described in Fig. 10, with the addition of 1 mM benzphetamine. The sample was gassed with CO for 5 min in special anaerobic cuvettes and then varying amounts of trypsin added. One min after addition of trypsin, 0.3 mM NADPH was added and the reduction of cytochrome P-450 measured at 450 minus 490 nm. Sodium dithionite was then added to determine the extent of chemically reducible cytochrome P-450.

statistical probability that adjacent molecules of hemoprotein will be no more than 80-100 angstroms away from each other if they are randomly distributed in the membrane.

BIPHASIC KINETICS OF CYTOCHROME P-450 REDUCTION

Studies of the kinetics of enzymatic reduction of cytochrome P-450 have

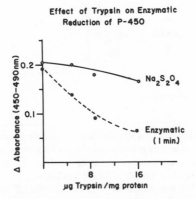

Fig. 13. The effect of varying trypsin concentrations on the extent of reduction of cytochrome P-450. A series of experiments were carried out as described in Fig. 12.

appeared in the literature from time to time (32,33). Most of these were performed with animals which had not been treated with drugs to induce cytochrome P-450 and its associated electron transport system and the reaction was quite slow. With the advent of low cost computers for the acquisition of stopped flow spectrophotometric data, this method has been adapted (34) to a dual wavelength stopped flow spectrophotometer for the examination of turbid samples. The type of data collected by the computer is shown in Fig. 14. Each of these experiments was performed under a CO atmosphere with particular care being taken to exclude oxygen from the reaction mixture. The reaction was initiated by the addition of NADPH and the reduction of cytochrome P-450 was observed by the increase in absorbance at 450 nm relative to 490 nm. The extent of enzymatic reduction was always greater than 95% of that observed in static dithionite reduction experiments. As can be seen in Fig. 14, there is an initial rapid burst of reduction followed by a much slower phase of reduction of cytochrome P-450. The inset of Fig. 14 shows the log plot of the first 90% of the reduction reaction. The time scale for this inset is the same as the rest of the figure. It can be clearly seen in this plot that the reaction is not first-order. What is difficult to determine from such a plot is the type of kinetic curve which will best fit the data.

For our kinetic analysis we decided to assume that the reaction was composed of two concurrent first-order processes: one fast, the other slow. The data was then fitted to the following equation by a nonlinear estimation procedure and

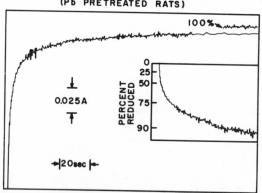

CYTOCHROME P-450 REDUCTION
(Pb PRETREATED RATS)

Fig. 14. The kinetics of cytochrome P-450 reduction. A sample of liver microsomes was diluted in a buffer mixture containing glucose and glucose oxidase as described in Fig. 10. After gassing with CO the suspension was placed in one syringe of the stopped-flow apparatus. An equal volume of CO gassed 0.6 mM NADPH was placed in the other syringe. The reduction of cytochrome P-450 was measured at 450 minus 490 nm and the data analyzed by computer evaluation as described in the text. The inset in the figure is the first order plot of the data.

the correlation coefficient for the fit was always greater than 0.999.

$$A_t = A_\infty [(1-F_S)(1-e^{-k_f t}) + F_S(1-e^{-k_s t})]$$

where A_t = absorbance at time t

A_∞ = absorbance at infinity

F_S = fraction of cytochrome P-450 reduced in the slow phase

k_f = rate constant for the fast phase

k_S = rate constant for the slow phase

A study of the temperature dependence of the reduction reaction was carried out to gain further information on the mobility of the electron transport carriers associated with the cytochrome P-450 reductase system. Fig. 15 shows an Arrhenius plot for the reduction of cytochrome P-450 in hepatic microsomes prepared from phenobarbital pretreated rats. The reaction was carried out in the presence of hexobarbital, a drug substrate of cytochrome P-450. As can be seen in Fig. 15, the temperature dependence of the fast phase of reduction of cytochrome P-450 is a single straight line with no break, while the kinetics of the slow phase clearly show a break at about 22°. The top line gives the fraction of cytochrome P-450 reacting in the fast phase. At the higher temperatures as much as 80% of the cytochrome P-450 is reduced in this portion, i.e. the fast phase. The energy of activation for this reaction can also be calculated from

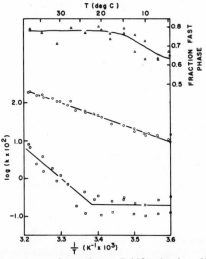

Fig. 15. Temperature dependence of cytochrome P-450 reduction. NADPH-dependent cytochrome P-450 reductase activity was measured at different temperatures with the dual-wavelength stopped-flow apparatus as described in Fig. 14. The values obtained for k_f and k_S, the rate constants for the fast and slow phases, respectively, are indicated by (-○-) and (-□-). The fraction of the cytochrome P-450 which is reduced, in the fast phase, is indicated by (-△-).

these curves and the E_a for the fast phase of reduction if 15.7 Kcal/mole, while the E_a for the slow phase is an astounding 38.1 Kcal/mole above the break point.

Comparable results were obtained whether the animal had been pretreated with a barbiturate, untreated or the cytochrome system modified by induction with the carcinogen, 3-methylcholanthrene. The slope of the curve and actual rate constants at any given temperature will vary somewhat with pretreatment procedure but the shape of the curves is invariant.

Thus this data is consistent with the model for the functional organization of the cytochrome P-450 reductase–cytochrome P-450 electron transport system. It is proposed that the fast phase represents the reduction of a cluster of cytochromes around a single reductase. These cytochromes P-450 can be randomly reduced by the reductase *without* the intervention of translational motion of the reductase or cytochrome P-450 through the membrane. This is shown by the lack of a break in the temperature dependence of cytochrome P-450 reduction. On the other hand, it is proposed that the slow phase of reduction represents the reduction of cytochrome P-450 molecules which are further removed from the reductase and their reduction requires some form of limited transport in the membrane as evidenced by the break in the temperature dependence curve.

DISCUSSION

The microsomal fraction of liver, obtained by differential centrifugation of homogenates, is composed of membrane vesicles derived from disruption of the endoplasmic reticulum. If one assumes the average diameter of such a vesicle is 0.2 μ, a simple calculation suggests that each vesicle contains about 20 molecules of flavoprotein reductase and about 500 molecules of cytochrome P-450. Calculations of protein mobility in membranes (35-37) suggest a diffusion constant of 1-5 x 10^{-9} cm^2 per second. Expressed a different way, a protein molecule would have the ability to migrate about 6000 angstroms in one second. Since liver microsomes contain 10% or greater of the protein as cytochrome P-450 a calculation of cytochrome P-450 distribution indicates that molecules of cytochrome P-450 are on the average no greater than 80-100 angstroms distant from each other. The observation that a significant portion of the cytochrome P-450 is rapidly reduced suggests that a major share of hemoprotein molecules are adjacent to a flavoprotein reductase molecule. The existance of a portion of the pool of cytochrome P-450 molecules, more slowly reduced in a temperature dependent reaction with a sharp phase transition around 20°, indicates molecules of hemoprotein more distant from the flavoprotein. Modification of the number of functional flavoprotein reductase molecules, either by inhibition with the mercurial, mersalyl, or by limited proteolytic digestion of microsomal membranes, markedly alters the proportion of cytochrome P-450 molecules rapidly reducible. Recently Yang (38) has challenged the interpretation of a rigid organization of electron transport carriers in the microsomal membrane. As with many biological problems, the question resolves to the absoluteness required to support a hypothesis. As described here, and discussed by Yang (38), given sufficient time and appropriate conditions (25 min and 37°) the further reduction of cytochrome P-450 molecules can be observed in the presence of an inhibitor of the flavoprotein, NADPH—cytochrome *c* reductase. The rates of protein mobility, however, would be at least orders of magnitude slower, under these conditions, than predicted from measurements reported elsewhere (35-37). Of course, one cannot exclude the possibility of rapid protein movement with an extremely high percentage of non-productive, abortive interactions which do not result in electron transfer. The challenge remains to better describe the nature of these interactions and the factors which regulate and control the participation of cytochrome P-450 in the oxidative metabolism of a wide variety of chemical compounds.

REFERENCES

1. Strittmatter, C.F. and Ball, E.G. (1952) *Proc. Nat. Acad. Sci. (U.S.A.) 38*, 19-25.
2. Omura, T., Sato, R., Cooper, D.Y., Rosenthal, O. and Estabrook, R.W. (1965) *Fed. Proc. 24*, 1181-1190.
3. Kamin, H., Masters, B.S.S., Gibson, Q.H. and Williams, C.H. (1965) *Fed. Proc. 24*, 1164-1172.
4. Estabrook, R.W., Cooper, D.Y. and Rosenthal, O. (1963) *Biochem. Zeit. 338*, 741-756.
5. Estabrook, R.W., Baron, J., Peterson, J.A. and Ishimura, Y. (1972) In *Biological Hydroxylation Mechanisms* (Boyd, G.S. and Smellie, R.M.S., eds.) Biochemical Society Symposium Number 34, pp. 159-185, Academic Press, New York.
6. Cohen, B.S. and Estabrook, R.W. (1971) *Arch. Biochem. Biophys. 143*, 46-53.
7. Hildebrandt, A. and Estabrook, R.W. (1971) *Arch. Biochem. Biophys. 143*, 66-79.
8. Ichikawa, Y. and Yamano, T. (1972) *J. Biochem. (Tokyo) 71*, 1053-1063.
9. Hoffstrom, I., Ellin, A., Orrenius, S., Backstrom, D. and Ehrenberg, A. (1972) *Biochem. Biophys. Res. Commun. 48*, 977-981.
10. Holloway, P.W., Peluffo, R. and Wakil, S.J. (1963) *Biochem. Biophys. Res. Commun. 12*, 300-304.
11. Oshino, N., Imai, Y. and Sato, R. (1966) *Biochim. Biophys. Acta 128*, 13-28.

12. Ziegler, D.M. and Mitchell, C.H. (1972) *Arch. Biochem. Biophys. 150*, 116-125.
13. Estabrook, R.W., Shigematsu, A. and Schenkman, J.B. (1970) *Adv. Enz. Reg. 8*, 121-130.
14. Conney, A.H. (1967) *Pharmacol. Rev. 19*, 317-366.
15. Eriksson, L.C. and Dallner, G. (1973) *FEBS Letters 29*, 351-354.
16. Remmer, H. and Merker, H.J. (1965) *Ann. N.Y. Acad. Sci. 123*, 79-97.
17. Estabrook, R.W., Franklin, M., Baron, J., Shigematsu, A. and Hildebrandt, A. (1971) In *Drugs and Cell Regulation* (Mihich, E., ed.) pp. 227-254, Academic Press, New York.
18. Claude, A. (1969) In *Microsomes and Drug Oxidations* (Gillette, J.R., Conney, A.H., Cosmides, G.J., Estabrook, R.W., Fouts, J.R. and Mannering, G.J., eds.) pp. 3-39, Academic Press, New York.
19. Omura, T., Siekevitz, P. and Palade, G.E. (1967) *J. Biol. Chem. 242*, 2389-2396.
20. Ito, A. and Sato, R. (1967) *J. Cell Biol. 40*, 179-189.
21. Kuriyama, Y., Omura, T., Siekevitz, P. and Palade, G.E. (1969) *J. Biol. Chem. 244*, 2017-2026.
22. Ito, A. and Sato, R. (1968) *J. Biol. Chem. 243*, 4922-4923.
23. Mihara, K. and Sato, R. (1972) *J. Biochem. (Tokyo) 71*, 725-735.
24. Spatz, L. and Strittmatter, P. (1971) *Proc. Nat. Acad. Sci. (U.S.A.) 68* 1042-1046.
25. Spatz, L. and Strittmatter, P. (1973) *J. Biol. Chem. 248*, 793-799.
26. Jonen, H., Werringloer, J. and Estabrook, R.W. Manuscript in preparation.
27. Estabrook, R.W., Matsubara, T., Mason, J.I, Werringloer, J. and Baron, J. (1972) In *Microsomes and Drug Oxidation* (Estabrook, R.W., Gillette, J.R. and Leibman, K.C., eds.) pp. 98-109, Williams and Wilkins, Baltimore.
28. Franklin, M.R. and Estabrook, R.W. (1971) *Arch. Biochem. Biophys. 143*, 318-329.
29. Phillips, A.H. and Langden, R.G. (1962) *J. Biol. Chem. 237*, 2652-2660.
30. Franklin, M.R. (1972) *Mol. Pharmacol. 8*, 711-721.
31. Masters, B.S.S., Williams, C.H. and Kamin, H. (1967) In *Methods in Enzymology* (Estabrook, R.W. and Pullman, M., eds.) Vol. X, pp. 565-574, Academic Press, New York.
32. Gigon, P.L., Gram, T.E. and Gillette, J.R. (1968) *Biochem. Biophys. Res. Commun. 31*, 558-562.
33. Diehl, H., Schadelin, J. and Ullrich, V. (1970) *Hoppe-Seyler's Z. Physiol. Chem. 351*, 1359-1371.
34. Peterson, J.A. and Mock, D.M. (in press) *Anal. Biochem.*
35. Poo, M.M. and Cone, R.A. (1973) *Exp. Eye Res. 17*, 503-510.
36. Poo, M.M. and Cone, R.A. (1974) *Nature 247*, 438-439.

37. Edidin, M. and Fambrough, D. (1973) *J. Cell Biol. 57*, 27-37.
38. Yang, C.S. (1975) *FEBS Letters 54*, 61-64.

MEMBRANE ASSOCIATION OF STEROID HYDROXYLASES
IN ADRENAL CORTEX MITOCHONDRIA

Tokuji Kimura, Hann-Ping Wang, Jau-Wen Chu
Perry F. Churchill and Jeff Parcells

Department of Chemistry
Wayne State University
Detroit, Michigan 48202

INTRODUCTION

The adrenal cortex steroid hydroxylases catalyze the biosynthesis of cortical steroid hormones from cholesterol. Steroid 17α-hydroxylase and 21-hydroxylase are located in the microsomal fraction, whereas steroid 11β-hydroxylase and cholesterol side-chain desmolase are located in the mitochondrial fraction. The electron transport systems from NADPH to molecular oxygen which are responsible for these hydroxylases have been extensively studied by us, by Dr. Estabrook's group, and by many other investigators. The mitochondrial system which distinctly differs from that of the microsomal system is as follows (1,2,3, 4).

NADPH → flavoprotein → iron-sulfur protein → P-450 → molecular oxygen

It should be emphasized here that the mitochondrial system requires a unique protein factor called adrenodoxin, whereas the microsomal system does not have such a protein.

A diaphorase called adrenodoxin reductase contains one mole of FAD per mole of protein. It has a molecular weight of 54,000 with a high content (54%) of hydrophobic amino acid residues. The Km's toward NADPH and NADH are 1.82 μM and 5.56 mM, respectively. The oxidation-reduction potential is -274 mV at pH 7.0. The iron-sulfur protein called adrenodoxin is a protein of the ferredoxin type. It contains 2 g atoms of iron and 2 moles of labile sulfur per mole of protein. The molecular weight is 12,500. It is an acidic protein with

447

11 and 18 residues of glutamic and asparatic acids, respectively. The oxidation-reduction potential is -272 mV at pH 7.0, and the number of electrons transfered by this protein is one. Cytochrome P-450, which is the terminal oxidase of the steroid hydroxylases, has not yet been purified from adrenal glands to a homogeneous state. It is assumed to be a hydrophobic protein with a molecular weight of about 50,000. The optical absorption properties of this heme protein are very similar to those of P-450$_{cam}$ and liver microsomal P-450.

We have previously shown that adrenodoxin reductase and adrenodoxin make a stoichiometric complex with a molar ratio of 1:1. The dissociation constant is 10^{-9} M in a low ionic strength medium. Many lines of experimental evidence strongly suggest that the steroid hydroxylases of adrenal mitochondria are associated with the inner membrane but not with the outer membrane.

In this report, we wish to present our views on the organization of these redox components in the mitochondrial membrane, and to discuss how the differences in structural organization reflect their biological functions.

RESULTS

Phospholipid composition of bovine adrenal cortex mitochondria (5)

Although it is known that the relative ratios of the major phospholipids in mitochondria of mammalian tissues are fairly constant, highly specialized tissues have vastly different ratios. Most mammalian mitochondria have the ratios of phosphatidyl ethanolamine:phosphatidyl choline:phosphatidyl inositol (PE: PC:PI) as 1:1:0.1, the flight muscle mitochondria of insects have the ratios of PE:PC:PI as 5:1:1. Adrenal cortex mitochondria are different from the mitochondria of other tissues. They possess unique vesicular structure and steroid hydroxylases with the unique electron transport system. On this basis, we decided to determine the phospholipid content in adrenal mitochondria. As presented in Table I, the salient features of adrenal cortex mitochondria are summarized as follows: (1) they contain a high phospholipid content per mg mitochondrial protein; and (2) they have a high content of PI and a low content of cardiolipin (DG). This distribution of phospholipids rather resembles that of microsomes. Coupled with the fact that adrenal cortex mitochondria have cytochrome P-450 which is a typical microsomal pigment of liver cells, it appears that the mitochondria of the adrenal cortex are of microsomal character.

Effects of phospholipids on steroid hydroxylase activities (6)

In order to gain some insight into the nature of membrane-association of the hydroxylase redox components, the effects of various phospholipids on the

TABLE I

Phospholipid Content of Bovine Adrenal Cortex Mitochondria

Phospholipids	Adrenal	Heart	Liver*
DG	14%	21%	17%
PE	40	36	35
PC	42	41	43
PI	4	2	–
PE/DG (moles/mole)	5.9	3.5	4.0
PC/DG (moles/mole)	6.1	3.0	5.0
Total phospholipid P/mg protein (nmoles/mg)	450	371	187

* The data are taken from Fleischer et al. (11).

Adrenal and heart mitochondria, approximately 30 mg protein, were extracted with chloroform:methanol (2:1, v/v) under a nitrogen atmosphere in the presence of an antioxidant. After removal of non-lipid materials, aliquots of the concentrated extract were taken for the determination of total phospholipid phosphorus and for phospholipid composition by two-dimensional thin-layer chromatography. The solvent systems were chloroform:methanol:28% aqueous ammonia (65:35:5) and chloroform:acetone:acetic acid:water (5:2:1:0.5). The spots were detected by charring. Heart mitochondria were prepared by the method of Crane et al. (10).

NADPH–DCIP, -cytochrome c reductase, and steroid 11β-hydroxylase activities were examined. It was previously known that the DCIP (2,6-dichloroindophenol) reductase reaction is catalyzed by the reductase alone, while the cytochrome c reductase reaction is catalyzed only in the presence of adrenodoxin. The 11β-hydroxylase activity was reconstituted from adrenodoxin reductase, adrenodoxin, and an extracted cytochrome P-450 particle preparation. These results are shown in Table II. These enzyme activities were determined in the presence of one mg of phospholipid per mg protein of the reductase. Among the four major phospholipids, PC and PE had a slight effect on these enzymatic activities, whereas acidic phospholipids such as cardiolipin (DG) displayed a marked inhibition for all three activities.

The titration curve of cardiolipin with respect to the DCIP reductase is shown in Fig. 1. When adrenodoxin was absent in the reaction mixture, a significant stimulation of the DCIP-reductase activity was observed. For the maximum

TABLE II

Effects of Phospholipids on Steroid 11β-Hydroxylase,
NADPH-DCIP Reductase, and NADPH-Cytochrome c Reductase Activities

Additions	11β-Hydroxylase	Cytochrome c reductase	DCIP reductase
none	100%	100%	100%
PC	---	99	98
PE	97	90	77
DG	65	30	20
PA	---	57	36
DG + 0.33 mM Ca	315	47	---
DG + 0.33 mM Mg	231	65	---
DG + 0.07 mM Mg	---	---	94
DG + 0.07 mM Ca	---	---	62
DG + 1.67 mM Mg	---	101	---

The steroid 11β-hydroxylase activity was determined essentially similar to the method of Kimura and Suzuki (3). The cytochrome c and DCIP reductase activities were measured as previously reported (8). Phospholipid was present in the reaction mixture at one mg/mg protein of adrenodoxin reductase. The turnover of cytochrome c reduction by the reductase and adrenodoxin (a slight excess relative to the reductase) was 650/min at 22°. The respective control experiments without the addition of phospholipid were taken as 100%.

stimulation, cardiolipin was required at the mole ratio of 10 with respect to the reductase protein. A similar enhancement of the activity was seen by the addition of linoleic acid instead of cardiolipin. The maximum stimulation occurred at a mole ratio of 110. Assuming that the molecular weights for cardiolipin and linoleic acid are 1550 and 280, respectively, the mole ratios were converted into the mass ratio of lipid to protein. The mass ratios necessary to obtain a maximum stimulation were calculated to be 0.3, and 0.5, respectively. At the higher concentrations of cardiolipin or linoleic acid, the activity gradually decreased as the mole ratio increased.

When adrenodoxin was present in the reaction mixture, there was neither activation nor inhibition up to a mole ratio of 10. Thereafter, the activity decreased as the mole ratio increased. To be emphasized here is the fact that in the presence of adrenodoxin, the DCIP-reductase activity is enhanced about 2-fold relative to the activity in its absence. Thus, the maximum activity obtained by cardiolipin or linoleic acid is comparable to that obtained by the addition of adrenodoxin. Since the enhanced DCIP reductase activity by adrenodoxin can-

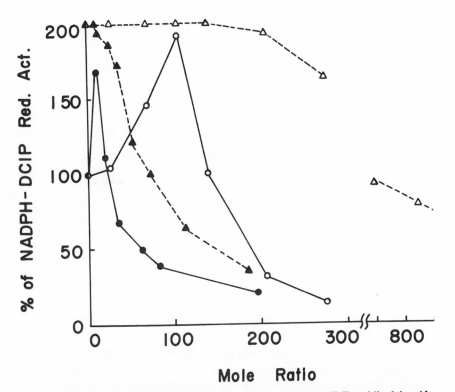

Fig. 1. Titration of NADPH-DCIP reductase activity by cardiolipin (DG) and linoleic acid in the presence or absence of adrenodoxin. ●: cardiolipin in the absence of adrenodoxin; ○: linoleic acid in the absence of adrenodoxin; ▲: cardiolipin in the presence of adrenodoxin; △: linoleic acid in the presence of adrenodoxin. The reaction mixture contained: adrenodoxin reductase, 1.68×10^{-7} M; adrenodoxin, when added, 1.68×10^{-7} M; DCIP, 1.9×10^{-5} M; lipids as indicated in 3.0 ml of 10 mM Tris-HCl buffer, pH 7.4. The control experiment with the reductase alone but without the addition of lipid was taken as 100%.

not be inhibited by the addition of antiserum against adrenodoxin, electrons from NADPH toward DCIP must flow through the reductase directly to DCIP regardless of the presence of adrenodoxin. Therefore, the activation by either adrenodoxin or cardiolipin must be due to changes in the reductase molecule. One can speculate that the hydrophobic portion of the adrenodoxin reductase may play an important role for the electron transfer reaction from reduced adrenodoxin to DCIP.

Fig. 2 shows the titration curves by cardiolipin for the NADPH-cytochrome *c* reductase (curve A), and the steroid 11β-hydroxylase (curve B) activities. Since these activities require the presence of adrenodoxin, there was no stimulatory effect by cardiolipin. Thereafter, these activities gradually decrease. The mole

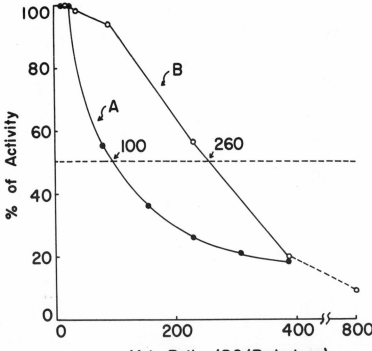

Fig. 2. Titration of NADPH-cytochrome c reductase (curve A) and steroid 11β-hydroxylase (curve B) activities by cardiolipin (DG). Control without any addition of lipid was considered as 100%. The steroid 11β-hydroxylase activity was determined by the method of Kimura and Suzuki (3). The cytochrome c reductase activity was measured by the method of Chu and Kimura (8).

ratios at a 50% inhibition were 100 and 260, respectively. These values are considerably higher than the value of 60 for the DCIP-reductase, suggesting that some of the cardiolipin binds to cytochrome c or P-450 molecules. It is emphasized again that since adrenodoxin is supposed to convert the reductase to a fully-active form by its binding, there should not occur any further activation by cardiolipin or linoleic acid for these activities.

From the experiments with cardiolipin it is concluded that: (1) a low level of cardiolipin stimulates the diaphorase activity; (2) in the presence of adrenodoxin this stimulation is not seen; (3) a high level of cardiolipin inhibits the diaphorase activity; and (4) cardiolipin appears to have no effect on adrenodoxin and P-450, although these proteins bind to the phospholipid.

Another interesting point is the effect of divalent cations. As shown in Table II, when 1.67 mM Ca^{2+} or Mg^{2+} is present together with cardiolipin in the reaction mixture, the DCIP and cytochrome c reductase activities are merely restored to the original level. The steroid 11β-hydroxylase activity is enhanced 2- to 3-fold by the addition of divalent cations. In the absence of cardiolipin, the activation was not significant, indicating that the activation requires both cation and cardiolipin. Since the enhancement is seen only for the steroid hydroxylase activity, but not for the DCIP and cytochrome *c* reductase activities, the site of the stimulation must be at the level of P-450. It is implicated that the Ca^{2+}-phospholipid-steroid complex serves as a better substrate for the steroid hydroxylation reaction. Further work in this regard is currently in progress in this laboratory.

Phospholipase C treatment of particulate cytochrome P-450 (5)

The crude preparation of cytochrome P-450 appears to be a membrane fragment. Therefore, we decided to examine the phospholipid composition of the particulate P-450 preparation.

Cytochrome P-450 particles were prepared by extracting previously sonicated adrenal cortex mitochondria with sodium cholate (0.5 mg/mg mitochondrial protein) at 4° with stirring for 60 min. The resulting P-450 preparation contained about 2 nmoles of P-450 heme/mg protein.

Phospholipid samples extracted from these particles were subjected to thin-layer chromatography. It was revealed that they contain PE and PC as major components while cardiolipin was not detected. The ratio of PE to PC was found to be about unity. The ratio of phospholipid phosphorus to P-450 was 200 nmoles/mole of P-450 heme. Although a possibility of redistribution of phospholipids during the preparation cannot be eliminated, the environment of P-450 appears to be devoid of cardiolipin.

Effects of phospholipase C pretreatment of the P-450 particles on the reconstituted 11β-hydroxylase activity was then examined. As shown in Fig. 3, the enzyme activity decreases as phospholipids are hydrolized by phospholipase C. These results are suggestive of the requirement of either PE, PC or both for the enzymatic activity.

Effects of temperature on steroid 11β-hydroxylase activity

The properties of membrane-bound functional proteins can be influenced by the physical state of membrane lipids. Breaks in the slopes of Arrhenius plots of membrane-associated enzyme reactions are interpreted as reflecting phase transitions of membrane lipids. In order to obtain some information about the influence of lipids on the steroid hydrolase, we have studied the effects of temperature by utilizing the freshly prepared adrenal cortex mitochondria.

As shown in Fig. 4, the Arrhenius plots of the steroid 11β-hydroxylase reaction

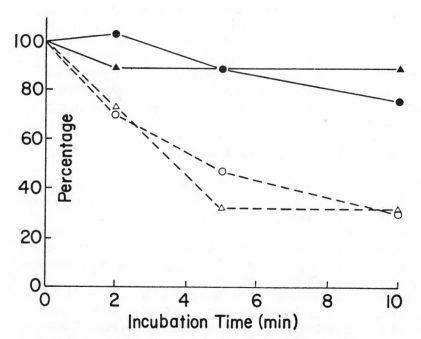

Fig. 3. Correlation between disappearance of phospholipid phosphorus and decrease in steroid 11β-hydroxylase activity. ○ : steroid 11β-hydroxylase activity with the pretreated P-450 particles by phospholipase C; ● : the same without the pretreatment; △ : phospholipid phosphorus remaining after the pretreatment; ▲ : the same without the pretreatment. 2.4 nmoles of the P-450 was incubated in a reaction mixture of 3.1 ml, which contained 0.2 units of phospholipase C (Bacillus cereus), 30 μmoles of phosphate buffer, pH 7.4, and 1 μmole of dithiothreitol. The reaction was carried out at 37° for the period of time indicated. Then, the reaction was terminated by adding 10 μmoles of EDTA. The pretreated P-450 was subjected to the steroid 11β-hydroxylase activity in the presence of adrenodoxin reductase, adrenodoxin, and NADPH. The rest of the sample was subjected to the determination of phospholipid phosphorus. Phospholipids were extracted from the samples by the method of Blogh and Dyer (12), and phosphorus was determined by the method of Bartlett (13).

have a distinct break at about 24° (the upper curve). The activation energies for the two slopes were calculated to be 5.5 and 19.2 kcal/mole. When the mitochondrial preparation was sonicated, the hydroxylase activity was significantly reduced, and the break became no longer distinct (the lower curve). The activation energy was 10.7 kcal/mole. The discontinuity in the Arrhenius plots observed in the intact mitochondria indicates that the lipid fluidity of the mitochondrial inner membrane is closely associated with the activity of the steroid hydroxylase.

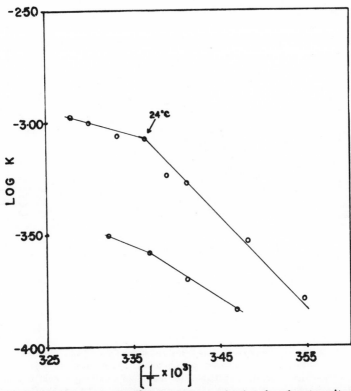

Fig. 4. Arrhenius plots of steroid 11β-hydroxylase reaction by adrenal cortex mitochondrial preparations. The steroid hydroxylation reaction was carried out at various temperatures. The activity assay was similar to the method previously described (3). Upper line: intact mitochondria; lower line: sonicated mitochondria.

Lipid peroxidation and P-450 (7)

As presented in Fig. 5, the rate of the degradation of P-450 at 0° was comparable to the rate of the formation of malondialdehyde by lipid peroxidation. Here, the amount of P-420 formed during the reaction was negligible, indicating that the heme moiety of P-450 had decomposed. This degradation reaction of P-450 was prevented by the addition of EDTA and stimulated by the addition of ferrous ions. These results show that, being similar to the case of liver microsomes, the degradation of P-450 of adrenal cortex mitochondria is closely associated with the lipid peroxidation reaction.

When chemiluminescence during the peroxidation reaction was monitored by the use of a liquid scintillation counter with off-mode of coincidence circuits, there were two phases of the chemiluminescence reaction. From measurements of the amounts of malondialdehyde formed, the later phase of chemilumines-

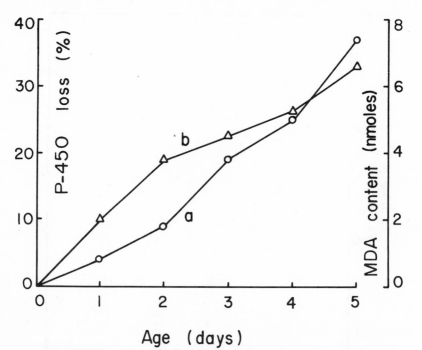

Fig. 5. Relationship between cytochrome P-450 and malondialdehyde contents during aging processes of isolated adrenal cortex mitochondria. 36 mg protein of bovine adrenal cortex mitochondrial preparation containing 0.8 nmole of P-450/mg protein were incubated at 0° in a total volume 36 ml of 10 mM phosphate buffer, pH 7.4. At the time indicated 2.0 ml of the sample was taken for the determination of P-450 (curve A) and 1.0 ml was taken for the determination of malondialdehyde (curve B).

cence appeared closely associated with the formation of malondialdehyde (Fig. 6). This chemiluminescence was not inhibited by either superoxide dismutase or the singlet oxygen quencher, DABCO (1,4-diazabicyclo-(2,2,2)-octane). Therefore, the involvement of superoxide or singlet oxygen in the iron-mediated peroxidation reaction appears to be unlikely. At present, the responsible chemical species for the luminescence is not defined.

Superoxide generation by adrenodoxin reductase and adrenodoxin

The superoxide anion radical is formed by the addition of one electron to an oxygen molecule. This radical is produced by the autoxidation of reduced flavoproteins, iron-sulfur proteins and heme proteins. Superoxide functions as a powerful oxidant and as a reductant for many biologically important substances. Upon dismutation of superoxide, it produces hydrogen peroxide which is a strong oxidant and, therefore, very toxic at high concentrations. Another pro-

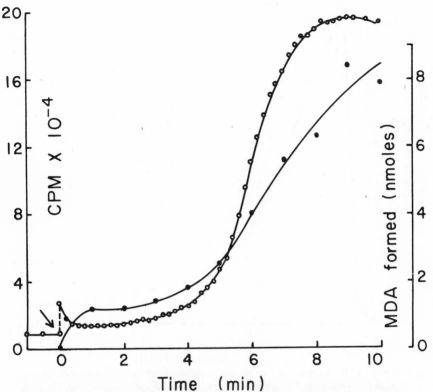

Fig. 6. Correlation between lipid peroxidation and chemiluminescence in ferrous ion-mediated reaction. For the determination of malondialdehyde (●), the reaction mixture had 1.0 mg mitochondrial protein, 10 mM phosphate buffer, pH 7.4, and 0.2 mM Fe^{2+} in a total volume of 1.0 ml. For the chemiluminescence measurements (○), the reaction mixture contained 10 mM phosphate buffer, pH 7.4, 1.0 mg protein of mitochondria and 0.2 mM Fe^{2+} in a total volume of 15.0 ml. The reactions were carried out at 22° and were initiated by the addition of Fe^{2+} (at arrow).

duct of the dismutation reaction is singlet molecular oxygen, which is quite harmful to living organisms due to its high reactivity. Additionally, when superoxide reacts with hydrogen peroxide, hydroxyl radical, an extremely strong oxidant, may be formed by a Haber-Weiss mechanism.

We have previously reported the formation of adrenochrome from epinephrine which is mediated by adrenodoxin reductase, adrenodoxin and NADPH (8). This superoxide generation reaction was absolutely dependent on the presence of adrenodoxin. The formation of superoxide by this system was completely inhibited by the addition of superoxide dismutase.

Fig. 7 indicates the effects of NaCl on the epinephrine co-oxidation during the course of NADPH oxidation. The rate of the formation of adrenochrome

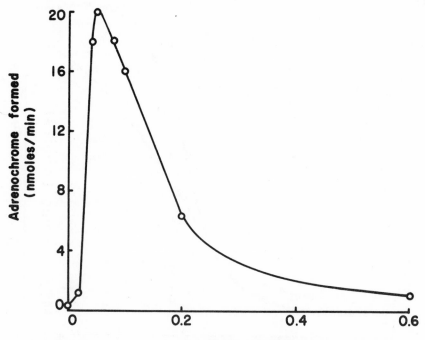

Fig. 7. Effects of NaCl concentrations on ephinephrine co-oxidation by adrenodoxin reductase and adrenodoxin. The reaction mixture contained: adrenodoxin reductase, 3.4 x 10^{-7} M; adrenodoxin, 3.9 x 10^{-7} M; epinephrine, 5.0 x 10^{-5} M; EDTA, 2.5 x 10^{-5} M; and NADPH, 17.5 x 10^{-5} M in 3.0 ml of 0.006 M phosphate buffer, pH 7.2. The ionic strength was adjusted by addition of NaCl. The assay was carried out at 22°.

was dramatically stimulated by a low concentration of NaCl. The maximal stimulation was seen at an ionic strength of 0.05. In the absence of adrenodoxin, the reductase alone did not generate any significant amount of superoxide in a wide range of salt concentrations.

Essentially similar effects were observed when NaCl was replaced by KCl or $MgCl_2$. Furthermore, polycations such as cytochrome c, histone and polyamines had the same stimulatory effects. The malondialdehyde formation from the mitochondrial lipids is significantly enhanced by the addition of cytochrome c. This stimulation could be achieved by either oxidized or reduced cytochrome c. Although cytochrome c reacts with superoxide effectively, the stimulatory effect by cytochrome c is not related to the oxidation-reduction cycle of this heme protein. Rather, cytochrome c acts as a polycation. Additionally, the malondialdehyde formation reaction was inhibited by the addition of superoxide dismutase. Therefore, the involvement of superoxide in this reaction might be the case. Yet, the active chemical species for the lipid peroxidation is not clear for sure. In view of the inhibitory effect of DABCO, the participation of singlet oxygen should be considered. In any event, polycations in a wide variety induce the superoxide generation reaction catalyzed by adrenodoxin reductase and adrenodoxin, resulting in the lipid peroxidation reaction and consequently the membrane degradation.

Tentative topography of membrane-associated steroid hydrolase

At present there is little known about the membrane topography of the steroid hydroxylases in adrenal cortex mitochondria. Nevertheless, from available data we can speculate their organization in the adrenal mitochondrial inner membrane. According to measurements of the P-450 and adrenodoxin contents, it was revealed that the P-450 and adrenodoxin contents are 1.85 and 2.04 nmoles/mg mitochondrial protein, respectively. These are in good agreement with the values reported by Simpson and Estabrook (9). We have previously observed that adrenodoxin reductase forms a complex with adrenodoxin in a molar ratio of 1:1. The dissociation constant is estimated as 10^{-9} M in a low ionic strength medium. The binding of adrenodoxin to P-450 appears to be much weaker than that of adrenodoxin to the reductase. The dissociation constant would be more than 10^{-6} M. From amino acid analysis, adrenodoxin reductase is a protein containing 46% of polar amino acid residues. Adrenodoxin is an acidic protein with 51% of polar amino acid residues. The amino acid sequence from the residue 21 to the residue 40 is, however, rich in hydrophobic amino acid residues. The forces between adrenodoxin and adrenodoxin reductase are known to be largely ionic (8). Yet the hydrophobic interaction would partially involve the formation of the catalytically active complex. As stated before, the effects of lipid on this reductase would support this implication. Based on the fact that the solubilization of the P-450 requires the addition of detergents, the P-450 is an intrinsic membrane protein which is buried in the membrane. Our membrane topography of the redox components of the steroid hydroxylases is schematically presented in Fig. 8. This membrane organization can be destroyed by at least two oxidative processes: one is iron-mediated lipid-peroxidation and the other is the degradation initiated by superoxide.

As a summary, adrenal cortex mitochondrial steroid hydroxylases which are responsible for the biosynthesis of steroid hormones are under weak influences of the membrane lipids, and their protein components form a cluster. Their mode of catalytic activity changes depending on the ionic strength of medium and at a certain concentration of cation superoxide is generated, leading to membrane degradation. Furthermore, as a consequence of the ACTH-stimulation

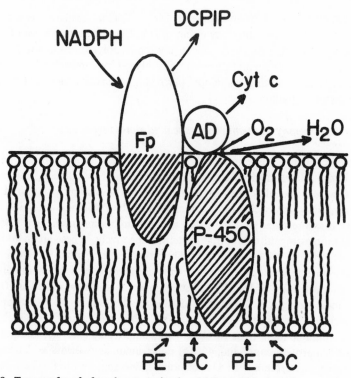

Fig. 8. Topography of adrenal cortex mitochondrial steroid hydroxylase complex.

toward adrenal cells, the redox components of the steroid hydroxylases change their levels, reflecting the change in the steroidogenic activity.

ADDENDUM

Adrenodoxin reductase was prepared by the method of Chu and Kimura (8) with slight modifications: the solutions used for the purification procedures contained 5 mM EDTA and a hydroxylapatite column separation was added at the last step of the purification procedure. The turnover number of the purified reductase was 650/min at 22° for the cytochrome c reductase activity in the presence of a slight excess amount of adrenodoxin. Adrenodoxin was purified by the method of Kimura (14). The ratio of absorbance at 414 nm to that at 276 nm was 0.86.

Asp, Asn, Leu, Gln, Lys, Ser, Arg, Thr and His are considered to be polar amino acids. The others are non-polar residues (15).

ACKNOWLEDGEMENTS

We express our sincere thanks to Professors T.T. Tchen and A. Paul Schaap for their collaboration in this study. Phospholipase C from *Bacillus cereus* was a kind gift of Professor R. Vagelos, and kind guidance for the determination of phospholipids was provided by Professor S. Fleischer.

Thanks are due to our previous colleagues, Dr. D.R. Pfeiffer and Dr. K. Goda, who participated in certain of the experiments reported here.

This study was supported by a research grant from the National Institutes of Health (AM 12713).

REFERENCES

1. Suzuki, K. and Kimura, T. (1965) *Biochem. Biophys. Res. Commun. 19*, 340-345.
2. Kimura, T. and Suzuki, K. (1965) *Biochem. Biophys. Res. Commun. 20*, 373-379.
3. Kimura, T. and Suzuki, K. (1967) *J. Biol. Chem. 242*, 485-491.
4. Omura, T., Sanders, E., Estabrook, R.W., Cooper, D.Y. and Rosenthal, O. (1966) *Arch. Biochem. Biophys. 117*, 660-673.
5. Wang, H.P., Pfeiffer, D.R., Kimura, T. and Tchen, T.T. (1974) *Biochem. Biophys. Res. Commun. 57*, 93-99.
6. Chu, J.W. and Kimura, T. To be published.
7. Wang, H.P. and Kimura, T. To be published.
8. Chu, J.W. and Kimura, T. (1973) *J. Biol. Chem. 248*, 5183-5187.
9. Simpson, E.R. and Estabrook, R.W. (1969) *Arch. Biochem. Biophys. 129*, 384-395.
10. Crane, F.L., Glenn, J.L. and Green, D.E. (1956) *Biochim. Biophys. Acta 22*, 475-487.
11. Fleischer, S., Rouser, G., Fleischer, B., Casu, A. and Kritchevsky, G. (1967) *J. Lipid Res. 8*, 170-180.
12. Blogh, E.G. and Dyer, W.J. (1959) *Can. J. Biochem. Physiol. 37*, 911-917.
13. Bartlett, G.R. (1959) *J. Biol. Chem. 234*, 466-468.
14. Kimura, T. (1968) In *Structure and Bonding* (Jørgensen, C.K., Neilands, J.B., Nyholm, R.S., Reiner, D. and Williams, R.J.P., eds.) Vol. 5, p. 1, Springer-Verlag, Berlin.
15. Capaldi, R.A. and Vanderkooi, G. (1972) *Proc. Nat. Acad. Sci. (U.S.A.) 69*, 930-932.

VIII

Summary of the Symposium

SUMMARY OF THE SYMPOSIUM

F.M. Huennekens

Department of Biochemistry
Scripps Clinic and Research Foundation
La Jolla, California 92037

> "Yea, the first morning of creation wrote,
> What the last dawn of reckoning shall read."
>
> —Omar Khayyam

> "Coupled information-storage (nucleic acid)
> and catalytic (enzyme protein) systems, under
> the influence of phase-boundary separations,
> ultimately gave rise to a cellular structure,
> encased in a boundary membrane."
>
> —M. Calvin

As the concluding speaker in this Symposium, it is my privilege to express, on behalf of all the participants, a deep sense of appreciation to our hosts for organizing this significant and timely conference on Membranes and for receiving us with the traditional Persian hospitality which has made the visit so pleasant. It was most impressive to have the meeting opened by a gracious message from His Imperial Majesty, The Shahanshah of Iran, which was presented by Mr. Assodollah Alam, Minister of the Imperial Court. Dr. Houshang Nahavandi, the distinguished Chancellor of the University of Tehran, added his greetings and told us about the recent commitment of the institution to the support of fundamental research and teaching in the field of molecular biology. In the coming years, we shall take pleasure vicariously in the benefits that our colleagues in Iran will receive from this enlightened policy. Our thanks are due also to: Dr. Khashayar Javaherian and the members of the Symposium Committee; Dr. Youssef Hatefi, who served as an advisor to the Committee and also supervised, together with Dr. L. Djavadi-Ohaniance, the publication of papers presented at the Symposium; Dr. Julius Kirchner who aided Dr. Hatefi with these activities; the International Union of Biochemistry and the International Union of Pure and

Applied Biophysics for their support; Dr. Philip Handler, President of the National Academy of Sciences (U.S.A.); and, finally, the speakers and session chairmen who so successfully communicated to the audience, especially the students, a sense of achievement and excitement in these frontiers of Membrane Structure and Function.

In planning this Symposium, the organizers realized that it was virtually impossible to cover in three days all facets of the bio-membrane field. Nevertheless, they felt that it was important to aim for the broadest possible mix of both topics and disciplines. In this age of specialization, as Hatefi pointed out in his opening remarks, workers in a particular area (e.g., the mitochondriacs) need to be exposed to the problems and the progress in other areas of membranology in order to better cope with their own system. And so the Symposium brought together a number of experts, each of whom presented the latest data and theories from his research group. In attempting to summarize this Symposium, with its formidable array of highly specialized and diverse contributions, Oppenheimer's observation that "We know too much for one man to know much" became painfully evident.

We might begin, however, by examining the general theme of the Symposium. Although the phrase "structure and function of membranes" is commonly used with the words in that order, it tends to place the cart before the horse. Both Green and Stoeckenius, early in the Symposium, pointed out that membranes exist because of their *functions* and that their *structures* evolved in a way that would support those functions. In primitive cells it is likely that membranes were only boundaries, designed to protect contents from the external environment in accordance with Orgel's maxim, "The operation of natural selection leaves little room for charity." When the first energy-yielding metabolic sequence (probably glycolysis) appeared, an opportunity was provided for membranes to acquire a more sophisticated function, namely energy-driven *transport* of metabolites. Because of their unique hydrophobic-hydrophilic milieu, membranes of subcellular organelles such as mitochondria and chloroplasts became the repository for oxidative phosphorylation and other complex *energy-transducing* systems. Later, as cells became differentiated and began to interact with each other (either by direct contact or by means of messenger molecules that could go from cell to cell), membrane surfaces elaborated a variety of recognition factors such as receptors and antigens. This evolutionary scenario, of course, is speculative, but it serves to focus attention upon three of the most important aspects of membrane function, viz. transport, energy transduction and receptors, which will form the basis for the subsequent discussion of highlights of the Symposium.

TRANSPORT

It is convenient to divide this aspect of membrane function into transport of: (a) ions; and (b) metabolites. The former process is bidirectional and rapid — characteristics which are well-suited to its involvement in nerve and muscle excitability, renal regulation of electrolyte balance and secretion in glands and epithelial structures. The wealth of information on the overall characteristics of ion transport in these systems has led to the formulation of several mechanisms which provide a basis for further experimentation. From the kinetics of Na^+ movement (ca. 10 nmoles/sec/cm^2) across the squid axon, Keynes has deduced that there are about 5×10^{10} channels per cm^2 and that each channel contains 3 charged "gating particles." Cessation of Na^+ transport is believed to be due to conformational changes in the channels. The channel hypothesis for ion transport was further supported by Läuger's description of the gramicidin-mediated movement of Na^+ across artificial lipid bilayer membranes. Consistent with Urry's hypothesis that only a gramicidin *dimer* (formed by head-to-head association of two molecules of the pentadecapeptide) would have the required structural characteristics, Läuger has found from electrical relaxation measurements that channel formation has a second-order dependency upon gramicidin concentration.

In addition to supporting a specific mechanism for Na^+ transport, the two preceding papers also illustrated the power of sophisticated biophysical instrumentation for the continuous monitoring of ion movement across membranes. Whether channels and gates are real structural entities, or merely convenient descriptive terms (like the famous "squiggle P" in oxidative phosphorylation), must await further resolution, characterization and reconstruction of the appropriate transport systems. An alternate mechanism, in which ions are moved through membranes via carriers (ionophores), has received support from the ability of certain low molecular weight compounds (e.g., valinomycin) and some naturally-occurring membrane components to facilitate the passage of ions through natural and synthetic membranes. Studies on the transport of cations and anions through mitochondrial membranes, which are discussed below in the section on energy transduction, generally invoke carrier mechanisms. It is conceivable that channel and ionophore mechanisms are not mutually exclusive but may have some as yet unexplained relationship to each other.

Transport of metabolites is slower than that of ions, and it involves essentially a unidirectional influx. To be sure, efflux or countertransport may also be demonstrable under special conditions, but these processes would not appear to be physiologically useful to the cell. Although some metabolites enter cells by diffusion, most appear to require specific carrier proteins — a stratagem which allows for regulation. During transit across the membrane, metabolites may, or may not, undergo structural transformation. Roseman's classical work on the transport of sugars into bacteria has been the prototype of a mechanism involving alteration of the metabolite. At the outer surface of the membrane each sugar binds to a specific carrier, and the resulting complex then moves to the inner surface where the sugar is phosphorylated by phosphoenolpyruvate and released into the cytoplasm.   Compelling evidence for a similar mechanism involving metabolite transformation during transport of amino acids in mammalian tissues was presented by Meister. The amino acid reacts initially with glutathione to form the γ-glutamyl amino acid and cysteinyl glycine; this step is catalyzed by a membrane-bound γ-glutamyl transpeptidase. After transit across the membrane, the γ-glutamyl amino acid is cleaved by a γ-glutamyl cyclotransferase to yield the free amino acid and 5-oxoproline. The amino acid is thus brought into the cell, while cysteinyl glycine and oxoproline are recycled to glutathione via a series of reactions in which 3 molecules of ATP are consumed. Specificity in this system may be accomplished, as it is in sugar transport, by a set of carrier proteins, one for each amino acid or group of amino acids; transpeptidation is restricted to the carrier-bound amino acids. If any of the enzymes in this sequence is also responsible for movement of the metabolites through the membrane, it, too, could be considered as a "carrier."

    Carrier-mediated systems that transport metabolites without modification were also covered at the Symposium. An extremely useful membrane vesicle system, obtained by treatment of *Escherichia coli* cells with lysozyme, has enabled Kaback's group to examine individual facets of the mechanism for transport of certain sugars and amino acids. Although the Roseman vectorial phosphorylation system is operative in *E. coli*, transport of most metabolites in the vesicles is driven by the oxidation of D-lactate. A D-lactate dehydrogenase (MW 75,000) has been purified to homogeneity from these vesicles and its position established in a respiratory chain linked to oxygen. The isolated dehydrogenase, Kaback reported, can be incorporated into vesicles prepared from dehydrogenase-deficient mutants, and this procedure restores the capacity for transport. Carrier proteins which undergo changes in affinity have been detected through the use of dansyl galactoside probes. These substrate analogs (which are not transported) show changes in fluorescence when bound to the putative carriers in the membrane; this effect is greatly enhanced when the vesicles have been energized by the addition of D-lactate. The value of bacterial systems for studying transport processes is also illustrated by the work of Dr. Gary Henderson in our laboratory. By extraction of lysozyme-treated cells of *Lactobacillus casei* with Triton X-100, he has been able to release a folate carrier protein which then can be purified to homogeneity by conventional procedures (MW 230,000; subunit MW 28,000). Transport of folate in this anaerobic organism is driven by glycolysis, but it is not yet clear whether energy from the latter is responsible for movement of the carrier protein, changes in its affinity, or both processes.

Carrier proteins also occur in the membranes of subcellular organelles. Klingenberg reviewed the isolation and properties of an ADP-ATP carrier (MW = 29,000) which was extracted by Triton X-100 from beef heart mitochondria and purified to homogeneity. This protein, too, shows high and low affinity states for its substrates. Carboxyatractylate and bongkrekate, inhibitors of oxidative phosphorylation that show a high affinity for the protein, have proved to be valuable probes for delineating the nature of the adenine nucleotide binding site, for examining conformational states of the protein, and for trapping the carrier at its stations on the outside and inside surfaces of the membrane.

From the above contributions, the following picture of carrier-mediated transport emerged: (1) Membranes contain a series of carrier proteins, each having an enzyme-like specificity for a substrate or class of substrates; (2) A carrier protein can exist in either a high or low affinity state with respect to interaction with its substrate — possible ways in which this transition might be regulated include conformational changes or subunit $\rightleftharpoons$ aggregate equilibria; (3) At the outer surface of the membrane, the carrier in its high affinity state binds the substrate; (4) The carrier-substrate complex moves across the membrane to the inner surface where the carrier is converted to the low affinity state and discharges its substrate. This type of formulation, which has some analogy to Michaelis-Menten kinetics, predicts that hydrophobic carrier proteins will be most amenable to extraction with agents like Triton when they contain bound substrate, since this is probably their state of maximum mobility in the membrane.

Knowledge of the constituents and mechanism of transport systems will lead inevitably to a better understanding of how these processes are regulated. It is regrettable that the exigencies of programming did not allow the inclusion of presentations directed specifically at the role of cyclic nucleotides in the regulation of transport. The concept that high levels of cyclic nucleotides suppress the transport of certain metabolites, and the consequences of this action upon cell replication, have been pointed out by a number of investigators, notably Holley and Tomkins. Relevant to this matter, Cuatrecasas reported that the deleterious effects of cholera toxin upon cellular transport processes stem from its stimulation of adenyl cyclase. A different mechanism for regulating transport was proposed by Rasmussen for the aldosterone-mediated flux of Na^+ in the amphibian urinary bladder. The hormone was found to stimulate the incorporation of acetate into long-chain, unsaturated fatty acids which, in turn, led to alterations in lipid composition of the membrane and changes in permeability. The effects of aldosterone on protein synthesis, originally thought to be the primary mode of action, may only accentuate lipid changes in membrane structure when the proteins in question are responsible for the enzymatic acylation and deacylation of phospholipids. The role of lipids in the structure and function of membranes was further stressed by Lynen who, in his remarks that opened the Symposium, called attention to the ease with which the lipid composition of bacterial membranes can be altered through the use of mutants and/ or manipulation of the growth medium. Another view of membrane lipids was presented by Van Deenen, who described the asymmetric distribution of specific lipids in the erythrocyte membrane. Phospholipases of varying specificity were used in conjunction with electron microscopy to demonstrate that phosphatidylcholines are located mainly on the outer layer of the membrane, while the ethanolamine- and serine-containing lipids are positioned on the inner surface.

ENERGY TRANSDUCTION

A major portion of the Symposium centered about the important, and still unsolved, problem of energy transduction in mitochondria and sub-mitochondrial particles. Here again, it was generally agreed that structure subserves function and, therefore, that structures should be scrutinized thoroughly for possible clues regarding the underlying function. The ultrastructure of mitochondria, however, received surprisingly little attention, although Green briefly presented an interesting new way of visualizing transducing complexes in which the tripartite units (headpiece, stalk and basepiece) are embedded in a series of ribbon-like structures that encircle the inner membrane. Treatment with lysolecithin dissociates these ribbon networks from the electron transfer complexes, allowing each component to be studied separately and paving the way for meaningful reconstitution experiments.

Several papers dealt with detailed structure-function relationships in particulate preparations derived from mitochondria. Extending previous work on the resolution of the mitochondrial electron transport chain which established the four standard complexes (I-IV), Hatefi and his colleagues have now isolated a new complex (V) that contains a portion of the oxidative phosphorylation apparatus, including a specific uncoupler binding site (see below). Further insight into the oxido-reduction mechanism of complex I (DPNH-coenzyme Q reductase) was also reported by this group. Oxidation of TPNH by complex I involves both transhydrogenation to DPN and a previously undetected pathway namely direct transfer of reducing power to iron-sulfur centers 2 and 3 of the chain.

At the other end of the mitochondrial electron transfer chain, cytochrome oxidase (complex IV in Hatefi's nomenclature) from yeast has been subjected to intensive structural analysis by Schatz. Electrophoresis on SDS-polyacrylamide resolves this complex into 7 polypeptides. Each polypeptide has been purified to homogeneity and is being subjected to detailed analysis. Specific inhibitors of protein synthesis have established that the three largest polypeptides, located largely in the core of the complex, are synthesized in the mitochondrion, while the four smaller polypeptides which form the shell of the complex are made on cytoplasmic ribosomes. Reconstitution of a functional cytochrome oxidase has now become a real possibility with the availability of these purified polypeptide components, but correct placement of the oxidoreduction groups remains a formidable obstacle. In related studies, Schatz has determined that cytochrome c_1 from yeast consists of a single polypeptide (MW 31,000) and that synthesis of this polypeptide, which occurs in the cytoplasm, is coordinately linked to heme synthesis.

The Mg^{++}-activated F_1-ATPase, a major component of the oxidative phosphorylation system of heart mitochondria, has also been analyzed in detail with respect to its polypeptide components. Penefsky reported that the solubilized ATPase (MW = 347,000) could be resolved by SDS-polyacrylamide electrophoresis into 5 polypeptides whose molecular weights were 54,000, 50,000, 33,000, 17,500 and 5700, respectively. Parenthetically, it should be noted that the uncoupler binding protein (MW 30,000) discovered by Hatefi and Hanstein in complex V is *not* a component of F_1. Each of the ATPase subunits has been obtained in homogeneous form and is thus available for further structural analysis (e.g., amino acid sequencing) and for reconstruction experiments. New information on the mechanism of the F_1-ATPase has been obtained through the use of aurovertin, an inhibitor of oxidative phosphorylation. Lardy described how changes in the fluorescence of this compound can be exploited to study its interaction with the solubilized ATPase. The enzyme exhibits a single binding site for the inhibitor in the absence or presence of ADP, and two binding sites of reduced affinity when ATP is present.

The Ca^{++}-activated ATPase from sarcoplasmic reticulum has been studied by Metcalfe for a different purpose, viz., to understand its relationship to membrane lipids. Since ATPase activity is lost when the number of lipid molecules surrounding the enzyme drops below a critical threshold (ca. 30), an opportunity is provided for examining the ability of various phospholipids to interact with the enzyme and regulate its activity. Instead of attempting to reconstitute completely lipid-free preparations, Metcalfe has utilized a conservative technique in which lipids in the membrane are exchanged with external lipids (including those with spin labels) in the presence of detergents. The lipid molecules are believed to form an annulus around the enzyme, sealing the latter into the membrane. The sarcoplasmic reticulum ATPase was further placed in context by Huxley in his comprehensive summary of muscle contraction. In the resting muscle, Ca^{++} concentration is very low; consequently, actin is unable to combine with the myosin crossbridges, and the actomyosin ATPase is not activated. The tropomyosin-troponin system further prevents the Ca^{++}-independent activation of the ATPase by actin. Upon receiving a nerve impulse,

Ca^{++} is released from the sarcoplasmic reticulum, and its combination with troponin allows the actin-myosin interaction (and muscle contraction) to occur. X-ray diffraction studies indicate that continuous strands of tropomyosin lie in grooves of the actin and somehow serve to transmit the effect of the spaced, globular troponin molecules to the actin.

The depth of understanding about the structural aspects of energy-transducing systems, especially mitochondria and submitochondrial particles, is not matched by a comparable degree of clarity regarding the mechanism by which the exergonic process of electron transport is coupled to the endergonic processes of ATP synthesis, ion-translocation and conformational changes. The major theories of oxidative phosphorylation, including its thermodynamic aspects, were clearly summarized by Slater, who served as chairman of one of the sessions. A variety of experimental techniques for the study of oxidative phosphorylation, and tentative mechanisms based upon the resulting data, were contributed by a number of the speakers. Griffiths, for example, drew attention to the power of a genetic approach. By using nuclear and mitochondrial mutants of yeast, he has uncovered a series of altered complexes III, IV and V. Mutants resistant to inhibitors such as oligomycin, venturicidin and triethyl tin are also available, and these may exhibit altered or deleted components when energy-coupling preparations are subjected to the types of analysis described above for cytochrome oxidase and the F_1-ATPase. New quantitative data on various exchange reactions catalyzed by mitochondria and sub-mitochondrial preparations have been obtained by Boyer, through the development of techniques that enable precise values to be obtained over short time periods. He reported that the Pi $\rightleftharpoons$ ATP and ATP $\rightleftharpoons$ HOH exchange reactions, but not the Pi $\rightleftharpoons$ HOH exchange, are inhibited by most uncouplers of oxidative phosphorylation; oligomycin, on the other hand, inhibits all three exchange reactions. This was interpreted to mean that Pi $\rightleftharpoons$ HOH exchange results from the reversible hydrolysis of tightly but noncovalently bound ATP. Boyer also offered the novel hypothesis that, in oxidative phosphorylation, energy from the electron transport system may be used, not primarily for the synthesis of ATP, but to promote a conformational change necessary for the release of preformed ATP.

Ion translocation in mitochondria continues to receive much attention, in part because it is a readily manipulable system. Lehninger reviewed the status of one of the more well-defined examples, viz. Ca^{++} transport, which has been studied extensively in his laboratory. At each of the three energy-coupling sites in the electron transport chain, careful measurements have revealed that two calcium ions are taken up (along with four molecules of the undissociated form of a weak acid). This ratio of 4 plus charges transported per $\sim$ is also observed with the Na^+-K^+-ATPase and the Ca^{++}-ATPase. Binding of Ca^{++} to the mitochondrial membrane, prior to translocation, may provide a measure of a putative Ca^{++}-carrier protein. Azzone has examined the capacity of mitochondria to bind and transport anions of various weak acids. Anion influx is accompanied by an increase in water content and concomitant swelling of the mitochondria which can be readily monitored by absorbance changes. Transport of anions is believed to occur by physical diffusion and to be facilitated by agents such as valinomycin and nigericin that make the membrane permeable to the cation component of the ion pair. The energy-linked extrusion of ions seems to depend upon conformational changes of the membrane resulting from the swelling process.

Mitchell's chemiosmotic hypothesis of oxidative phosphorylation was, as always, a central theme in the discussions of ion translocation. It makes the basic assumptions that the mitochondrial membrane is impermeable to protons, that at each of the three energy-coupling sites two protons are taken up inside the mitochondrion (to compensate for the 2 e^- present) and two other protons are ejected into the medium, that the carriers at the coupling sites are situated vectorially in the membrane in order to facilitate establishment of the proton gradient, and that the proton gradient drives ATP synthesis. Papa has examined this hypothesis with respect to the coupling sites in complexes III and IV. In the latter case, he concluded that cytochrome oxidase is, indeed, a transmembrane complex with the a_3 heme component reacting with oxygen at the matrix side of the inner membrane, while the a heme reacts with cytochrome c at the outer side.

The chloroplast system has been skillfully exploited by Avron to test the ability of pH gradients and membrane potentials to generate ATP. pH gradients, which can be measured very accurately via quenching of the fluorescence of 9-aminoacridine, was achieved with the aid of internal buffers while membrane potentials were produced by the rapid injection of a K^+-free chloroplast suspension into a medium containing K^+ plus valinomycin. Both pH gradients and membrane potentials could be used to drive ATP synthesis, but the energy associated with the movement of two protons was too low to support synthesis of an ATP molecule. In order to resolve this dilemma, Avron suggested, among other possibilities, that proton gradients and membrane potentials may be storage devices in equilibrium with some energy-rich intermediate that actually drives ATP synthesis.

Purple membranes of halobacteria also contain a light-driven proton pump which has been studied extensively by Stoeckenius with respect to both the structure and the mechanism of this energy-transducing system. The pigment, bacteriorhodopsin, is seen by X-ray diffraction measurements at 7 Å resolution to be located within hexagonal lattices in the rather rigid membrane. Protons are taken up and released by a light-driven cycle in which the pigment (linked to the protein via a Schiff base) passes through a series of intermediates having different pK values.

Some comfort could be derived from the progress, noted above, that had been made in understanding the mechanisms of the individual processes of electron transport, ATP synthesis, ion translocation and conformational changes. But two baffling questions remain. How are these individual processes coupled to each other and, during oxidative phosphorylation, which processes are primary and which are secondary? Despite the continued failure to detect the postulated compounds, there was still considerable support for the venerable hypothesis that, at the three coupling sites in the electron transport chain, energy-rich intermediates of defined chemical structure are generated and that this is followed by a series of essentially isoenergetic transformations culminating in the synthesis of ATP. In this view, conformational changes and ion translocation are seen as necessary, but secondary, consequences of the above chemical reactions. Lynen's classical discovery in 1951 that the ethereal "active acetate" was in reality S-acetyl coenzyme A, and the simple yet ingenious manner in which the energy-rich intermediates, 1,3-diphosphoglycerate and phosphoenolpyruvate, are generated in glycolysis, reassure the faithful that comparable energy-rich compounds will eventually be detected as intermediates in oxidative phosphorylation. Alternatively, these or comparable intermediates may soon be uncovered by current studies on transport processes, particularly if the plasma membrane proves to be more amenable to resolution than the mitochondrial membrane. If the chemical intermediate theory is correct, the mode of action of uncoupling agents may provide the key to the puzzle. 2-Azido-4-nitrophenol, an uncoupler of oxidative phosphorylation that contains a photoaffinity group, has been used by Hatefi and Hanstein to identify the protein which contains an "uncoupler site." It may be possible to use this purified material in conjunction with other components to construct a system that will reproduce the uncoupling phenomenon. Arsenate "uncouples" glycolysis by virtue of its ability to exchange with the energy-rich 1-phosphoryl group of 1,3-diphosphoglycerate, and the resulting 1-arseno derivative decomposes spontaneously to yield the energy-poor 3-phosphoglycerate. If the analogy is valid, one might ask what type of energy-rich intermediate is likely to interact with a nitrophenol and what end-product should be sought? Equally intriguing is the question of what process normally occurs at the site where uncoupling can be made to take place?

In contrast to the chemical, chemiosmotic or conformational mechanisms which were invoked in connection with the specific problems described above, a broader, unifying principle of energy transduction, termed the "paired moving charge (PMC) theory," was proposed by Green. Coupling of two processes (e.g., electron transfer and ion transport) is predicated upon the creation of paired charges at the physically separated centers; for the example chosen, H^+

and e^- would originate from one center while a cation-anion pair would be generated elsewhere. In each instance, charge separation would need to be maintained by a fail-safe mechanism. For example, H^+ and e^- might be separated by the electron transfer chain itself (e.g., reoxidation of a reduced flavin could result in the electron and the proton being transferred to an iron-sulfur center and water, respectively) while ionophores could sequester cations from anions, or vice versa. These paired charges are then envisioned as moving in concert through the membrane. Electrons and protons would "move" as described above, but cations and anions would be carried by specific ionophoroproteins, each of which would consist of a protein moiety (activating protein) and a prosthetic group (ionophore). The loaded ionophoroproteins would move along a prescribed course defined by specific "track" proteins whose molecular weights are ca. 30,000. Finally, elimination of the charges would complete the operation.

The PMC theory has some features in common with the chemiosmotic hypothesis, but it differs in the following important aspects: (a) charge separation is always paired; (b) coupling is designed to avoid rather than to create a membrane potential; (c) the flow of cations is directly coupled to the flow of electrons; and (d) coupling does not require an intact membrane. Experimental support for the PMC theory includes the recent isolation from mitochondria of a series of small molecular weight ionophores, several of which have been characterized as derivatives of octodecadienoic acid. "Activating proteins" associated with these ionophores have molecular weights of about 10,000.

Although they do not appear to be involved in energy transduction, cytochrome P-450 systems will be included in this section because they have some things in common with the mitochondrial electron transfer chain. The four papers that covered various aspects of this subject were preceded by an excellent overview from Gunsalus, who served as chairman of the session. The overall characteristics of this system are well-established, particularly in mammalian microsomes and certain bacteria: Reduced pyridine nucleotides are linked, via a flavoprotein or iron-sulfur protein, to cytochrome P-450, and the reduced form of the latter brings about the reduction of oxygen to water and the concomitant hydroxylation of various acceptors. Drawing from the work of his laboratory with highly purified components of the *Pseudomonas putida* monoxygenase system, Gunsalus noted that various proposed mechanisms involving oxidized, reduced and liganded forms of P-450 must conform to thermodynamic constraints imposed by the redox potentials of the system. This theme was amplified by Estabrook who proposed a detailed cyclic mechanism in which the heme of P-450 combines successively with substrate and O_2, and the ternary complex undergoes electronic rearrangement culminating in the release of water and the hydroxylated substrate. Absorbance spectra and EPR measurements have been employed to determine the nature of various intermediates in this cycle and the kinetics of their appearance and disapparance. An intriguing mechanistic problem in the above oxidoreduction sequence centers about the pathway of the two electrons needed to reduce O_2 to OH^+ (the probable hydroxylating species) and OH^- (the precursor of water). The reduced pyridine nucleotide donates a hydride ion which is subsequently broken up into two electrons, but it is not clear whether both of the latter flow directly to P-450 (and thence to oxygen) or whether one goes to the heme and the other to oxygen.

A liver microsomal system capable of hydroxylating fatty acids has been studied in detail by Coon's group. Components include the P-450, a TPNH-dependent reductase (MW 79,000) that contains equal amounts of FAD and FMN, and phosphocholine. From the liver microsomes of rats fed phenobarbital, a P-450 was purified to homogeneity (MW 49,000) and shown to contain an oligosaccharide. Other agents, however, induced different forms of P-450 that could be distinguished by their absorbance spectra, molecular weight and response to antibodies. Comparable progress with the P-450 system from adrenal cortex that hydroxylates steroids in the 11β- position was reported by Kimura. Here the electron transport system consists of the P-450, an iron-sulfur protein (adrenodoxin), and a flavoprotein (TPNH-adrenodoxin reductase). The latter two components form a tight 1:1 complex, but the binary complex shows little tendency to associate with the P-450. At least two different P-450 cytochromes have been observed in adrenal microsomes; these can be distinguished by their relative sensitivity to phospholipase or to iron-induced lipid peroxidation.

An interesting variation on the P-450 story, namely its involvement in chemical carcinogenesis, was discussed by Ernster. Polycyclic hydrocarbons become carcinogenic only after being activated via epoxide formation catalyzed by P-450 systems. The epoxide, or some more reactive compound derived from it, interacts with DNA which, in turn, leads to the transformation of the normal cell into a malignant cell. Working with lung tissue, a primary site for chemically-induced tumors (as well as secondary metastases), Ernster has partially purified and characterized a benzpyrene-induced P-450 monooxygenase and has begun to examine the reaction sequence and intermediates formed during activation of the hydrocarbon.

RECEPTORS

Four speakers combined to give a lucid picture of the current status of several membrane receptors. Cuatrecasas described the purification of cholera toxin and its resolution into an active (MW 36,000) and an inactive (MW

66,000) subunit. When the toxin is labeled with a fluorescent dye, its interaction with liver membranes or isolated cells (e.g., fat cells) can be demonstrated.
^{125}I-Labeled toxin was utilized to show that binding of the material actually
involves interaction with a specific ganglioside contained in the membrane (ca.
10^4 per fat cell) — a fact which prompted Cuatrecasas to note that the term
"receptor" needs to be defined, in each instance, in terms of a specific chemical
entity. Membranes can also spontaneously incorporate the purified ganglioside,
thereby increasing the amount of toxin that can be bound. Unlike some hormones, whose receptors are cytoplasmic proteins that conduct the bound
hormone to the nucleus, the toxin-ganglioside complex moves laterally in the
membrane and stimulates adenyl cyclase; this is believed to be the basis for
the deleterious effects of the toxin in clinical cholera.

DeRobertis reviewed the broad study that has been carried out in his laboratory on a variety of receptor proteins in excitable membranes. Receptors
specific for indoleamines, amino acids and cholinergic and opiate drugs have
been solubilized by treatment of the membranes with organic solvents and
detergents and further purified by conventional chromatography or, in some
instances, by affinity chromatography. Drug-receptor interaction has been
studied by various techniques, including light-scattering, fluorescence polarization, electron microscopy and X-ray diffraction. Purified receptor proteins
introduced into artificial lipid membranes provided a model system which
exhibited drug-sensitive ion transport. DeRobertis envisions the receptor macromolecules forming a channel through the membrane; "gating" is believed to be
controlled by conformational changes in the channel resulting from the binding
of agonists and antagonists.

Various aspects of the acetylcholine receptor (also called the "cholinergic"
or "nicotinic" receptor) were discussed by Changeux and Reich. In the former
presentation, the receptor was first examined under physiological conditions,
i.e., in the electroplaque. Three states were defined: resting (impermeable to
ions and stabilized by antagonists); active (permeable to ions and having an
affinity for agonists); and desensitized (impermeable to ions although able to
bind agonists). Anaesthetics are believed to induce the desensitized state. In
membrane fractions the protomer containing the receptor and the ionophore is
seen by electron microscopy (particularly when using the freeze-etch technique)
to consist of ring-like particles. Homogenization of the membrane fragments
in the presence of Ca^{++} yields closed vesicles or microsacs in which transport of
$^{22}Na^+$ is stimulated by carbamylcholine and inhibited by tubocurarine.

Reich and his colleagues have utilized non-ionic detergents to solubilize the
cholinergic receptor from the electric organ of the eel, *Electrophorus*. Binding
of ^{3}H-labeled cobra neurotoxin formed the basis for a quantitative assay of the
receptor during its separation from acetylcholine esterase and purification to
homogeneity; assuming 1:1 binding, the molecular weight of the receptor was

estimated to be 90,000. The receptor contained a high proportion of hydrophobic amino acids but no tryptophan. Equilibrium binding studies gave a value of ca. 10^{-9} M for the dissociation constant of the receptor-toxin complex. Measurement of the rate constants revealed that the forward reaction (complex formation) was slow and temperature-dependent, and that the reverse reaction was even slower and temperature-independent. Consistent with the kinetics, a biphasic mechanism was proposed involving an intermediate receptor-toxin complex which is eventually transformed into a slower-dissociating complex. In the presence of hexamethonium, a cholinergic ligand, the dissociation rate of the slower complex becomes equal to that of the faster complex. Binding constants for a number of agonists and antagonists were determined by allowing these compounds to compete with the toxin for the receptor. The data obtained correlate well with previous information gained from *in vivo* pharmacological studies.

Thus, the meeting closed, as it had opened, on a high note. With their work completed, the participants gathered slides and notes, and looked forward to the enjoyment of seeing the columns of Persepolis, the roses of Shiraz and the mosques of Isfahan. But as they departed, many undoubtedly asked themselves, "What were the *significant* accomplishments of this Symposium?" Certainly this question surfaced continually during the preparation of the above summary, and the latter thus represents an attempt to select a few highlights of the presentations and to put them into some kind of perspective relative to the directions that the membrane field is taking. For the omission of results that some speakers might consider to be the most important aspects of their presentations, and for any inadvertent misinterpretations of what was said, the reviewer can only apologize and hope that these lapses will be rectified by the speakers' own written accounts which appear elsewhere in this volume.

The question of significance is not a trivial one, especially since Symposia have become commonplace events in our highly developed and compartmentalized scientific world. However, it seemed to this reviewer, a veteran attendee at such affairs, that the present Symposium did have real significance, not only because of the numerous scientific advances that were reported, but also because it brought together a remarkable collection of investigators noted for their sound scholarship, provocative ideas and lucid presentations. One could not help but be impressed by the skillful manner by which these individuals blended together sophisticated physical measurements such as electron microscopy, light scattering, electron paramagnetic resonance (in conjunction with spin labels), and electrical conductance for delineating membrane properties with sensitive analytical procedures for identifying membrane components. It is clear that a wide variety of fundamental and challenging problems are finally amenable to study, largely because the proper tools are available. But the future is even brighter. Fernandez-Moran, one of the pioneers in the use of electron microscopy to study membranes, provided the participants with an exciting preview of the powerful new techniques that are just appearing on the horizon. And yet, conversely, it was refreshing to see that some of the most significant advances reported at this Symposium had been obtained by investigators using traditional equipment and simple procedures.

To close on a personal note – Dr. Hatefi, Dr. Kirchner and myself have the good fortune to live in La Jolla, California, a burgeoning scientific community which is also blessed with a Mediterranean setting and climate. As you might expect, surfing is a major activity there, and one of our colleagues has a picture in his office showing a surfer in fine form sliding down the front face of a giant wave. What the young man does not see, however, is the curl of the wave looming overhead with its promise of even greater acceleration. The caption under the picture reads "It's been a great ride so far, but the best is yet to come!" I think that perhaps the same could be said for the field of membranology.